Trees and Forests of Tropical Asia

TREES AND FORESTS OF TROPICAL ASIA

Exploring Tapovan

Peter Ashton and David Lee

The University of Chicago Press

The University of Chicago Press, Chicago 60637
© 2022 by The University of Chicago

Published 2022
Printed in the United States of America

31 30 29 28 27 26 25 24 23 22   1 2 3 4 5

ISBN-13: 978-0-226-53555-5 (cloth)
ISBN-13: 978-0-226-53569-2 (paper)
ISBN-13: 978-0-226-53572-2 (e-book)
DOI: https://doi.org/10.7208/chicago/9780226535722.001.0001

Library of Congress Cataloging-in-Publication Data

Names: Ashton, Peter S., author. | Lee, David Webster, 1942– author.
Title: Trees and forests of tropical Asia : exploring Tapovan / Peter Ashton and David Lee.
Description: Chicago : University of Chicago Press, 2022. | Includes index.
Identifiers: LCCN 2021017524 | ISBN 9780226535555 (cloth) | ISBN 9780226535692 (paperback) | ISBN 9780226535722 (ebook)
Subjects: LCSH: Forests and forestry—Asia. | Forests and forestry—Tropics. | Forest ecology—Asia. | Forest ecology—Tropics.
Classification: LCC SD219 .A886 2022 | DDC 634.9095—dc23
LC record available at https://lccn.loc.gov/2021017524

♾ This paper meets the requirements of ANSI/NISO Z39.48-1992 (Permanence of Paper).

We dedicate this book to the students of tropical Asia,
the future naturalists and silviculturists who
may make the restoration of magnificent forests
in tropical Asia a reality,

and to Mary and Carol, who helped us on
our journey every step of the way.

Contents

Preface

I am a firm believer that without speculation there is no good and original observation.

—*Charles Darwin (in a letter to Alfred Russel Wallace at Ternate, 22 December 1857)*

This book embarks the reader on an expedition into leafy, clammy, forested landscapes as they still are. It is a natural history of the tropical forests of Asia in its broadest and deepest sense. It is a book for all who love nature and forests, and a book for exploring, inquisitive minds. It analyzes the diversity of species (particularly trees) within these forests, and it suggests the main reasons for this diversity. It also discusses the human roles in modifying the forest landscapes, from early history through European hegemony, independence, and current pressures, and concludes with a cautious assessment of the likelihood of the survival of these forests in the future.

Trees and Forests of Tropical Asia: Exploring Tapovan is also a summary of a much larger work, *On the Forests of Tropical Asia* (London: Royal Botanic Gardens, Kew, 2014), by Peter Ashton. It is the first text on the forests of a whole tropical biogeographic region—a synthesis in which the results of contemporary research are set in context by deep field experience during more than half a century of all the region's forest formations and nearly all its nation states. *Tapovan* is the Sanskrit word for "forest wisdom," where knowledge is attained through *tapasya* (or inner struggle), fitting because of the historical cultural influence of Indian civilization and the book's integration of often onerous field research throughout the region.

Purpose and contents. Trees and Forests of Tropical Asia is a complete natural history of tropical Asian forests. It covers the forest types in the broad longitudinal arc from India east to New Guinea. The book begins with tropical Asia's long geological history and its influence on past and present climates. That sets the stage for a discussion of soils and the plants of the Asian tropics. A chapter on the structure of the lowland forests of everwet regions, and another chapter on forests in seasonally dry climates follow. Chapter 9 is a comparison of species compositions in tropical lowland forests, and chapter 10 is dedicated to montane forests of the entire region. Chapters 11 and 12 discuss animals as mobile links that mediate natural selection in forest trees, first considering pollination and breeding systems, and then seed dispersal, herbivory, diseases, and mycorrhizae. Chapter 13 reviews the phylogeographic influences (again in the context of geological history) on the distribution of species, and general species diversity in tropical Asia. Chapter 14 discusses the biggest evolutionary problem in tropical forests: the question of high species diversity on a local scale.

The final section of the book is the history of humans in the forest environment. It begins with a chapter on early human history, including influences of early civilizations on forest use and deforestation. Chapter 16 relates the long history of European influence in the region, first in exploration and purchasing commodities, and then in colonial administration. Chapter 17 is a history of forest use in postcolonial countries throughout the region. The final chapter deals with the future of forests in tropical Asia, factors in their destruction, and hope for changed policies and public attitudes. For each of the chapters, references citing sections summarized from the original large monograph, as well as more recent research, are provided in the notes at the end of each chapter.

Three appendixes will aid the less-specialized reader. The first is a geological history diagram, with major eras and epochs, allowing the placement of the times in the text (in millions of years ago, abbreviated Ma) into geological history. The second is a description of ForestGEO (formerly the Center for Tropical Forest Science, CTFS), emphasizing the locations of plots used most frequently in this book. The third is a description of recommended sites for student and eco-tourist visits, including their special features, means of access, and availability of accommodations. Chapter and illustration notes provide documentation.

The audience. This book is intended for a general audience, with minimal advanced training in biology or ecology required. It is suitable for undergraduates in those fields. It will also appeal to those interested in tropical Asia as a region, including—beyond those in botany and ecology—development, political and physical geography, political science, history, economics, anthropology, and sociology. For travelers to the region, the book provides needed understanding of the vegetation and commoner trees where they visit. It will also be valuable to anyone interested in all tropical forests, for the unique features of the Asian tropics described in this book will add to their understanding of other such forests in equatorial Africa or the Neotropics. The book is well illustrated with color photographs and diagrams.

About the authors. Peter Ashton has studied these forests, particularly in the field throughout tropical Asia, for more than sixty years. Peter was born in England and, during childhood, developed a love for the human-shaped landscapes and pockets of nature of the island. He became an expert in butterflies, taking advantage of the arrival of rare species from the neglected fields of wartime Normandy. He took a passion for natural history with him to Cambridge University in 1954, where he was a student at St. John's College. During a college expedition to the Santarem region of the Brazilian Amazon (p. 90), his first experience of the tropics, he developed his lifelong interest in the biology of trees of the tropical forests. He was advised by the eminent tropical botanist E. J. H. Corner to first get practical experience by starting his career in the forest itself. Peter joined the Forest Department of the Sultanate of Brunei, as forest botanist, in 1957 (fig. P.1). He explored the ecology of the rich forests of northern Borneo and studied the systematics of its most important canopy tree family, the Dipterocarpaceae. Soon after his arrival, he married, and Mary's and Peter's three children were later born on the island of Borneo. During his field work, Peter was visited by Corner, who witnessed his bite by a venomous snake, which Peter barely survived! He developed a close relationship with his Iban field guides (fig. P.1, top). His respect for and goodwill toward the indigenous people of the Asian tropics was a hallmark of his entire career.

After completing a PhD degree in 1962, leading to publications on the vegetation and soils of the forests of Brunei, and on the systematics of the dipterocarps, he performed similar work as forest botanist for the government of Sarawak. He continued his work on forest ecology and dipterocarp systematics, applying the keen eye for patterns that he developed during his childhood sojourns in English nature, and drawing inspiration from the insights of his Iban assistants.

In 1966, the University of Aberdeen hired several faculty members with expertise in tropical biology, to develop a relationship with the University

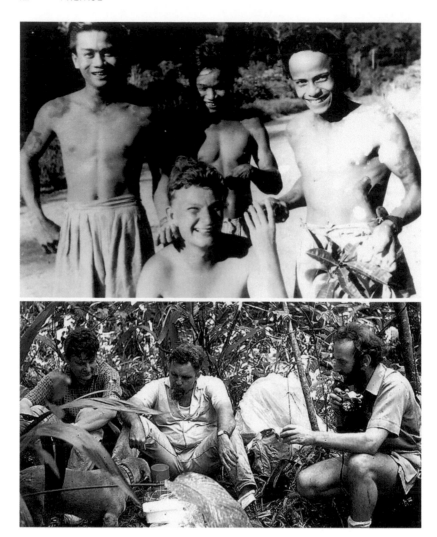

Fig. P.1 Teachers. *Top*, Peter in Brunei with friends and teachers: (L to R) Naban, Ladi, and Asah, 1957 (T. W.). *Bottom*, David's forest teachers on Gunung Ulu Kali, chasing the tropical citrus (*C. halimii*), West Malaysia, 1973: (L to R) Peter (rain forest ecology), Ben Stone (floristics), and Brian Lowry (phytochemistry); Poochie in foreground (D. L.).

of Malaya, and Peter joined the university. He then established a multi-disciplinary study of the reproductive biology and population genetics of Malayan rainforest trees, which began as David Lee arrived independently to work at the University of Malaya in 1973. They quickly became friends and collaborators (fig. P.1, bottom).

Fig. P.2 David and Peter (by Carol), Carol and Mary (by David) at the Ashtons' home in Chiswick, London, celebrating forty-two years of friendship, April 2015.

David had first been exposed to the tropics during a long trip to the South Pacific in 1964. He began to study the functional ecology of plants in tropical forests while at the University of Malaya. David and Carol left Malaysia at the end of 1976. They visited the Ashtons in Aberdeen during the return trip and continued to meet during the next forty years (fig. P.2). Soon thereafter, Peter was selected as the director of the Arnold Arboretum and became a Professor at Harvard University in 1978. The Lees lived in France, working with Francis Hallé at the Université Montpellier. They returned to the United States, and David began work at Florida International University, in Miami, in 1980. He continued research in functional ecology, addressing such topics as the nature and function of leaves of tropical plants, leaf optics, understory light climates, and developmental plasticity in the forest understory.

At the Arnold Arboretum, Peter and Mary invigorated an august but stale institution. He reestablished the arboretum's link with tropical Asia and investigated the questions raised by his earlier years of field research on the ecology and plant systematics of tropical Asian forests. He traveled extensively and developed friendships with scientists in the region. In 1981 he established a partnership with Stephen Hubbell (p. 255), creating large research plots in the Asian tropics. This partnership led to the formation of the Center for Tropical Forest Science (CTFS, p. 429). Peter's research, and that of numerous outstanding graduate students, along with his leadership in helping to establish and lead the CTFS and his long-term efforts to protect tropical forests through better management, led to his receiving many awards, most importantly the Japan Prize, for Science and Technology toward Harmonious Coexistence between Humanity and Nature (2007). More than sixty years after that first visit to the rainforests of the Amazon, Peter continues to visit the tropics, especially in Asia, engaging with scientists and stimulating research and sound forest management.

David continued his research on the functional ecology of tropical plants at Florida International University until his retirement in 2009. He received the Botanical Society of America's Distinguished Fellow award in 2019, in recognition for his service to the field and his research in the Asian tropics.

Preserving the legacy of this long period of research, Peter wrote the monumental monograph *On the Forests of Tropical Asia* (hereafter referred to as *OTFTA*). It is a huge book, 9" × 11" and eight hundred pages. Although only fully understandable by a small number of specialists and graduate students working in the region, it is a gold mine of information of use to a much wider audience. Although every effort is being made to make the book available to workers in the countries described by it, through reduced pricing and an electronic version free of cost, a need to produce a shorter version of the book for a wider audience was seen, and Lee volunteered to help make it happen. This is the book, primarily a summary of the ideas of a master tropical ecologist, aided by an old friend sharing an interest in tropical Asian forests, but a different research focus to complement Peter's interests. Except for chapter 8, it is Peter's voice in the opening segment of each chapter; in chapter 8, it is David's.

While condensing the densely presented information of *OTFTA*, we have added introductory material to make the later chapters more understandable. We use page numbers to cross-reference definitions in the early chapters. We also add relevant research summaries (and citations) of work published since 2012. Although the number of geographical place names has been greatly reduced, readers still may want to access *Google Earth* to find them. We have simplified references to time periods primarily by using the years before present (as Ma = millions of years ago and Ka = thousands of years ago) instead of using geological periods (the equivalents described in appendix A).

The original work was documented with 2,289 notes, citing research from 2,587 articles, plus websites and other sources of information. Here, the first endnote for each chapter generally provides page intervals from which the text of *OTFTA* was condensed and revised, and the notes in that original page range will cover much of the documentation for the text. We have reduced the number of citations and references by referring the reader to relevant sections of *OTFTA*, which can be downloaded as a PDF document from the internet free of cost from the following websites:

https://bibleandbookcenter.com/read/on-the-forests-of-tropical-asia/
  and
https://alishabooks1.blogspot.com/2020/04/dowload-on-forests-of
  -tropical-asia.html.

The notes in the present work cite sources that are particularly important, for which the authors have been mentioned in the text, as well as acknowledgments for illustrations. Most of the notes cite recent research, for 2015 and later. Up to six authors of articles and books are listed by convention here; more are listed by first author only. When a work is cited more than once in a chapter, only the first author's name and a short form of the title are given.

My nascent curiosity about nature was first attracted to butterflies. They ignited an interest in pattern detection, which was then aesthetic and sensual rather than intellectual but became a theme that has been central to my work as a field biologist. Much later, as a freshman undergraduate, I first visited the tropical rainforest during a summer vacation with two friends, near Monte Alegre, Brazil, along the great river. That visit inspired my fascination with tropical trees and forests, and with the apparent tapestry of variation among their species communities within their landscapes, in relation to geology, soil, and water. Later, on first gazing up beneath a mighty columnar kapur (Dryobalanops) stand, their branches fan-vaulting into vast domed crowns, I realized I must spend my life studying Asian rainforests. — P. A.

1

THE ASIAN TROPICS

The tropics evoke diversity—both biological and cultural—and we focus on the diversity of tropical Asian trees and forests. To describe their natural history, we need to start with the tropics in general. We first discuss the general climatic conditions of the tropics, what produces them, and what factors determine their extent north and south of the equator. Then we discuss differences among the tropical regions, the Neotropics, Africa, and Asia, emphasizing the forests of the Asian tropics (fig. 1.1).[1]

Energy and the Tropics

Radiation from the sun strikes the earth at a near constant 1.33 kilowatts per square meter ($kW\ m^{-2}$). A watt is an amount of energy delivered over time (power, or joule s^{-1}), familiar to us in our monthly utility bill (in kW hours). Even in clear, cloudless atmosphere, some of that radiation is reflected or absorbed, so that around 1 $kW\ m^{-2}$ reaches the surface of the earth. The equatorial regions are slightly closer to the sun than the high latitudes, but the energy increases are insignificant. All parts of the planet receive the same duration of sunlight per year, an average of 12 hours. Since the earth's orbit around the sun is slightly elliptic, there is a slight increase of irradiance, about 3.3%, when the earth is closest, in January, and 3.3% less in July. None of these factors are important in explaining the large increase in energy arriving in the equatorial belt (fig. 1.2).

The important difference is the average angle of sunlight to the zenith. This varies with latitude during the day. The seasonal variation, as at midday,

Fig. 1.1 Forest in everwet tropical Sabah (H. H.).

also varies with latitude. As the earth moves around the sun, the tilt of its plane of rotation, at 23.35° latitude, exposes different latitudes to different zenith angles. Those near the equator are close to zero, and those further north and south increase, depending upon the time of year. As the angle increases, the 1 m² shaft of radiation is spread over a large area, varying with the angle's cosine. That irradiance decreases with latitude and is greatest near the equator at all times of the year, although latitudes near the Tropic of Cancer are slightly greater at summer solstice (table 1.1). Such calcula-

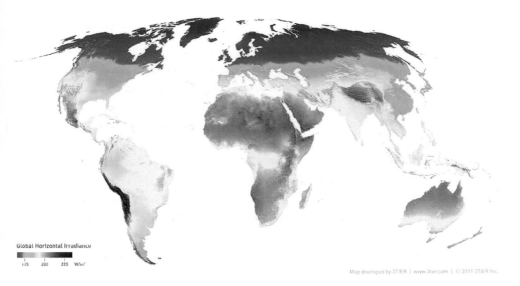

Fig. 1.2 Map of mean annual radiation in different regions and latitudes, units in Wm⁻², by www.3Tier.com. Lower radiation in India is due to the cloudy summer monsoon.

Table 1.1 Estimates of irradiance at midday (solar zenith) at different latitudes and times of the year, plus means

Location	Latitude ° N	Irradiance per day as kW-hours/m²				
		21-Dec	21-Mar	21-Jun	21-Sep	Mean
Nome, Alaska	64.5	0.2	4.6	11.5	4.6	3.1
North Petherton, UK	51.1	2.0	6.4	11.6	6.4	6.6
Miami, Florida	25.8	6.3	9.3	11.2	9.3	9
Singapore	1.3	9.7	10.3	9.2	10.3	9.9

Source: Thayer Watkins, www.sjsu.edu/faculty/watkins/newpages.htm.

tions permit estimations of the annual total radiation at a site, which also can be calculated from satellite observations (fig. 1.2). These values do not consider the effects of clouds, which reduce the direct solar radiation but increase the diffuse sky radiation to a lesser extent. Since cloud cover is often thick along the equator, higher irradiance is generally seen in the tropics, but at some distance north and south of the equator.

The Hadley Cells

The climatic equator is the center of the belt of tropical weather. It moves north in north-temperate summers and south in north-temperate winters, as the earth, tilted on its axis in relation to the sun, receives solar energy differentially over its surface during its annual cycle. The tropical climatic zone consists of two separate wind-circulation systems, termed Hadley cells (named for the English lawyer who first described them in the 18th century), north and south of the climatic equator (fig. 1.3). Earth's rotation, with greatest spin at the geographical equator, constantly draws tropical surface winds from east to west. They originate at the dry margins of the tropics and pick up speed as they move diagonally toward the climatic equator, where they slow and cease. That wet climatic-equatorial region is known as the intertropical convergence zone (fig. 1.4). At their maximum speed over the oceans, these winds are known as the north-east and south-east

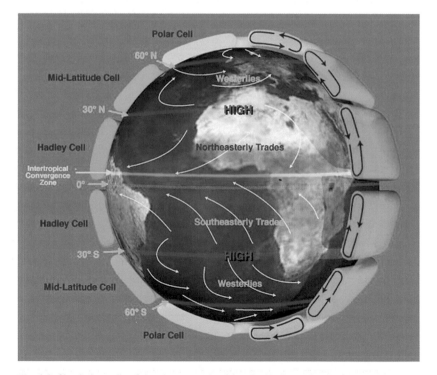

Fig. 1.3 Circulation cells of the planet, emphasizing the Hadley cells, the intertropical convergence zone, and the directions of trade winds (Courtesy of NASA).

Fig. 1.4 The intertropical convergence zone seen as the band of clouds over the western Pacific, from a satellite photograph (Courtesy of NASA).

trade winds, and they are the most constant winds on Earth. Where they move over water or forest, the moving air accumulates moisture. As the winds eventually slow, the increasingly warm and humid air at low altitude becomes unstable beneath cooler heavier air above. Thunderclouds form, especially where there are mountains in the path of the wind, or near the equator, where even low peaks draw the moist, warm air up their slopes like giant wicks. The lowland air thus rises, cooling and losing its moisture, at first through condensation to cloud, and then as rain. The heat of condensation adds to the air's warmth; it rises further, where it finally cools and is returned as high-altitude dry wind to the margins of the tropical climate zone. There it descends again, now desiccated and warmer, creating deserts. Lands to leeward of the trade winds are among the most predictably dry on Earth.

The Limits of the Tropics

Geographically, the tropics are defined as that part of Earth's surface which receives the sun vertically overhead at least once a year; that is the region between 23.35° north and south latitudes. Of the three continental tropical regions, Asia is the smallest. It further differs in being oriented east-west rather than north-south, encompassing a greater distance of 9,100 km. It alone is both terrestrially and tectonically fragmented, comprising both the southern part of a continent, abutted on its north by the world's highest mountain range, and the world's largest archipelago, which includes two of the planet's largest islands. It includes, through the Malay archipelago, known to botanists as Malesia, the only extensive terrestrial region under continuously wet tropical climate (see fig. 4.1).

Ecologically, these latitudes do coincide with the limits of tropical forests, but only roughly. Thus, it is convenient, but not at all accurate, to consider the edges of the tropics as the Tropics of Cancer or Capricorn. A simple examination of the ecosystems at different latitudes indicates that tropical regions may establish well beyond those latitudes (fig. 1.5), and their forests are discussed in chapter 10 (p. 193). The limit of an 18° C mean annual temperature in the coldest month has been considered as the thermal limit of the tropics, but it misses the most important limiting weather factor in the lowlands: freezing weather at rare intervals. For instance, Miami is not truly tropical—despite the efforts at advertising its tropical ambience. It is subject to rare freezing weather—the last hard freeze occurred in 1990.

Nowhere are these limits less precise than in continental Asia, owing to the winter shelter from cold, dry, continental northerly winds offered by the Himalaya and Burma mountains and the warmth imparted by the ocean current that originates in the bottled-up South China Sea and hugs the coast northward. Here, deciduous tropical sal (*Shorea robusta*) forests extend northward to 32° N in the western Central Himalaya. Mangroves (Rhizophoraceae) reach far north along the East Asian coast to Kyushu in southern Japan at nearly the same latitude as Tel Aviv, Israel. Wet seasonal dipterocarp forests reach 28° N in northeast India and Burma, following the limits of the southwest summer monsoon winds, which are drawn north by the low summer pressure created by the high Tibetan Plateau. Specifically, the limits of these lowland forests coincide with the limits of ground frosts and ice storms in moist climates (see fig 4.1).

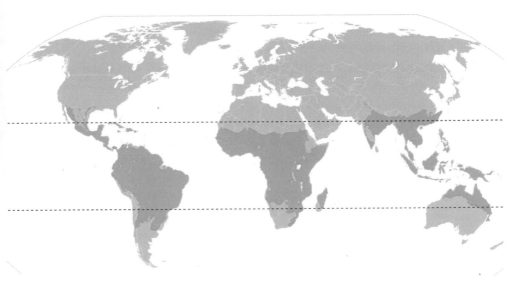

Fig. 1.5 Tropical terrestrial biomes. The tropical terrestrial biome (green) is all tropical mesic ecoregions (see Jos Barlow et al., "The future of hyperdiverse tropical ecosystems," *Nature* 569 [2018]: 517–29; Eric Dinerstein et al., "An ecoregion-based approach to protecting half the terrestrial realm," *Bioscience* 67 [2017]: 534–45). These ecoregions span 82% of the 50 million km² of land between 23.5° N and 23.5° S but extend into the subtropics in some areas, following the frost line shown in fig. 2.1. For plants, the extensions north and south are reduced by killing frosts (C. C.).

A Day in the Tropics

In the tropics along coasts, precipitation ultimately originates from the sea according to a daily pattern (fig. 1.6). During the night, land cools while ocean water retains heat; consequently, columnar and anvil-shaped cumulonimbus clouds arise offshore. In the morning, the land heats; by midmorning, it is warmer than the sea. Humid warm air rises through the heavy cool air above, forming clouds over land. This rising warm air is at first replaced by moist air drawn inland from the relatively cooler sea, causing diurnal onshore breezes. Coastal regions of the intertropical convergence zone therefore lack prevailing winds.

Inland, mornings in the humid tropics are typically cloudless, although fog may lie in valleys or over plains if there was rain the previous day. This fog, usually less than 200 m thick, rises and dissipates by 9:30 a.m. By 10:00 a.m. the first small clouds appear, clustering over high ground. These are at first shallow, but by midday they begin to spread over the plains and to bil-

Fig. 1.6 Photographs of a typical day in the tropics, North Borneo. *Top left*, early afternoon clouds, Matang, Sarawak (H. H.); *top right*, heavy afternoon rains, Gunung Mulu, Northwest Sarawak (P. A.); *bottom left*, evening fog in Maliau Valley, Sabah, after rains (H. H.); and *bottom right*, clouds hugging the Matang Summit, following rain (H. H.).

low upward, signaling the beginning of displacement of the hot humid air of the lowlands by the cooler heavier air above. Humid lowland air, forming the flat cloud base that precedes rain, condenses at the same range of altitudes under the convergence zone throughout the tropics. Over the plains and smaller hills, daily clouds form a level base at ~1,200 m. This base may descend as low as ~800 m near the sea and on coastal mountains, as air rising over warming inland surfaces draws cooler moister breezes off the sea. By early afternoon, major cloud accumulations in the convergence zone have risen to heights exceeding 10,000 m, and precipitation starts, often accompanied by thunder. The rain is heavy but of relatively short duration, and much of it is lost as runoff. In late afternoon it is noticeably cooler. Frequently, clouds dissipate, leaving a sunny late afternoon. Trails of mist rise from the forest canopy when hot moist air, previously trapped, rises into the cooling air above (see fig. 1.1).

The Seasonally Dry Tropics

When sufficient moisture is available for at least eight months (i.e., a minimum monthly precipitation of 100 mm), the classical tropical evergreen rainforests occur. These moisture conditions are most often met in the intertropical convergence zone, where moist rising columns of air produce frequent rains. However, various climatic factors may reduce rainfall in certain months, and forests there have different phenology and structure. This usually occurs in tropical Asia as a single dry season and long rainy season, because tropical Asian lands are aligned to the north, so that only the southern trades bring moisture off the ocean. However, two rainy seasons and two dry seasons do occur in the few places where the sea lies both to their northeast and southeast. They also prevail in much of Africa and the Neotropics, wherever rain is brought by both north and south trades. The prevailing seasonal winds of the Asian tropics are known as *monsoons* (Arabic, *mawim*: a season). The southern monsoons veer southwest because they are drawn north of the equator by the Tibetan Plateau's summer low pressure, becoming affected by the Coriolis force (p. 56). They arrive in May–June and carry heavy precipitation in parts of South Asia, Indo-Burma, southeast China and Japan. The northeast monsoon arrives in October–November. Its winds are dry in most areas, but they are moist when they pass over the ocean before arriving on land. These strong seasonal winds and accompanying rain make the Asian tropics distinctive, because the patterns of precipitation vary greatly along the long east-west extent of the region.

The El Niño Southern Oscillation (ENSO)

The Pacific Ocean to its east lends Asia a greater range of yearly variation in rainfall than other wet tropical regions, excepting Neotropical coasts also facing that ocean. There, a climatic phenomenon, El Niño Southern Oscillation (El Niño, ENSO), influences all of the earth's climate. Regular supra-annual famines caused by the cessation of the Humboldt Current down the Peruvian coast (leading to warming of the sea surface and disappearance of fish stocks) and accompanied by exceptional coastal rain, have been known to local fishermen since ancient times, but it was not until 1893 that a meteorologist in India, Charles Todd, observed that droughts and famines in India coincided with droughts in Australia. Gilbert Walker, then director-general of Indian Meteorological Services, recognized their connection with

Peruvian storms across the Pacific Ocean and coined the phrase "Southern Oscillation." As with cyclonic storms originating under the influence of the trade winds, small differences between surface air and sea temperatures in the windless, equatorial central-east Pacific can cause variation in the evaporative power of the air and the speed of the trade winds. Here, though, there is a coupled atmospheric and oceanic oscillation across the tropical Pacific Ocean, centered south of the equator. In the atmosphere, this is manifested as an east-west seesaw of surface pressure and related patterns of wind, temperature, cloud, and rain. In the ocean below, there is an interdependent east-west flip-flop of the location and depth of warm and cool bodies of water. During El Niño events, unusual high pressure develops in the western Pacific. Trade winds weaken, and moisture from the ocean is diminished, coming solely from daily onshore breezes. Rain moves to the central and eastern Pacific, where the weakened trades eliminate upwelling of the deep, cool, nutrient-rich water of the Humboldt Current (fig. 1.7). During the longer intervening La Niña periods, the easterly trades strengthen to full force, restoring the upwelling of deep water in the eastern Pacific. Pressure rises in the east and declines in the western Pacific, completing the customary high year-round rainfall.

In most years, therefore, the constant flow of the trade winds mediates both the currents of the tropical oceans and the concentration of rainfall over the daily warming land surfaces. During El Niño events, rainfall around the western Pacific slackens, becoming localized and dominated by daily air movement. Regions distant from coastal mountains experience droughts of varying intensity. Brush, coastal peat, logged forests, and parched grasslands become susceptible to fire. Equatorial regions with mountains close to the coast and facing away from the Pacific, such as north-west Borneo and western Sumatra, are least affected. Among these are some of the most floristically rich forest communities and landscapes in the world.

El Niño events may persist for more than a year. The variability of the wet monsoon is also correlated;[2] continental areas experience tropical Asia's longest droughts. El Niño affects the biology of the forest more intensely, and specifically, in tropical Asia than elsewhere.

The Asian Tropics

Of all the regions in the tropical zone, that of Asia is by far the smallest. It is also distinct from the Neotropical and African regions in several ways. The

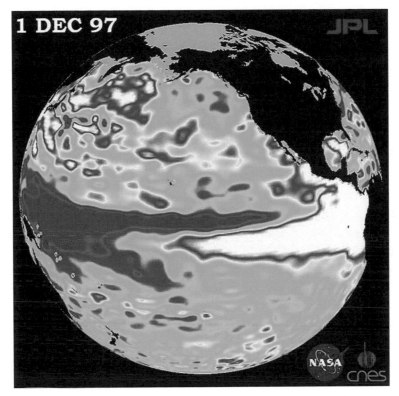

1 DEC 97

Fig. **1.7** An El Niño event in 1997–98 documented by a composite of satellite imagery using false color to show ocean temperatures; elevated temperatures are shown by white and red (eastern Pacific and tropical Asia), and reduced temperatures by blue and purple (Courtesy of NASA).

latter two areas primarily comprise extensive continental land masses and, for the most part, are dominated by the drainage of major rivers, the Amazon and the Congo. The Asian tropics are made up of the two Laurasian southern extensions of the Indian subcontinent and Indo-Burma, plus a multitude of islands, a few large ones that are part of continental blocks (tectonic plates), but also many small, oceanic ones (see fig. 1.5). New Guinea, Borneo, and Sumatra are the second, third, and fifth largest islands on the planet, respectively. These different landforms are the results of a very complex geological history, especially compared to the other tropical regions. They also support a great diversity of climates. The Himalayan range, extending eastward into western China, protects an extensive area from freezing winter weather, and tropical forests extend far northward. No such protection is provided in the

Neotropics and Africa. From east to west, the Asian tropics span a distance of 9,100 km, and the western third extends well north of the Tropic of Cancer. The monsoon winds of the region influence regional climates, which vary greatly in the length of seasonal drought conditions—more so than other tropical regions. The ENSO also modifies weather patterns in parts of the Asian tropics, making rainfall less predictable in some areas.

All these tropical regions support a disproportionate amount of biodiversity for their area, more than half of Earth's total plant diversity in 12% of the land area. Was this diversity produced by higher rates of speciation? Molecular evidence is lacking.[3] For their area, the Asian tropics probably contain a greater diversity of plants than the other tropical regions, with about the same total tree diversity as the Neotropics on a third of the area. Ferry Slik,[4] along with 325 co-authors (including Peter Ashton), estimated the tree species diversity among tropical regions (table 1.2). Their calculations were based on species prevalence in long-term plots (particularly the ForestGEO 25–50 ha plots; see appendix B). Although these plots do not capture all the tree diversity (perhaps even less in the Neotropics), they allow a more objective comparison among the tropical regions—and the Asian tropics contain by far the greatest diversity in terms of area. However, the Asian tropics have been surveyed for diversity for a much longer time, even though ecological research in the Neotropics is generally more advanced.

These three regions have traditionally been seen as phylogenetically distinct, although Africa and the Neotropics are related, and Asia also has affinities to parts of East Africa and Madagascar, though it is otherwise quite distinct.[5] We never use the term *jungle* to describe any of these forests. This term was derived from the Hindi/Sanskrit *jangal*, which actually describes waste or scrub-land, and its use was popularized by Rudyard Kipling in the 19th century.[6]

Table 1.2 Areas and estimates of vascular plant and tree species from the three tropical regions

Tropical Regions	Neotropics	Africa	Asia	All tropics	World
Area (million km²)	9.18	5.27	3.11	17.56	148.33
Vascular plants	98,000	24,000	100,000	213,000	390,000
Trees	24,580	5,984	24,819	53,345	98,000

Source: Rafaël Govaerts, "How many species of seed plants are there?" *Taxon* 50 (2001): 1085–90; Lucas N. Joppa, David L. Roberts, and Stuart L. Pimm, "How many species of flowering plants are there?" *Proceedings of the Royal Society B* 278 (2010): 554–59.

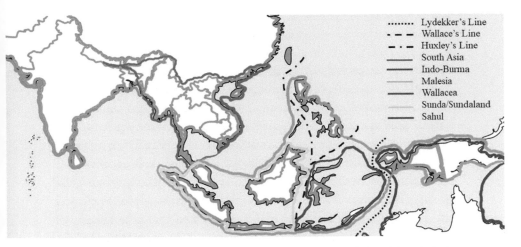

Fig. 1.8 Map of tropical Asia, showing country boundaries and boundaries of the principal political and geographical subregions discussed in this book. Black lines are deep water troughs that have isolated regions from each other (D. L.).

There is a lot of geography in this book; readers can locate unfamiliar mountain ranges and river basins via Google Earth. However, political and geographical boundaries define subregions not used in global maps, and they are used in this book to describe patterns of biotic diversity and the distributions of different forest types. Deep troughs divide regions and prevented land connections during Pleistocene sea-level changes, as shown in figure 1.8, and we refer readers to this map to help in discussions elsewhere in the book.

Tropical Diversity

One of the longest-running and biggest questions in ecology is, Why is there a gradient of increasing species diversity among most groups toward the equator? Alternatively, if we think of high species diversity under the conditions of optimal moisture, temperature, and sunlight, Why does that diversity diminish toward the poles? Robert Whittaker, in an important paper on the issue of species diversity in the vegetation of the Siskiyou Mountains,[7] provided the initial context for the partitioning of diversity that enabled more refined study by other researchers. He considered the total species number for the entire region as its gamma (γ) diversity. He also recognized two other types of diversity in the forest: beta (β) diversity, which is the number of habitat types (as alpine tundra, riparian zones, and coniferous forests); and alpha (α) diversity, which is the number of species in each of those habitats. The gamma diversity will then be the summation of the alpha diversities in all of the habitats, or $\gamma = \Sigma \ \beta \cdot \alpha$.

A variety of techniques are used to assess the diversity of a single habitat. Just the total number of species present is an important estimate: species richness, or S_t. Species diversity indices take into consideration the total number of species and their relative frequencies. A common such index is Fisher's alpha (Fisher's α), which we occasionally use in later chapters.

This book seeks to explain the patterns of species distributions, particularly trees, in the forests of the Asian tropics. It begins with a survey of the forest types of the region (chapter 2), and goes on to examine the geological history of the region (chapter 3), its climatic history (chapter 4), and soil diversity (chapter 5). A focus of the book is to understand how high species diversity has accumulated in the Asian tropics. One explanation for that diversity is the rich assemblage of forest types that have appeared, under the influence of geology and past climate, described in chapters 7–10. Those forests constitute the β diversity described above. The species that make up those forests were derived from ancestors that moved to the region from different regions and at different times, as described in chapters 9 and 13. Some of those trees and plants are truly distinctive for Asian forests, as described in chapter 6. Within those forest types, high species diversities become established under local conditions; this is the α diversity discussed above, which is extensively treated in chapters 9 and 14. The roles of animals in influencing species biology and evolution are described in chapters 11 and 12. The net result is the enormous diversity of tree species collectively growing in Asian forests—the γ diversity.

The Importance of Natural History

One may wonder, in our times of population growth and global climate change (see chapter 18), as every organism is being affected by human activity, if there is anything "natural" and therefore any natural history left. Yet, we have much early data collected when the human impact was still small, and there are still tropical forests of different types remaining, often in very inaccessible places, that are little affected by human activity, past or present. These forests can be studied through natural historic studies, with site visits, lists of organisms, and so forth. Those initial studies give rise to speculation about why the organisms are there and how they interact with other species. This inspires us to formulate hypotheses testable by the techniques of forest ecology: small plots (and then larger ones) censused over long periods, growing plants under environmental control, and measuring their physiological activity with field instruments. The bulk of the information in this book comes from natural historical observations that have led to rigorous research. The revolution in bioinformatics (collecting and analyzing very large data sets) has strongly affected contemporary research, and we now see papers that present data spread among different forest sites from hundreds of authors, reminiscent of papers in experimental physics. This has provoked even more questions, mentioned in the text, that may foster further research.

Today tropical forests are retreating to ever smaller remnants. Policymakers focus more intently on the increasingly difficult challenge of accommodating the needs of still-expanding forest communities and of the inexperienced immigrants to the residual forest lands that they barely understand. Meanwhile, science and the universities are channeling young professionals into increasingly specialized fields of endeavor. Yet we must still first depend on the hypotheses of those natural historians (generalists, synthesizers, and communicators) who seek to comprehend these tropical forests, their plants and animals, and how they interact, survive, grow, and die.

Notes

1. Ashton, Peter, *On the Forests of Tropical Asia*, 29–40; cited hereafter as *OTFTA*. See also Corlett, Richard T., *The Ecology of Tropical East Asia*, 2nd ed. (Oxford, UK: Oxford University Press, 2014); Ashton, P. S., and H. Zhou, "The tropical-subtropical evergreen forest transition in East Asia: An exploration," *Plant Diversity* 42 (2021): 255–80.

2. Gloor, E., et al., "Tropical land carbon cycle responses to 2015/16 El Niño as recorded by atmospheric greenhouse gas and remote sensing data," *Philosophical Transactions of the Royal Society B* 373 (2018): http://dx.doi.org/10.1098/rstb.2017.0302; Malhi, Y., L. Rowland, L. E. O. C. Aragaō, and R. A. Fisher, "New insights into the variability of the tropical land carbon cycle from the El Niño of 2015/2016," *Philosophical Transactions of the Royal Society B* 373 (2018): http://dx.doi.org/10.1098/rstb.2017.0298.

3. Igea, Javier, and Andrew J. Tanentzap, "Angiosperm speciation cools down in the tropics," *Ecology Letters* 23 (2020): 692–700.

4. Slik, J. W. Ferry, et al. (~325 co-authors), "An estimate of the number of tropical tree species," *Proceedings of the National Academy of Sciences U.S.* 112 (2015): 7472–77; Govaerts, Rafaël, "How many species of seed plants are there?" *Taxon* 50 (2001): 1085–90; Joppa, Lucas N., David L. Roberts, and Stuart L. Pimm, "How many species of flowering plants are there?" *Proceedings of the Royal Society B* 278 (2010): 554–59.

5. Slik, J. W. Ferry, Janet Franklin, et al. (~184 co-authors), "Phylogenetic classification of the world's tropical forests," *Proceedings of the National Academy of Sciences U.S.* 115 (2018): 1837–42.

6. Dove, Michael R., "The dialectical history of 'jungle' in Pakistan: An examination of the relationship between nature and culture," *Journal of Anthropological Research* 48 (1992): 231–51.

7. Whittaker, Robert H., "Vegetation of the Siskiyou Mountains, Oregon and California," *Ecological Monographs* 30 (1960): 279–338.

*In November 1958 I took the Golden Blowpipe
Express from Singapore to Bangkok to attend a Pacific
Science Congress. At Tampin, the Main Range rises to
the east, and I was struck by the majesty of the great
trees following the ridges, and by the density of their
huge, domed, grey-green crowns. On the slopes, the forest canopy was different—broken, rich green, the emergent crowns scattered or in clumps on spurs and mostly
of smaller diameter. Much further north, we changed
trains at Kuala Krai, not far from the Thai frontier. As
the sun went down, we began passing through hills on
whose ridges the grey-green crowns were absent, the
emergents more scattered and less clearly raised above
the rest of the canopy, while the slopes appeared more
disturbed. Upon waking the following morning, I was
startled to find myself in a different world. The humid
heat and the cloudy afternoon Malayan sky had been
replaced by a cool, dry, golden haze, the sun already
reflecting off steep craggy hills with short, sparse woodland that was beginning to change color like a New
England autumn: this forest was clearly deciduous. All
was buff, gold, umber, and terracotta.* — P. A.

2

FOREST IN THE LANDSCAPE

Many Asian train routes traverse dramatic changes in climates and forest types, as from everwet to seasonally dry deciduous forest. Describing these forests is an exercise of ecological and economic importance, since important timber species may be absent or present among them, and they require different approaches to their silviculture. Efforts in this direction originated at least a hundred and fifty years ago, with research along the roughly 9,100 km longitudinal range of the Asian tropics. The different languages and colonial histories of the eighteen countries along this distance make assembling the research literature a formidable task.

Asian Tropical Forests

This chapter summarizes the distribution of the major forest types. H. G. Champion presented the first regional classification of forests types for former British India (Pakistan, India, Bangladesh, and Burma, but not Sri Lanka) and Thailand.[1] It is still commonly adopted and works well, especially for India, and such research has continued, particularly with the cartographic work at the French Institute in Pondicherry. Forest types can be described by the climates in which they establish, their structure and function, and the tree species present. Unfortunately, Champion claimed correlations between his main regional forest formations and mean annual rainfall, an assumption that is clearly false. Table 2.1 outlines an updated classification, which roughly correlates with the number of dry months (p. 51).

Champion's major forest classes have been accepted, with names changed

Table 2.1 Diagnostic characteristics of the major forest formations of tropical Asia. For climate and stature, the numbers in parentheses are the extremes (often due to soils and human disturbance) of the main range.

Forest formation	Mixed dipterocarp forests (MDF)	Seasonal evergreen dipterocarp forests	Semi-evergreen forests	Tall deciduous forests	Short deciduous forests	Thorn woodland	Semi-evergreen notophyll forests
Climate: months with < 100 mm rain (chapter 4)	0	2–4.9	(2–) 5.0–6.5 (–8)	5.0–6.5 (–8)	6.6–8.5	More than 8.5	Two monsoons
Stature (chapters 8, 9)	Canopy (20–) 45 (–80) m	Canopy (20–) 35 (–50) m	Canopy 20–35 m	Canopy 20–35 m	Canopy < 20 m	Canopy > 10 m with occasional emergents	Canopy mostly < 20 m, some emergents < 35 m
Structure (chapters 8, 9)	3 strata: emergents above main canopy	3 strata: emergents barely above main canopy	2 strata: main canopy and subcanopy	2 strata: main canopy and subcanopy	2 strata: subcanopy frequently shrubby	1 shrubby canopy stratum	2 strata: main canopy and subcanopy
Foliage phenology (chapters 8, 9)	Canopy and subcanopy overwhelmingly evergreen	Canopy and subcanopy overwhelmingly evergreen; some late successional deciduous	Canopy deciduous and evergreen mixed; subcanopy evergreen except in succession	Canopy overwhelmingly deciduous; subcanopy partly evergreen	All woody plants deciduous	All woody plants deciduous	Canopy mostly evergreen; subcanopy evergreen except in succession
Reproductive phenology (chapters 11, 12)	Supra-annual, high interspecific synchrony	Annual with occasional intense years; low interspecific synchrony	Annual with occasional intense years; no interspecific synchrony	Annual with occasional intense years; no interspecific synchrony	Annual with occasional intense years; no interspecific synchrony	Annual with occasional intense years; no interspecific synchrony	Annual with occasional intense years; no interspecific synchrony

to provide more generalizable descriptions, to encompass the Asian tropics in general.[2] British India had no everwet climates for Champion to study, so the forest types of the everwet Asian tropics are added. Worldwide, most lowland tropical forests occur in climates with rainfall that seasonally is less than the expected rate of evapotranspiration, which we generalize here as 100 mm a month, though this may be an underestimate, because meteorological data rounded to this figure are more easily available. This influences their phenology and floristic composition, strongly differentiating them from tropical forests of everwet climates.

The forest formations (forests of everwet climates, seasonal evergreen forests, semi-evergreen forests, tall and short deciduous forests, thorn woodland) whose distributions are mapped (fig. 2.1), differ at a *regional* scale in structure, dynamics, phenology, and floristics. They are correlated with rainfall seasonality and, nowadays, human influence (e.g., in the current almost complete overlap in the distribution of semi-evergreen and tall deciduous forests; see chapter 8). The subdivisions of these formations differ little in structure, physiognomy, and what relatively little is known of their comparative dynamics (about which relatively little is known); they mainly relate to historical biogeography, hence also to tree *species* composition (see chapter 9).

Mixed dipterocarp forest (MDF) and its associated forest types occur in three isolated blocks: a tiny, ~10,000 km[2] area in lowland southwest Sri Lanka; the vast lowlands of western Malesia—Sundaland and the Philippines—which are the main focus of this book; and in everwet regions of Wallacea and New Guinea (fig. 2.2). These blocks are isolated by forests of seasonal climates in continental Asia, and by the Malesian sea passage of Makassar Straits, recognized by biogeographers as Wallace's Line (pp. 13, 289), which separates Asian from Australasian forests. These forests are distinguished by the dominance of a single tree family, the Dipterocarpaceae (p. 104).

Seasonal evergreen dipterocarp forest and other forests comprise similarly separated blocks in Peninsular India west of the Western Ghats (there is almost none in Sri Lanka), in a discontinuous block from eastern India through Indo-Burma, and again east of Wallace's Line (fig. 2.3). From the eastern Himalayan foothills, the eastern Indian states, and Burma north of ~27° N, to Thailand, Laos, and Vietnam north of ~19° N, a northern seasonal evergreen dipterocarp forest formation, floristically distinct but similar in structure and physiognomy (p. 91) to the southern seasonal evergreen forest to its south, coincides with copious nocturnal winter precipitation as

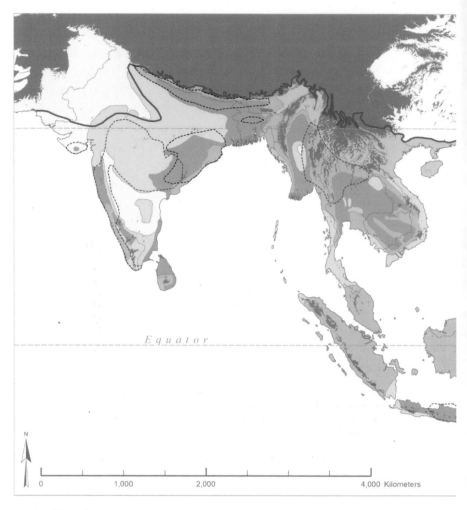

Fig. 2.1 The major zonal forest formations of tropical Asia. Boundaries are approximate, on the basis of herbarium specimens and field observations of Peter Ashton (Jeff Blossom).

dew and, toward the coast of the China sea, wet temperate winter northeast-erlies off the Pacific Ocean.

Semi-evergreen forests and tall deciduous forests currently share simi-lar distributions, owing to degradation of semi-evergreen forests by human-induced fire and cattle browsing (see chapter 8). Semi-evergreen forests prevail on moister valley sites; tall deciduous forests exist on drier, sandy or ridge-top sites (fig. 2.4). Toward the moister end of their ranges, the tall deciduous forest is more restricted. However, toward regions with shorter

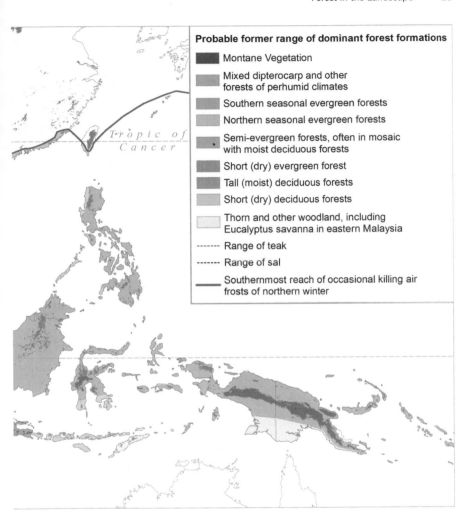

Probable former range of dominant forest formations

- Montane Vegetation
- Mixed dipterocarp and other forests of perhumid climates
- Southern seasonal evergreen forests
- Northern seasonal evergreen forests
- Semi-evergreen forests, often in mosaic with moist deciduous forests
- Short (dry) evergreen forest
- Tall (moist) deciduous forests
- Short (dry) deciduous forests
- Thorn and other woodland, including Eucalyptus savanna in eastern Malaysia
- ------- Range of teak
- ------- Range of sal
- ——— Southernmost reach of occasional killing air frosts of northern winter

monsoons, semi-evergreen forest disappears or is confined to riparian habitats. These formations prevail as a biogeographically uniform southern semi-evergreen forest and tall deciduous forest down the eastern foothills of the Western Ghats, their western foothills north of ~11°, and the northern hills of the Eastern Ghats (fig. 2.5). A floristically distinct further southern block is distributed from Bangladesh and the eastern Indian states eastward, from Burma and Thailand north of the Malay-Thai peninsula to the Indo-Chinese nations. Another floristically distinct northern block ranges from

Fig. 2.2 Tropical everwet forests, MDF. *Top left,* MDF forest profile, track to Labi village, Brunei in 1958. Pale foreground crown *Dipterocarpus crinitus, Shorea curtisii* to left behind it (P. A.); *top center,* detail of crowns on ridge slope, Lambir, Sarawak (© Christian Zeigler); *top right,* canopy density and uniformity of heights, with dipterocarps and young copper foliage of *Koompassia excelsa,* east Sabah (H. H.); *bottom left,* MDF at Sinharaja, Sri Lanka (M. A.); *bottom center,* coastal hill forest, grey crowns of *Shorea curtisii*, Penang, Peninsular Malaysia (J. O.); *bottom right,* trunk of *S. curtisii*, Lambir National Park, Sarawak (S. D.).

the central Himalayan foothills west to Central Nepal and eastward to the Burma-China frontier north of ~27° N, as well as through Thailand, Laos, and Vietnam north of ~25° N to the tropical margin in southernmost Yunnan and Guangxi, where it replaces northern seasonal evergreen forests on drier ridges and slopes, and as a secondary succession following repeated swidden. These forests therefore share the climate and much of the range of the northern seasonal evergreen dipterocarp forest.

The two continental deciduous forests are floristically differentiated west and east of an evergreen forest corridor running down the Indian-Burmese frontier: Chatterjee's Partition (p. 290). The eastern section is

Fig. 2.3 Seasonal evergreen forests. *Top left*, southern seasonal evergreen dipterocarp forest, Machinchang, Langkawi, Malaysia (S. L.); *top right*, *Calophyllum polyanthum*, Khao Ban Thad Wildlife Sanctuary, Thailand (H. H.); *bottom left*, northern seasonal evergreen dipterocarp forest, Mengla, Xishuangbanna, China (D. L.); *bottom center*, trunk of *Terminalia myriocarpa*, Mengla, (D. L.); *bottom right*, *T. myriocarpa* in flower, Phuntsholing, Bhutan (H. H.).

much richer in species. Tall deciduous forest includes several genera that are rare or absent in short deciduous forest. Short deciduous forest dominates the northern Deccan Plateau of Peninsular India, extending far south into Tamil Nadu; its climate is that of the Gangetic Plain south into Bangladesh, now cultivated but for the sands and gravels of the terai terraces bordering the Himalayan foothills (fig. 2.5). Whereas tall deciduous forests are among the most pervasive in Indo-Burma, short deciduous forests are confined there to the upper plains of the Irrawaddy immediately south of Mandalay, and probably northeast Thailand and adjacent Laos, where they seem to have been entirely degraded. This formation shares only magnifi-

Fig. 2.4 Semi-evergreen forests. *Top left*, semi-evergreen dipterocarp forest, *Dipterocarpus turbinatus* with deciduous associates, Chiangdao, northwest Thailand; *top right, Ficus strictus*, Huai Kha Khaeng Wildlife Sanctuary, Thailand; *bottom left*, semi-evergreen dipterocarp forest, Khorat Plateau, Thailand, pale crowns: *Shorea henryana*; dark crowns: *Hopea ferrea*; *bottom right, Dipterocarpus turbinatus* with *Dendrocalamus strictus* understory, Doi Inthanon, northern Thailand (P. A.; the rest are H. H.).

cent, yellow-flowered *Cochlospermum religiosum* as its characteristic species, on either side of the transition.

Thorn woodlands are absent in Indo-Burma but for a limited area in the northern Irrawaddy plain (fig. 2.6). There are two areas in India, a northwestern block, abutting the Thar Desert, and the tropical climatic frontier

Fig. 2.5 Deciduous forests. *Top left*, short deciduous forest, Bandhipur National Park, India; *top center*, tall deciduous sal forest with evergreen understory, Simlipal National Park, India (sal is 30 m tall); *middle left*, Tall deciduous sal forest, with evergreen understory, Buxa National Park, West Bengal; *right*, deciduous dipterocarp forest, *Shorea siamensis* in background, riparian semi-evergreen forest in foreground, Northwest Thailand (P. A., all others H. H.); *bottom left* and *center*, short deciduous teak forest, March and November, Mudumalai Wildlife Sanctuary, India.

immediately to its northwest. Beyond, temperate thorn woodland extends into the plains of Pakistan and Mediterranean woodlands; there is also a southern peninsular block.

Semi-evergreen notophyll forests (p. 150) are now confined to the southeastern coastal hills and raised beaches of Peninsular India, and to adjacent northern and eastern Sri Lanka, where they remain substantially intact or have long regenerated, following abandonment of the massive irrigation program in the 16th century.

Besides the regional formations summarized here, there are subsidiary forest types specific to more limited soils or geology, notably *kerangas*, swamp forests, and forests on limestone and ophiolites. These represent variants of their prevailing regional forest formations in their structure and dynamics, and they draw their flora from the climate zone and region in which each occurs (figs. 2.7 and 2.8, discussed in chapter 9). In coastal re-

Fig. 2.6 Thorn woodlands and savanna. *Top left*, thorn woodland, Moyar Gorge, Karnataka, India, late March (H. H.); *top right*, thorn woodland, Tamil Nadu (P. A.); *bottom left*, semi-evergreen notophyll forest, Giritale, Sri Lanka (H. H.); *bottom right*, savanna, Mudumalai Wildlife Sanctuary, India (H. H.).

gions subject to cyclonic storms, as in the Philippines and Taiwan, *typhoon forests* occur, their canopies shaped by winds.

Floristic communities *within* the several forest formations vary according to nutrients and soil moisture (see chapter 9).

Landscapes

Variations in climate, topography, and geology limit the distributions of the forest types described in this chapter. Often, such variations produce landscapes with forest types adjacent to each other, adding to the biodiversity, natural history, and beauty of small regions. In northern Borneo, under an everwet climate, a fascinating mosaic of forest types has become established

Fig. 2.7 Other tropical everwet forests. *Top left*, riparian forest along an alluvium bank, Lower Kinabatangan, Sabah (P. A.); *top center*, flood limit and river at Khao Ban Thad, Thailand (H. H.); *top right*, canopy of typhoon-stunted lowland forest, Nanjeng Shan, Taiwan (S. D.); *bottom left*, kerangas, Bako National Park, Sarawak (P. A.); *left center*, wax-coated young leaves of *Cotylelobium burckii*, kerangas at Bako (P. A.); *right center*, Belait-Baram peat swamp forest, Sarawak (A. C.); *bottom right*, mature peat swamp forest profile, Seria, Brunei (P. A.).

Fig. 2.8 Tropical everwet forests on ultramafic and limestone. *Top left*, ultramafic main range on Palawan (P. A.); *top center*, 30 m karst needles, Mulu National Park, Sarawak (© Tim Laman); *top right*, high karst and forest, Mulu (H. H.); *bottom*, ultramafic soil profile, Palawan (P. A).

(see fig. 9.1). In Asia, each of these forest types presents different structures that play with light and shade and offer unique aesthetic experiences. This is most striking in the occasionally majestic stands in mixed dipterocarp forests (MDF; p. 146), but occurs in other types too, including riverine forests.

In season, the whitewater rivers of the Sunda uplands were glorious, with the orange or maroon-red flowers of *Saraca*: each species had its riparian habitat, with the orange-red young leaves of *Syzygium* and the coppery-winged fruit of *Dipterocarpus oblongifolius*, kesugoi, ensurai, and neram brushing overhead as one toiled upstream. In Borneo and the Philippines you find the tall, pale-boled, *Swintonia acuta* crowned high above with pale blue-green. In Brunei, and a few other places, you can still experience them, cooled by the daily downdraft off the mountains.

Notes

1. Champion, H. G., "A preliminary survey of the forest types of India," *Indian Forest Research*, new series, *Silviculture* 1 (1936); Champion, H. G., and S. K. Seth, *A Revised Survey of the Forest Types of India* (Nasik: Government of India Press, 1968).

2. *OTFTA*, chap. 2, 95–153; chap. 3, 183–239.

A summer field experience with my Cambridge professors piqued my interest in the importance of geology for appreciating plants within communities and landscapes, but it was not until I started to explore the forested landscape of Brunei that I grasped its deep significance for forest ecology and historical geography. The resident geologist of the North Borneo Geological Survey there was G. E. "Wilf" Wilford, a superb field investigator whom I accompanied on expeditions. We visited the Belait River headwaters, a complex mosaic of swamps, Pleistocene raised beaches, and the narrow Miocene sandstone rim of the Belait syncline that follows the frontier to the south. What better place to learn, in long evening discussions, of the ceaseless rise, fall, and folding of the mountainous regions and the erosional plains of that northwestern Borneo landscape. And what a contrast with those ancient, rounded Malayan hills, which have been above the ocean now for 150 million years! These histories and landscapes must surely have influenced the forest's story. — P. A.

3

GEOLOGY

The diverse forests of tropical Asia cover a tumultuous geologic history.[1] Charles Hutchison described these origins as "composite," as a sometimes contentious dialogue between ancient history and geology.[2] The contemporary product of this slow but relentless and ongoing dialogue is a uniquely fine-grained mosaic of topographies, soils, and habitats. Though the contemporary distribution of organisms, including plants, can mostly be explained by their places of origin and past opportunities for migration, the distribution of plants and forest types is also determined by the distribution of habitats—that is, soils and climate—to which the organisms are adapted. Geological history has determined the way in which landscapes have been created. The origin, history, and composition of rocks determine their rates and processes of weathering and porosity, which generate the diversity and patterns of the soils. These combine with the processes of uplift, buckling, and erosion to create topography, which influences climate. Each of the tropical land masses is unique in its geological history and climate, and consequently in its landscapes and the patterns of the forest cover, as well as in the geography of its fauna and flora. However, tropical Asia's history is unusually complex.

A Geological History of Tropical Asia

History and plate tectonics. A dramatic shift in our understanding of earth history was first suggested about a century ago by Alfred Wegener, who claimed that the continents formed granitic plates with stable coastlines, which had shifted their positions over time in expanding oceans. During the past sixty years, his idea has been buttressed by our understanding of convective currents in the earth's magma, or molten core, and the emergence, cooling, separation, and migration as undersea rock strata astride mid-ocean ridges, which form the basaltic oceanic plates. Evidence for plate tectonics has developed from fossils, from similar geology (rocks and strata) observed in lands distant from each other, from the dating of rocks, and from paleomagnetic signals in volcanic rocks. Continental plates have migrated and often collided since their formation some 4.6 billion years ago, but we only have a detailed knowledge of their movements during the past 300 million years. At that time, the present continents were joined in a vast, single continent, Pangaea, which sat astride the equator. This supercontinent had been formed from other continents that had collided after an earlier migration cycle. The fragmentation of Pangaea and the subsequent migration of the present continents are important for our understanding of the geography and diversification of the flowering plants, which originated 150 Ma.

Around 275 Ma, east-west trending Pangaea began to split into two east-west continental blocks ranging north and south of the equator (fig. 3.1). These blocks remained joined by the Americas in the west, so that as the split between them gradually deepened it created first a gulf, then a great sea, the Tethys. The result was a continent of two mighty lobes, Laurasia to the north and Gondwana to the south, joined in the west by a unitary America. Evolution followed separate lines on each lobe. That southern lobe was named after the Gonds, the aboriginal and still surviving Dravidian people of Central India, when ancient Indian plant fossils were unexpectedly discovered in Antarctic rocks.

Almost from the beginning, coastal pieces known as *terranes* broke from Gondwana and drifted northward. The first chunk separated from the coast of what is now northern Australia during the early Carboniferous period (~350 Ma), when it was positioned at 25° S. That split started before the major split of Pangaea had even begun. This terrane, followed by others, drifted north as the Tethys opened, colliding and then fusing with the southeastern coast of the northern (Laurasian) lobe more than 200 Ma. These terranes parted from Gondwana far too early to have carried flowering plants with them. Together, these chunks of the ancient Gondwanan land mass currently constitute the Tibetan Plateau, southern China, Burma, Indo-China, the mountains running south from Yunnan along the Burmese-Thai frontier south through Peninsular Malaysia into East Sumatra, and a triangular block

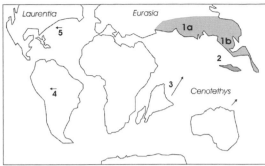

Fig. 3.1 Snapshots of plate tectonics in tropical Asia. (We emphasize geological time in millions of years ago [Ma] and occasionally mention a major period over a long length of time. The approximation of dates is indicated by a tilde [~]. The geological time scale is diagrammed in appendix A.) *Top left*, break-up of Pangaea 200 Ma. 1–3, terranes moving from NW Australian Gondwanaland toward South Laurasia; 4, Laurasia-Gondwana split continues; 5, Gondwanaland beginning to split. *Top right*, Break-up of Gondwanaland 200 Ma. 1, Gondwanaland terranes form high Asia (a) and form China and Sundaland Shelf (b); 2, northwest drift of younger terranes; 3, India and Madagascar move from SE Africa and split apart; 4, South America splits from Africa; and 5, North America splits from Eurasia. *Bottom left*, migration of Greater India. 1, 150 Ma, NE movement from Africa; 2, 90 Ma, Madagascar left behind, and India drifts rapidly NE; 3, 55 Ma, Indian plate collides with Eurasia, showing modern exposure of plate (a) and buried portion of plate; 4, Gondwanan and Asian plates collide, rocks metamorphose, and the Himalayan Range is formed; 5, eastward pressure into Sundaland; and 6, westward pressure into Baluchistan (all, I. B.).

with one side running down the western coast of Borneo and the others tapering to Borneo's eastern coast (fig. 3.1). Ophiolite extrusions (mantle and magma) mark the current positions of the sutures between them. The Southeast Asian continental blocks moved southward 5–7° in the late Cretaceous period to reach their present latitudes. They merged to form the currently partially submerged Sunda continental tectonic mass, which includes Peninsular Malaysia, the islands of Borneo, Java, and Sumatra, and the seas and islands between (Sundaland).

The first major rift of the southern (Gondwanan) lobe divided South Asia, Madagascar and Australia from Africa and South America, though Australia did not separate from Gondwana at this stage. The South Asian subcontinent (Pakistan, India, and Sri Lanka) first separated from the northern coast of Gondwana, where it had been sandwiched between what is now the northwestern Australian coast and southeastern Africa. It then broke off from Australia some time during the Jurassic period (160–200 Ma)—that is, before the origin of flowering plants—and then moved northward across the Tethys Sea along a course parallel to the coast of Africa but at an uncertain distance from it. By 100 Ma (fig. 3.1), it was remote from Australia. The Seychelles, a continental fragment, split off ~60 Ma and became increasingly isolated.

By ~66 Ma, drifting continents began to collide once again. Oceanic tectonic plates consist of basaltic rocks, which are heavier than the granitic cores of continental plates. When an oceanic plate collides with a continental plate, it is subducted beneath the continental coast, thereby raising mountains. As the huge weight of continental rocks thrusts the edge of the ocean plate down beneath Earth's mantle, its basalt melts back into magma. The resulting pressure is periodically released by volcanoes. The surfaces of the mountains thus formed are heavily influenced by volcanic lava and ash. Sometimes even the mantle, which is the layer of hot and semi-plastic but recognizably crystalline rock between the crust and the liquid magmatic core, is extruded. The resulting ophiolitic, ultramafic rocks often yield soils containing potentially toxic concentrations of nickel, cadmium, cobalt, chromium and copper, to which plants have adapted as characteristic floras rich in endemics.

Paleogeographers call the Indian subcontinent "Noah's Ark" because it carried so many taxa from the southern to the northern hemisphere (fig. 3.1). The subcontinent's submerged continental shelf made first contact with the southern margin of Laurasia 45 Ma, but land contact came 5 Ma later. The collision is now thought to have occurred (with continuing debate on the details) at the present latitude of the southern Laurasian coast, the Himalayan foothills, at 28–35° N and within the southern edge of the temperate Hadley cell, in a seasonally wet climate. Whereas Laurasia stood firm, the Indian subcontinent continued pushing northeast from the spread of the Indian sea-floor plate to its south, at a steady northward velocity of 7.5 cm/yr, causing earthquakes from Afghanistan and Nepal to southern China, and scattering island arcs as it collided to the east in the young Bay of Bengal.

Following the early stages of the Indian collision (~45 Ma), the trajectory of the Indian plate shifted northeast, starting ~35 Ma, and continues now. The Indian Gondwana plate squeezed the sediments which had been accumulating along the shelves of the northern and southern continental blocks into huge parallel folds–mountain ranges, now the Himalayas—curving southeastward into Myanmar, the Andaman Islands, and the Thai-Burmese frontier ranges, and initiated the other north/south ranges of Indo-Burma to their east. The heavy basaltic ocean plate to the east of the Indian subcontinent in the Bay of Bengal subducted obliquely north-northeastward, together with the Australasian oceanic plate to its immediate south, producing earthquake-induced tsunamis and raising the still-active "ring of fire" volcanoes of Sumatra and Java. Subsequently, the more northerly trajectory of the Australasian plate caused the anti-clockwise rotation of Java and the Lesser Sundas relative to Sumatra, around the Krakatau axis, thereby closing the Indonesian throughflow between the Indian and the Pacific oceans and initiating the East Asian monsoon (chapter 4). Sumatra has rotated clockwise less than 30° to its present position.

The Indian collision spun the bulk of northern Indo-Burma clockwise ~20º. Even before, Indo-Burma was massively faulted, creating shear-zones along the Red River and Thai-Burma Faults. The regional pattern of earth quakes and volcanoes indicates that pressure between some of the Southeast Asian plates continued (fig. 3.2), although the main movement ceased during the past ~20 Ma. Deep north/south and northwest/southeast rifts also developed across Sundaland, and this region also developed a pronounced topography of parallel ridges and valleys resulting from squeezing along fault lines. Many of the valleys formed deep freshwater lakes, which then filled in with organic-rich muds. Following subsequent burial, these eventually produced most of the region's hydrocarbons. The easternmost of these rifts resulted in the separation of southwest Sulawesi from the main part of Borneo and the formation of the Makassar Straits, creating part of the biogeographical barrier of Wallace's Line (p. 289). Later, in the early Oligocene epoch (~32 Ma), a deep north/south rift opened up in the Indo-China block, creating the South China Sea and subduction zones to the east, west, and south. A small oceanic plate, the Lucania Platform, now underlies the Neogene sediments south of northwest Borneo.

The Philippines, an arc of volcanic islands, were pushed northward by the spread of the Pacific and Philippine Sea oceanic plates arriving from the southeast during the Cenozoic era. They have become a zone of folded mountains, squeezed by subduction on both east and western flanks (fig. 3.3). By 15 Ma, Mindoro and Palawan approached Sundaland, the opening of the South China Sea was arrested, and these areas were joined with Borneo ~10 Ma. Only the Zamboanga region of southwest Mindanao has an

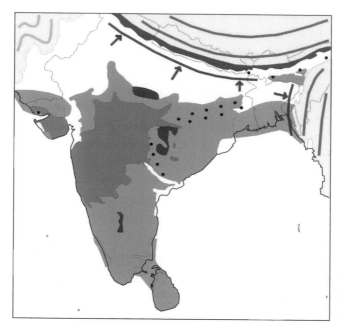

Fig. 3.2 The surface geology and tectonic features of South Asia (I. B.).

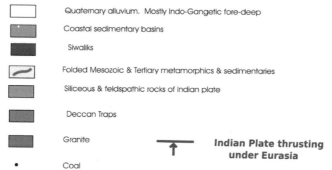

Quaternary alluvium. Mostly Indo-Gangetic fore-deep

Coastal sedimentary basins

Siwaliks

Folded Mesozoic & Tertiary metamorphics & sedimentaries

Siliceous & feldspathic rocks of Indian plate

Deccan Traps

Granite

Coal

Indian Plate thrusting under Eurasia

Australasian continental origin. It remained isolated all through the early Tertiary, not joining the archipelago until the middle or late Miocene epoch. Taiwan, on the northern side of the opening South China Sea, remained firmly attached to the Asian continental shelf and thus apart from the Philippine archipelago. The continuing northward shift of the Australian plate affected the Philippines 3–5 Ma, shouldering Luzon out to the north and west, and rotating the Philippine archipelago clockwise. The Philippines and Taiwan are thus closer now than at any time during the Cenozoic.

The volcanic rocks in northwestern Borneo mark the impact of the last

Current tectonic struture
of SE Asia

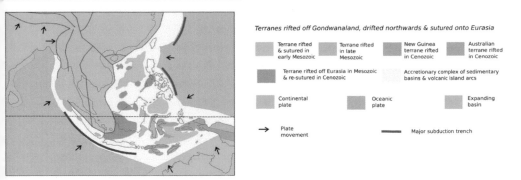

Terranes rifted off Gondwanaland, drifted northwards & sutured onto Eurasia

Terrane rifted & sutured in early Mesozoic	Terrane rifted in late Mesozoic	New Guinea terrane rifted in Cenozoic	Australian terrane rifted in Cenozoic

Terrane rifted off Eurasia in Mesozoic & re-sutured in Cenozoic   Accretionary complex of sedimentary basins & volcanic island arcs

Continental plate   Oceanic plate   Expanding basin

→ Plate movement   —— Major subduction trench

Fig. 3.3 The surface geology and tectonic features of tropical Asia (I. B.).

major southward thrust of the South China Sea plate. Subduction of the South China Sea plate beneath the southwestern Philippines and Borneo-Palawan has also ceased, but subduction of the Pacific plate beneath the eastern Philippine coast continues. Similarly, both sides of the southern Philippines and Wallacea are being pushed downward, as Australia continues to raft northward. The South China Sea rift was quiet in middle Miocene times.

The Australian continental plate separated from the Gondwanan mother-continent ~75 Ma. Australia has since drifted northward throughout the Cenozoic era. It pulled a jumble of continental fragments with it; these, along with a series of volcanic arcs, now comprise the eastern Indonesian islands, grouped by biogeographers as Wallacea (fig. 3.3; p. 289). The Australasian-derived Gondwana fragments meet the Sibumasu Southeast Asian coast at what is now Central Sulawesi. Thus, east and southeast Sulawesi originated from the Australian Gondwanan continental plate, but southwest Sulawesi shares the geology of southwest Borneo, having separated from it following the formation of the Makassar Straits in the middle Eocene epoch.

Thus, the Cretaceous period witnessed the comprehensive fragmentation of the southern Gondwana supercontinent. The Gondwanan core became isolated as the Antarctic. Africa and South America are fragments of Gondwana, which at first drifted north across the equator together as western Gondwana. Flowering plants probably started to diversify in western Gondwana, and—in contrast with the Asian tropics—must have been well advanced on these two continents before they eventually split apart and became distantly separated.

Therefore, although tropical Asia now lies almost entirely north of the equator, none of its land mass is geologically part of the original, late Jurassic northern continent of Laurasia, although its southern regions had started well within tropical climates and latitudes.

In contrast to the African and the American tropics, the Asian and Australasian tropics comprise an amalgam of many separate tectonic components. They collided at different times, both before and during the age of mammals and flowering plants. Then they may have transported species of one continent across the seas to another. Throughout much of the Cenozoic era–and into the present—these blocks have continued their collisions, forming new amalgams and raising great mountain ranges along their margins, while buckling parts of their hinterlands beneath shallow seas.

Regional Geology

The Gondwanan blocks that comprise the Asian tropics have stable cores of ancient rocks and surfaces with buckled terrain along those edges where collisions continue. These stable cores, although originally continental granite, have been variously melted, eroded into sediments, and crushed. In this way their rocks, albeit predominantly siliceous like granite, have been transformed into metamorphic forms or eroded and redeposited into sediments, which may themselves be metamorphosed. Many of these changes took place well before the origins of our contemporary flora, but they greatly influenced present plant distributions through soil formation.

South Asia. The Indian subcontinent—namely, all of South Asia south of the Himalayan summits—is by far the largest single Gondwanan continental block in Asia, and it has had by far the greatest climatic and plant-geographical influence on contemporary conditions. It comprises the largest expanse in tropical Asia on Precambrian and Cambrian siliceous, tectonic continental rocks and associated sediments, much of which has been metamorphosed (fig. 3.4). India's transit north from Gondwana produced a massive eruption of lava, evident now in one of the greatest extrusions of basaltic lava the world has known. This eruption occurred 65 Ma, coinciding with the extinction of the dinosaurs, well before the collision with Laurasia. These flows formed the Deccan Traps: horizontally bedded tablelands in and around the northern Western Ghats (see fig. 3.2).

The Himalayas, which began rising ~23 Ma, attained their present heights as late as 12 Ma (fig. 3.4). The subarctic Tibetan Plateau probably rose above 3,000 m following the Indian collision, only later to bulge and

Fig. 3.4 Tropical Asian landscapes, showing influences of geological history. *Top left*, Miocene sandstone overlaying shale at Bukit Lambir, Sarawak (H. H.); *top right*, metamorphic khondalite rocks, northeast Sri Lanka (H. H.); *middle right*, base of the Himalaya, Buxa National Park, West Bengal; *bottom left*, limestone karst, Kedah, Peninsular Malaysia (P. W.); and *bottom right*, Dhauladar range rising abruptly above the Kangra Valley in northwest India, a vertical rise from 700 to 5600 m (D. L.).

fold behind the Himalaya to form a so-called third pole, producing the Indian, or southwest, monsoon. These heavy rains promoted the accumulation of sediments filling the bed of the former Tethys Sea at the southern foot of the rising mountains.

The rise of the Himalayas may have contributed to the compensatory upwarping of the Indian peninsular mountains. These ranges are highest toward the south in Sri Lanka, which is part of the Indian continental shield, and in the west, such that much of the peninsula is an eastward-shelving and eroding tableland at ~900 m elevation. The narrow western coastal plain abuts a scarp, the Western Ghats, 1,600 km long and averaging more than 1,000 m elevation, with some peaks exceeding 2,500 m.

The obduction (emplacement of oceanic strata at the continental margin) of Indian Gondwana ("Greater India") started before the first surface-level contact between the southern and northern continents. A wide, fragmented belt of lignite fields from Gujarat east to Orissa was laid down in shallow vegetated lagoons around the northern Gondwanan continental shelf, which may have been dotted by a string of islands upthrust by the obduction. These swampy lagoons, formed at a time when their region was at equatorial latitudes in an everwet climate, preserved the pollen, resin, and other botanical evidence of the flora that was originally derived from equatorial east Africa when the northernmost Indian plate itself was equatorial. This rain-forest belt has continued to shift southward and survives now in southwest Sri Lanka. The alluvial sediments of the Indus-Ganges plains are a vestige of the Tethys Sea, persisting as an effective ecological barrier to southward migration of the wet mountain floras of South Asia and the Himalayas.

The Gondwana-Laurasia suture in the eastern Himalayas follows the obduction and subduction line, running southward through Nagaland and turning sharply southwest beneath the Indo-Burmese Patkai frontier range. It then continues south, following the frontier ranges, before descending beneath the Andaman Sea, which only opened 10 Ma. It finally continues along the deep to the west of the Sumatran outer islands and to the south of the Lesser Sundas before terminating south of Timor at the Australian continental shelf.

Indo-Burma. Pressure between the East Asian and Indian plates also produced subductions and obductions between the East Asian Gondwanan block fragments (fig. 3.2). This is marked by lines of isolated extinct volcanoes behind the contact-zone mountains on the overthrust side. Where intercontinental block movements are still active, they produce volcanoes and earthquakes along slip lines. Such lines of volcanoes occur east of the contact line between the Burmese and Indo-Burmalayan plates, rising between the Chindwin and Irawaddy rivers, where they culminate in the sacred Popa mountain (1,518 m). Other extinct volcanoes occur west of the contact line between Indo-Burma and the Malay Peninsula and Indonesia, as well as along the Red River valley.

Today, the Burma continental block receives the full impact of India's continued northeastward thrust. The obduction zone, where the continental plate of thrusting India continues to collide with that of Indo-Burma, is to the west of the Indo-Burmese frontier ranges and just east of the Ganges delta, thereby lifting the suture zone and exposing its ophiolite in places along the crest of the range. Deep water sediments—including some limestone— are exposed at higher elevations. In the western foothills, a broad band of Miocene sediments and some limestone karst formed in a shallow, saline environment. The main valley of the Irrawaddy and Chindwin rivers is a Cenozoic basin representing, to the west of the rivers, the northern part of the inter-arc trough, which descends southward into the Andaman Sea and between Sumatra and its western outer islands. The basin is filled with Miocene to Quaternary sediments. Whereas sandstone predominates along much of the western slopes of these Indo-Burmese ranges, clays dominate the eastern slopes.

To the east, from Yunnan south into Peninsular Thailand, is the principal north/south mountain system of Indo-Burma, the Sino-Burman Range, which divides Burma from Thailand. Parallel to it in the south, but 50 km to the east, the Peninsular Malaysian main range begins. These ranges were uplifted in the Permian period by the collision of the East Asian blocks with Laurasia. They may have remained above sea level and within the humid tropics at least since the genesis of flowering plants,[3] at which time the earliest sediments overlying the southern peninsular hills were deposited.

These ranges consist of Paleozoic-Mesozoic basement rocks, including the main granite exposures in Indo-Burma, flanked by extensive Paleozoic-Triassic sediments in which clay minerals and extensive Permian-Triassic limestone predominate. Although continental deposition occurred toward the north, the extensive limestone implies shallow and sediment-free waters (in a dry climate). Mudstones and shales, with associated karst, prevail in the mountains of northeast Burma, northern Thailand, Laos, northern Vietnam, and northward into tropical South China flanking the Red River shear zone, where karst mountains (below 2,000 m) extend far into southeastern China.

The Annamite range south to Dalat is Mesozoic granite, uplifted by westward subduction of the South China Sea plate. The range rises mostly to 700–1,200 m, with some peaks reaching 2,700 m, and it extends into

Laos and northeastern Cambodia. Extensive late Jurassic to early Cretaceous coarse sediments resulting from massive erosion are also found west
of the range, extending southeast to the Khorat Plateau. A sandstone range,
exceeding 1,000 m in the Cardamom Mountains along the western Cambodian coast, follows the southwestern scarp of the Khorat Plateau (~500 m),
forming the southwest flank of the Mekong valley.

The mountains and valleys of the southern part of the Sibamasu Terrane
(Burma, West Thailand, the Malay Peninsula, and Sumatra) could have been
corridors for migration of both lowland and montane humid tropical floras,
whereas its drier intervening plains, continuously isolated from one another
physically, and intermittently by climate, were barriers. The hill ranges have
similarly been barriers to dry-climate floristic elements.

Sumatra and adjacent islands. The west coast of Sumatra was uplifted
by the subduction of the Indian Ocean plate (see fig. 3.3). This produced a
trough off the west coast and a high volcanic mountain range, the Barisan,
which has been rising since the early Miocene epoch. It is predominantly
andesitic—neutral, though often rich in calcium. The Barisan is flanked by
Permian, fine sedimentary rocks, some converted into metamorphic phyl-
lite. Massive karst limestone is locally present, with some granites near the
range. Adjacent islands, the Riau, Banka and Belitung, are also composed of
granites and Triassic sediments similar to Peninsular Malaysia. To the east
are sediments accumulated in three basins, which also formed petroleum
deposits.

Peninsular Malaysia. This peninsula is an ancient land mass of exceptional stability, which in the north may have started to rise above the sea surface during the Jurassic period, and which has remained within 9° latitude of
its present position ever since. Although composed of granite from the main
range westward and of Triassic sediments to the east, its ancient, rounded
physiography gives the peninsula great uniformity, with isolated outcrops
of limestone (fig. 3.4). The east coast has recent sandy terraces deposited
during the highest sea levels.

Borneo. This is a large and geologically very diverse island. Its northern mountains are the southeastern prow of the Asian tectonic continent,
formed 12 to 60 Ma, again uplifted by the South China Sea oceanic plate and
the westward thrusts of the Philippine island arc and the Australian plate.
Basement granite from the Sibumasu terrane (p. 39) is exposed mostly in
the central ranges (reaching 2,278 m at Gunung Raya) and in mountains
running west to the coast.

Borneo had a seasonal tropical climate after the northern tropical temperatures in the Oligocene epoch. That everwet equatorial climate probably was limited to a lowland plain along the western coast from southern Burma to Java. However, by 20 Ma the earlier massifs of the Sunda lands had eroded into extensive featureless plains that, although buckled by further uplift since the mid-Miocene epoch, became the cradle, with a mostly everwet climate, of the present lowland, dipterocarp rainforest. At the heart of equatorial Sundaland, Borneo contains a continental core to the west with a series of sedimentary basins along its northwestern and eastern coasts, as well as continental basins in the center and south (figs. 3.3 and 3.4). Northwest Borneo is a massive sedimentary basin with scattered volcanic acidic and basic outcrops, initiated in the early Cenozoic as the South China Sea opened. This complex geological history has produced a diversity of sedimentary rocks, from very coarse (sandstones) to fine texture (shales), and from soft to very hard.

In the northeast there was earlier southward movement of the South China Sea bed and, later, collision with the Australasian continental outliers of northern Sulawesi and the eastward-trending Philippine island arc. Since the lower Miocene, these have produced complex ophiolitic (ultramafic) mantle exposures. These are associated with the 5 Ma late emergence of the granite summit zone of Kinabalu (4,094 m). They are also exposed at lower elevations extending through Palawan. Kinabalu is the tallest mountain between the Himalaya and New Guinea, the next tallest being the recent volcanoes Kerinci (Sumatra) at 3,805 m and Rinjani (Lombok) at 3,776 m. Borneo is thus distinguished by an extraordinary diversity of rock substrates, including, in the northwest, the most extensive exposure of sandstones and other siliceous substrates in Sundaland.

The Philippine archipelago. Palawan is the westernmost region in the archipelago. It alone rests on the Sunda Shelf, composed of limestone in the north and south, sediments in the north, but otherwise largely ultramafic ophiolite. The Philippines are otherwise a volcanic island arc: isolated mountainous islands dominated throughout by basic volcanic sediments. Limestone karst exists locally in Luzon and elsewhere, while ophiolite is found especially in lands immediately to the west of the major subduction zone, which borders the islands' eastern edge, and has created the Sierra Madre ranges.

Java, Bali, and the eastern archipelagos. Although Java and Bali are on the Sunda continental shelf and are of Quaternary origin, they share a widespread volcanic and mostly andesitic surface lithology with the Lesser Sunda Islands to their east; ash has universally influenced their soils. East of Borneo and Bali, from Sulawesi to New Guinea and from Lombok to Timor, are archipelagos of islands with different histories, cast up by the relentless northward transit of Australia (fig. 3.3).

Wallacea and New Guinea. Continental plates have remarkably sharp edges; their submarine shelves are generally narrow, dropping to the surrounding oceanic plates deep below. Along the western to southern coasts from the Andaman Islands to Timor, and down the east coast of the Philippine Islands, the edge of the shelf drops sharply into deep ocean trenches. A deep trough follows the strait between Borneo and Sulawesi. The rifting of these Makassar Straits began in the middle Eocene epoch. Extending south and through the Sunda Islands between Bali and Lombok, it is known to biogeographers as Wallace's Line, after the naturalist Alfred Russel Wallace, who first documented the extraordinary change in animal species across it (pp. 13, 289); the vegetation also changes.

Sulawesi is partly volcanic or ophiolitic. It includes a Sibumasu fragment, its southwestern arm. Important for plant geography, that fragment has long been traversed by a section of the Makassar Strait. Makassar is the most ruggedly montane of the islands between Borneo and New Guinea. East of Sulawesi, clay-rich sediments and karst limestone are associated with widespread volcanic rocks, including ultramafics. Continental Australasian Gondwanic rocks are exposed in Sumbawa and Sumba in the Lesser Sundas, and in West Timor. Ophiolitic rocks, implying the presence of deep-sea fragments, occur in Buton and Obi, while Buru-Seram-Ambon are considered a further micro-continent detached from Australia. However, most of the Wallacean islands are basic volcanic. The sandstone and granite ridges common in much of tropical Asia are rare or absent in Wallacea.

New Guinea is the only eastern Malesian island large enough to support a rich, indigenous tree flora. At least in its eastern part it seems to have arisen as late as the early Miocene epoch, through collision of the northward-thrusting Australian continent and the Pacific oceanic floor. The northern part has been much folded, together with metamorphosed volcanic intrusions. It remains volcanically active but was earlier a major depositional basin. Western New Guinea arose later, beginning in middle Miocene times; but the New Guinea mountains mainly uplifted in the Pliocene and Pleistocene epochs. The massive karst limestone on their southern flanks betrays a pre-uplift shallow sea, free of sediment. Subsequent extensive deposition created the southern foothills. New Guinea is the only island east of Wallace's Line with sandy, Pleistocene raised beaches, which occur along the foot of the western snowy mountains.

Geology and Natural History

Whereas tropical mainland Asia and Australasia comprise the ancient rocks and surfaces of stable continental blocks, the archipelago in between bears the imprint of massive collisions and upheavals along the still active "ring of fire," the zone of volcanoes and earthquakes extending from northern Sumatra eastward to New Guinea and beyond. In the western part of this Malesian archipelago, the edge of the Asian continental shelf has been continuously uplifted and buckled by collisions with the Indian Gondwanan fragment and Bay of Bengal oceanic plate from the west, the Philippine island arcs from the east, and the Australasian Gondwanan continent from the south. The contrast between the volcanic, mountainous Indonesian islands fringing the Indian Ocean and the infertile, sedimentary condition of Borneo to the north, which rests close to the edge of a stable continental plate, evokes the contrast between the Andes and the Guyana Highlands of South America (though these have been stable for very much longer).

Whereas the ocean constitutes a barrier, or at least a filter, to plant migration, mighty tectonic events have created a terrestrial setting of enormous complexity for the evolution and intermittent spread and restriction of plants throughout the Asian tropics, a setting that is in stark contrast to the relative stability of the other two major tropical continents. Persistent tectonic movement has also resulted in a diversity of surface topographies and of soils far greater than those in Africa and on a much finer spatial scale than those in the South American lowlands.

Peninsular India excepted, the siliceous rocks of tropical Asia form relatively small ecological islands in a sea of clay-dominated landscapes—either in coastal basins or exposed along ridges and plateaus of harder granite or sandstone within landscapes of predominantly fine, clay-yielding sediments (fig. 3.5). Limestone karst, originating in clear seas, may also be exposed as islands associated with fine sedimentary rocks. The same is true of acid volcanic extrusions, especially rhyolite and intrusive ultramafics.

The period of the last 20 million years—especially the last 1.8 million— might appear as a brief footnote in this long geological history, but it is disproportionately important to the present distribution of forests. New

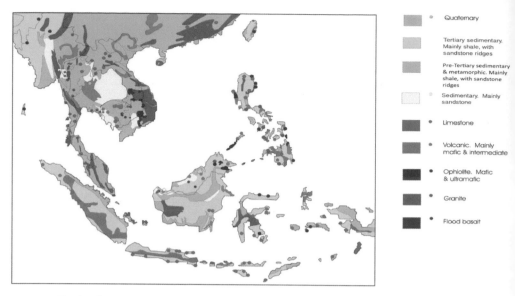

Legend:
- Quaternary
- Tertiary sedimentary. Mainly shale, with sandstone ridges
- Pre-Tertiary sedimentary & metamorphic. Mainly shale, with sandstone ridges
- Sedimentary. Mainly sandstone
- Limestone
- Volcanic. Mainly mafic & intermediate
- Ophiolite. Mafic & ultramafic
- Granite
- Flood basalt

Fig. 3.5 The contemporary surface geology of East Asia and Malesian Australasia (I. B.).

landscapes formed by the continuing uplift of sedimentary basins and by volcanoes at plate margins produce rich soils and diverse plant communities, and are home to most of the concentrations of agriculture and culture in tropical Asia. In addition, sea-level changes have altered runoff and siltation, producing conditions for human settlement.

Notes

1. This is a condensation of *OTFTA*, sections 1.1.1—1.2.9, pp. 13–25.

2. Hutchison, Charles S., *The Geological Evolution of Southeast Asia*, Oxford Monographs in Geology and Geophysics, 13 (Oxford, UK: Clarendon Press, 1989); Hall, R., "Reconstructing Cenozoic Southeast Asia," in *Tectonic Evolution of Southeast Asia*, ed. R. Hall and D. Blundell, Geological Society Special Publication 106 (London: Geological Society, 1996), 153–84.

3. Fu, Quiang, et al., "An unexpected noncarpellate epigynous flower from the Jurassic of China," *eLife* (2018): DOI.10.7554/eLife.38827.

I arrived in Phnom Penh at the beginning of November to advise on developing silviculture and botany in the Khmer forest service. It had been raining, lightly but continuously, and I was worrying about the prospect of prolonged fieldwork in the Cardamom and Elephant Mountains. It was therefore with surprise and delight that, on the morning of the tenth, I stepped out of the Hôtel de la Poste under a glorious, cloudless sky and walked to my jeep in the cool, crisp air of early temperate autumn. Leaving town on the Khmer-Soviet Highway to Sihanoukville, I saw a wall of purple thunderclouds to the south: the retreating monsoon. Phnom Penh was to receive no more rain for three months. — P. A.

4

CLIMATES

Climates are long-term patterns of weather that persist during many years.[1] The differences between weather patterns and climate are thus arbitrary or determined by arcane statistical analysis. Over geological time, shifts in climate are evident in the changes in geological proxies for mean temperature and in the fossil record. In tropical Asia, global changes as well as regional geological events have influenced the present complex distribution of climates (fig. 4.1) and vegetation.

Mapping climate distribution is important for understanding the distribution of tropical forests. Mapping requires a quantitative assessment of climate that considers the interplay between precipitation and temperature. Forests in the tropics can transpire and evaporate 130–170 mm of rainwater per month during windy periods (1,500–2,000 mm per year), but most release less because of the effects of droughts and deciduousness, and the saturated atmosphere of the wet monsoon in the seasonal tropics. Under tropical temperatures, periods of less than 100 mm of rain each month soon lead to water stress, producing drought. The number of dry days required to produce a drought varies with windiness, soil type, topography, and the prevailing vegetation. Increased temperature produces more evaporation and transpiration, which reduces the amount of water available for tree growth and forest establishment. One month is an arbitrary and approximate defining threshold—detectable from meteorological rainfall data—for drought. Köppen defined tropical climate classes by the mean rainfall of the driest month combined with the mean temperature of the coldest month. The systems of Holdridge, Gaussen, Walter, and others used different indices of

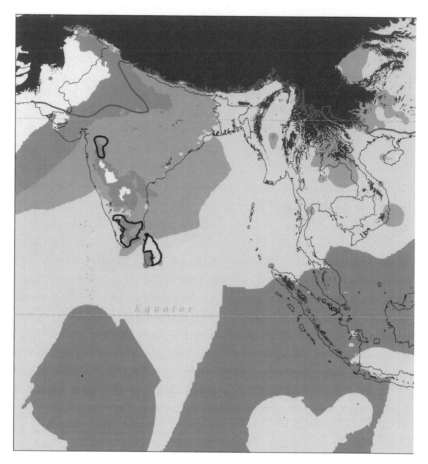

Fig. 4.1 A climate map of tropical Asia. Colors indicate the seasonality of precipitation in consecutive months with at least 100 mm precipitation. Blue frost line indicates the northern limit of the tropics. Areas with red boundaries have two dry seasons (Jeff Blossom).

seasonal moisture availability to define climate distribution. None of these methods have produced vegetation maps that correlate well with the actual distribution of forest types in tropical Asia. Perhaps the classification of Schmidt and Ferguson comes the closest.[2]

A problem with such attempts at climate characterization is that occasional extreme events may override average conditions. Exceptionally heavy rainfall may increase rainfall totals but be lost from the ecosystem by runoff. Extreme rainfall events are a feature of tropical climates. Their power

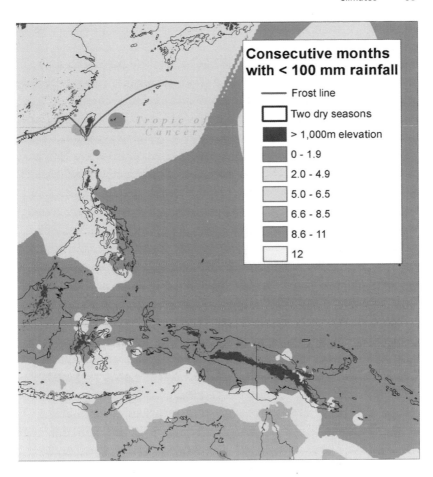

contributes to geological processes through erosion and sedimentation, and they are important in the dynamics of tropical soils.

Extremely rare drought events (one every five or more decades) may kill more vulnerable species and change forest composition, but not be revealed by mean precipitation data. This may be particularly true in the everwet (perhumid) regions. The margins of the tropics (and limits of tropical vegetation on high mountains) are delineated less by some mean temperature, even in the coldest month, than by a single limiting factor in the physical environment, an exceptional killing frost (see chapter 10 on the transition from tropical to temperate forests in East Asia).

The Everwet Tropics

The equatorial archipelago extends 6,000 km from Sumatra (including Peninsular Malaysia) through the Philippines and Malesia to the Pacific. Most of these lands and associated seas—except for a few areas influenced by dry, southeast trades off Australia and dry northeast trades over the western part of Luzon's Sierra Madre—are under the most extensive, continuously wet (everwet) and calm tropical climate in the world: they are known as the "lands below the wind" (see fig. 4.1). Their climate is defined by the absence of predictable annual periods of water stress and mean monthly rainfall in excess of expected evapotranspiration throughout the year. Everwet lowland tropical climates are restricted to within 8° latitude of the equator, except on oceanic islands and coastal regions fronting mountains with two monsoons. The pervasiveness of the everwet climate in the Far East is due to the open water around the Malesian archipelago, and particularly to the influence of the world's widest ocean to the east. This region is also within the intertropical convergence zone (p. 4) throughout the year. In the equatorial and everwet tropics, temperatures are uniformly warm and vary little with the seasons and during the day. Singapore averages an annual temperature of 27.2° C, and varies over a mere 1° C range in mean monthly temperature, whereas the daily range is ~7° C. Its minimum recorded night temperature is 17° C.

Rainfall here is mostly from thunderclouds (cumulonimbus), which form when daily heating of the land draws moist breezes off the ocean. Onshore breezes, warmer than the cool air above, become unstable and rise to replace it. This process is enhanced by mountains, which act like giant wicks, drawing the moist air up their slopes. The air cools as it rises, and its moisture condenses into clouds (figs. 4.2 and 4.3). Such an everwet climate does not exist in lowland Africa; in South America, it is confined to a relatively small area of the upper Amazon near the Andean foothills, the Choco of the northwestern coast, and narrow coastal regions of Central America that face the Caribbean. This climate supports the most species-rich terrestrial ecosystems known. They are generally associated with leached infertile soils, rapid growth and mortality, and distinct species compositions.

There are two small everwet regions in tropical Asia. Owing to the mountains to its east, the southwestern quarter of Sri Lanka receives the brunt of both the southwest monsoon and a somewhat wet northeast monsoon, along with orographic rainfall during the rest of the year. The hills of central Vietnam and the Annamite Range behind them receive a wet, north-

Fig. 4.2 Images of climates in the Asian Tropics. *Top left*, northeast monsoon sweeping over Kuching, Sarawak, Dec. 2006 (H. H.); *top right*, dry season cloud base, Rinjani Volcano, Lombok, July 1980 (P. A.); *bottom*, Tansa Valley, Maharashtra, India, *left*, March 1985, hot dry period before SW monsoon; *right*, August 1985, SW monsoon. That year ~2,500 mm of precipitation fell between 7 June and 10 October (D. L.).

east winter monsoon combined with frequent orographic rains during the rest of the year. These are unique patches in otherwise seasonal regions, and potential refuges for the hyper-rich floras of everwet climates.

Seasonally Dry Climates

The wind systems of the tropics are the trades: northeast, north of the equatorial convergence zone in the winter, when the zone moves south; and southeast, south of the convergence zone, when this zone moves north in the summer. These seasonal winds produce *three* wet monsoon systems, though the northeast one is not generally termed *monsoon* in English parlance (p. 9).

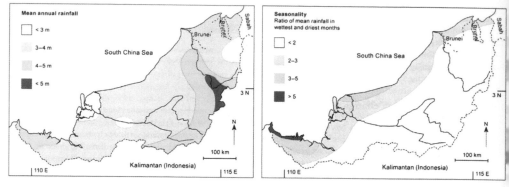

Fig. 4.3 Rainfall variation in everwet Sarawak, northwest Borneo. *Left*, mean annual rainfall, and *right*, seasonality of rainfall, as a ratio in the wettest and driest months. Rainfall in this everwet equatorial region is mainly orographic and diurnal, although west Sarawak receives the northeast monsoon (I. B.; see Ian Baillie *Further Studies on the Occurrence of Drought in Sarawak* [Kuching: Sarawak Forest Department Soils Survey, Research Section F.7, 1972]).

The Indian southwest monsoon. The climate system of tropical continental Asia is unique because it is distorted, extending far to the north beyond the geographical limit of the tropics. This is almost entirely due to the Tibetan Plateau, with its northern location and east/west shape. This vast region of highlands north of the tropics remains cool and under low pressure during the wet summer months and under high pressure during the snowy winter. Following the spring equinox, the intertropical convergence zone moves north of the equator in the Indian Ocean. As the sea surface warms, the atmosphere over it absorbs moisture like a sponge. In April the air starts to cross the west coast of southernmost India. Land and air temperatures increase. Fed by onshore breezes, the energy balance in the air over the land becomes greater than that over sea, radically altering atmospheric circulation. The transfer of energy to a contrasting terrestrial environment combines with the strength of the low-pressure system over Tibet. This draws the southwest trade winds with their ocean-derived humidity up and beyond the northern margin of the geographical tropics between May and November to 35° N in the upper slopes of the northwest Himalaya and to 45° N up the western edge of the Pacific.

The onset of the Indian monsoon along the coast is sudden and, being controlled by the seasonal movement of the sun, occurs relatively predictably between 20 April and 10 May (p. 144). Once the trades cross the geographical equator, they veer to the northeast from the earth's spin (the Coriolis force). Reaching Peninsular India, the monsoon winds divide. One branch chases up the Western Ghats, pouring out its rain onto their western slopes in quantities reaching 3 m per month at Mahabaleshwar. Passing over the hills to the east and north, they exhaust what moisture remains. Thus, much of continental South Asia receives less rain over a shorter season, which increasingly varies in intensity and time of onset as it moves north (fig. 4.2). Jodhpur (at 16° N in Rajasthan) has a mean annual temperature of 26° C, but a minimum recorded temperature of −2.2° C and a maximum of 45° C; annual precipitation is 315 mm, mostly falling in July and August. The annual range increases at higher latitudes and with the length of the dry season, increasing the intensity of summer droughts when the rain fails.

The other branch of the Indian monsoon drops even more rain onto the mountains from Sumatra north up the Bay of Bengal to northeast India and the eastern Himalaya (fig. 4.2). It produces driving rains, flooding the lower Ganges valley and drenching the hills of Odisha (Orissa). Cherrapunji, in the mountains of northeast India, averages 13 m of rain over its nine wet months. The Indian summer monsoon off the Indian Ocean also reaches east to cover Burma, but only covers Thailand slightly, and hardly penetrates Malesia.

The East Asian summer monsoon climate penetrates well beyond the tropics of Indochina in Asia east of the Tibetan Plateau, regularly reaching north to the east/west barrier of the Qinling mountains in temperate north-central China, continuing up the coast into southernmost Korea and to Japan. This temperate East Asian monsoon climate is where Chinese tea, native to warm temperate evergreen forest, can be reliably grown. In the absence of a Tibetan plateau as a "third pole," and with most land equatorial, average wind speeds and maximum rainfall are less; however, this monsoon is strengthened by the restriction and consequent heating of the Indonesian ocean current throughflow and by the South China Sea. Thus, the rains are derived from the South China Sea and warmed by the restricted Indonesian throughflow (fig. 4.3).

The Northeast Asian northeast winter "monsoon." The high Himalayas and northern Burmese mountains protect lands to their south, which are frost-free and tropical to 32° N in the west and 28° N in the east; but frost-free lowland climates do not occur north of 23° N to their east in South China. During the remaining part of the year, when winter descends on the northern hemisphere, northeast trade winds dominate the temperate continental Asian tropics. These winds are dry and cool where they originate over the continent, as they do everywhere except in the Far East. But the East Asian tropics, from eastern coastal Indochina, peninsular Malaysia, northwest Borneo, and the eastern coastal Philippines to New Guinea receive warm wet, northeasterly trade winds off the Pacific Ocean and the South China Sea during the northern temperate winter months of December-March (fig. 4.2).

Variations in rainfall seasonality. The duration of the wet monsoon over a given location generally decreases as it moves north, and less precipitation reduces its humidity. As it ascends mountain slopes, its orographically cooled air adds precipitation. The wet monsoon is generally shortest and most variable in the plains of Pakistan and parts of northwest India, as well as in the valley of the Irrawaddy, where the winds have already dropped most of their rain in the mountains of southern India or coastal Burma. It is longest in the Far East, where it follows the many north/south trending valleys of northeast Burma and the South China mountains. It drenches coastlines backed by mountains, including southwestern Sri Lanka, the west coast of Peninsular India, and northern coastal Burma with adjacent Bangladesh. Similarly, mountainous coasts facing the Pacific or the warm South China Sea, including the Philippines east of the Sierra Madre, the Vietnamese coast beneath the Annamite range, eastern coastal Peninsular Malaysia, northwestern Borneo, and the northern lowlands of New Guinea all receive exceptionally wet northeast monsoons. At lower latitudes, some rain may fall during the dry season from intermittent clouds.

The length and intensity of the dry season determines the nature of the vegetation and the species that survive and compete. It was occasional catastrophic drought in India that led to the discovery of the global El Niño Southern Oscillation pattern (p. 11). Where the dry season is long, the degree of annual variation in its duration—and the amount of rainfall received during the intervening monsoon—become increasingly critical in defining the potential for vegetation. In contrast, the dry season may last only one or two months in the most southwestern part of Peninsular India, southern Peninsular Thailand, and northwest Peninsular Malaysia, but it is annually dependable. However, occasional intense droughts may occur, even where the wet monsoon is generally long (p. 160).

Areas that receive both a wet southwest and northeast monsoon may approach double annual wet and dry seasons. They receive one as a true rainy season, and the other as intermittent storms. Thus, southeast Peninsular India and eastern Sri Lanka receive a full northeast monsoon, as well as a weak

Fig. 4.4 Global tropical cyclone tracks between 1985 and 2005, indicating the areas where tropical cyclones usually develop (NASA).

southwest monsoon that has already dropped most of its rain on the high mountains to their southwest. In southeast Vietnam it is the reverse.

Cyclonic storms. Catastrophic cyclonic storms (called *cyclones* in the Bay of Bengal and *typhoons* in the Far East) affect most of the coasts and coastal hills that face late summer wet monsoon winds in the wet seasonal tropics. These regions include the central Philippine islands, Vietnam, southern and central Japan, the eastern coastal hills of the Bay of Bengal, the Andaman Islands, and the area from the Chittagong Hill tracts of southern Bangladesh to Peninsular Thailand (fig. 4.4). More local cyclonic storms can occur south of both belts and along the western slopes of the Western Ghats and southwest Sri Lanka. Such storms can cause landscape-scale tree-throw. They vary in intensity and frequency, affecting stands along the eastern coast of northern Luzon as often as once a decade, and less often in the northern plains of Thailand—no more than once or twice a century.

Even in everwet climates, intense local turbulence, often preceding thunderstorms, can cause blowdowns of hundreds of hectares of forest. Individual stands, though, rarely receive more than one such event in several centuries, except in peat swamps and along floodplains, where rooting is shallow, and the forest is consequently less stable.

Fire. Periodic dry-season fires occur in regions with annual dry seasons of more than five continuous months (chapter 8). Fires that devastate human-modified rainforest are widespread nowadays during drought, and they have become an almost annual occurrence in some regions. Lightning has always been a major cause of canopy-tree mortality, particularly on peaks and ridges, where it occasionally initiates ground fires. The recent scourge of fire in logged and degraded everwet forests during El Niño droughts is not naturally caused. The use of fire by our hominin ancestors for cooking and driving game in the seasonal tropics has been inferred from evidence as old as 1.8 million years.

The Limits of the Tropics

The limits of the tropics are apparently determined by the distribution—in a complex pattern—of air frosts during certain nights of the year, principally during the dry winter monsoon (see fig. 4.1). Along the southern flanks of the central and eastern Himalayas, annual frosts occur above ~2,000 m. In South China, frost remains above ~1,500 m on south-facing slopes to ~23° N but may occasionally occur lower down when cold air pools in the plains. Even Guangzhou, at 23.08° N and not far inland, occasionally receives ground frosts, while inland, frost-free, south-facing pockets occur far northward. Frost defines a rough boundary between temperate and tropical lowland climates. It is therefore surprising that there is so little systematic observation of its occurrence or effects on vegetation, either in the lowlands or on tropical mountains.

With increase in elevation, gas expands and cools. This rate of change, the adiabatic lapse rate, varies with temperature and relative humidity from saturated air at 0.5° C/100 m to double that for dry air. In the northwestern Himalayas, winter frost occurs in the lowlands from at least 30° N 78° E northwestward, but apparently not severely or frequently enough to eliminate lowland tropical forests, which extend there to 32° N. Night freezes occur regularly above ~3,800 m on equatorial mountains. This cold, heavy air descends slopes and valleys to lower altitudes and "ponds" during dry weather in wide, gentle valley bottoms. Such freezes have been recorded as low as 1,500 m in New Guinea (~6° S), 1,700 m in Sri Lanka (~7° N) and the Western Ghats (10°–16° N)), and 950 m on Meghasani (21° N) in the more seasonally dry Eastern Ghats. On the other hand, such ponding has been noted only down to 2,500 m beneath the summit of Doi Inthanon (Thailand, 18°30′ N), and to 1,500 m in the rice stubble of the foothill paddies of Bhutan (27° N), which is at the tropical margin but where valleys are sheltered from the cool, northeasterly wind by the Himalayas (see fig. 10.2).

The tropical equatorial monsoon cloud base is universally at 1,000– 1,200 m, except on major mountains, where it is higher, and along the coast, where it is lower. On Mt. Kinabalu, North Borneo, diurnal fog within the forest canopy becomes pervasive at 2,200 m, but it is lower on the widespread lower Malesian ranges, down even to 800 m on coastal peaks. On the largest mountain massifs of the Asian tropics, including the Himalayas, the cloud base itself rises to 3,000 m or higher in the cool, dry season and during droughts, and to similar altitudes in the rain shadows of the core of New Guinea mountains. At subalpine and alpine elevations above ~3,000 m, cumulonimbus clouds are separated by cloudless aerial canyons. Therefore, both intermittent drought and dry season frost become increasingly frequent between long periods of fog: a demanding climate for plants.

The History of Asian Tropical Climates

Past climates and geology influence each other, and both influence forest distributions. The formation of the Himalayas dramatically strengthened the southwest monsoon, and the heavy rainfall subsequently added sediments to the shallow trough at its base and to the Bay of Bengal. By 23 Ma, the northward movement of the Australian plate had restricted the westward current, the Indonesian throughflow, increasing the strength of the northeast monsoon and increasing sedimentation along continental margins (fig. 4.5). Global climate changes, likely accompanying shifts in atmospheric composition (especially in carbon dioxide and methane), affected ice storage and thermal expansion of water, and produced some dramatic changes in sea levels during the last 75 million years. Sea levels affected sedimentation in the shallow basins adjacent to land.

Climates can be inferred from changes in palaeotemperature over time, and there are various proxies used, such as oxygen stable isotope ratios and fossil evidence from plants (including leaf margins and stomatal density) and foraminifera. Regional and local terrestrial climates can also be inferred

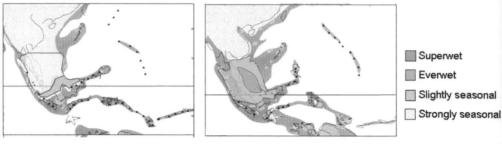

Fig. 4.5 Climate change in tropical Southeast Asia, at 38 and 20 Ma, showing the shift toward an everwet core in lowland Sundaland (I. B., after R. Morley; see Robert Morley, *Origin and Evolution of Tropical Rain Forests* [Chichester, UK: Wiley, 1999]).

from the identification of plant fossils from the late Cretaceous period onward, but reasonable precision is possible only from the Oligocene epoch, when most modern taxa (at the family and genus levels) can be distinguished.

Early climate change. We have a good account of the history of Cenozoic climate change in the Asian tropics. Past climates can be inferred from tectonic plate movement (p. 42). Although tropical climates have extended above their current latitudes at times in the geological past, the Hadley cells (p. 4) and, therefore, wind systems have remained constant at their current latitudes. Everwet tropical climates have always remained within 8° of the equator. The Indian block separated before the origin of flowering plants from Gondwana and Africa. It originated in a marginally tropical but mostly warm temperate, hot and humid climate. Then it moved north through currently desert latitudes before entering a seasonal wet—and then everwet—equatorial climate as it bypassed east Africa, where it picked up much of the tree flora that currently characterizes Asian rainforests. The broad northern margin of the Indian plate then reentered a seasonal climate before colliding with Laurasia, while its equatorial, rain-forested belt shifted south and currently survives only in southwestern Sri Lanka, at 6°–7° N.

The Eocene was the last epoch when tropical temperatures widely extended to 35° N or more, and wet tropical regions were more extensive than at present. The Indian plate had butted into the southern Laurasian coast, then as now located near the Tropic of Cancer. At that time, the contact zone between India and Laurasia—with the Tethys Sea retreating to the east—was under a seasonal tropical climate, although the Himalayas were no more than a line of hills. The climate was hotter than at present, and the hills received some rain off the retreating Tethys to the east, by now sheltered from the Pacific by the Southeast Asian peninsula.

This climate changed drastically during the Cenozoic era. By the Paleocene epoch, humid tropical and near-tropical warm temperate climates had reached high latitudes in both northern and southern hemispheres, but from 35 Ma, these climates have receded. Seasonal tropical climates penetrated to lower latitudes, and this change would have been accelerated along the India-Laurasia suture line by India's continuing northward thrust. A corridor of seasonal wet tropical climate between South Asia and the Far East likely existed at the time of contact and at least until 35 Ma, and a monsoonal climate existed in India from 55 to 34 Ma.[3] Evidence from sediments along the base of the Himalaya suggests that, for ~35–20 Ma, that region had become semi-arid, while global climates remained hot, and regional climates were generally wet. Humid tropical climates expanded again during the late Oligocene and continued until the thermal maximum during the middle Miocene. These climates remained seasonal in the lands between India and the Far East. The Himalayas and the Tibetan tableland emerged later, from 24–22 Ma. Climate became wetter but remained seasonal; the final and highest Himalayan uplift, along with the full monsoon climate pattern, occurred later. An everwet equatorial tropical climate was already confined to central peninsular India by the time of the Laurasian collision and continued to move south as India thrust north It left the peninsula altogether ~10 Ma, becoming restricted to SW Sri Lanka. Afterward, global climates again cooled until ~13 Ma, when warm and cool periods began to alternate on a cycle of 41 thousand years, correlated with a wobble in Earth's axis in relation to the sun.

Central and southern Europe were humid and tropical/subtropical during the Eocene epoch. However, this region was separated at the time from Asia (including India) by a broad, north/south, epicontinental strait, the Turgai. This strait closed at the end of the Eocene, as the Indian plate collided to its southeast. A migration route could then have opened for seasonally wet tropical climate species between southwest Eurasia and the eastward-receding Laurasian shores of the Tethys. This may have persisted intermittently at least until the Himalayas started to rise ~23 Ma. In the absence of a low-pressure zone like the current Tibetan Plateau to its north, it is likely that rainfall would have become increasingly seasonal as the Indian block pushed north. By the Miocene epoch, the Middle Eastern coastline was located well north of the equator in what had become a seasonal and

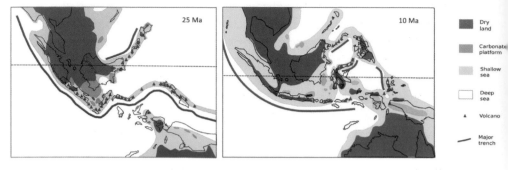

Fig. 4.6 The northward rafting of Australasia produced conditions for the everwet climate of Sundaland by partially blocking the westward flow of the Java current. *Left*, 25 Ma, and *right*, 10 Ma (I. B., after R. Hall; see R. Hall, "Reconstructing Cenozoic Southeast Asia," in *Tectonic Evolution of Southeast Asia*, ed. R. Hall and D. Blundell, Geological Society Special Publication 106 [London: Geological Society, 1996], 153–84).

even semi-arid tropical climate, while the lowlands south of the northwestern Himalayas became temperate.

The broad Sunda region was mostly drier or seasonally wet during much of the Oligocene epoch, except for a western lowland coastal ribbon from southern Burma through Sumatra to Java, where it broadened. At this time the northern part was a widespread upland area. As the ocean flow through the Malesian archipelago was constricted by the collision of the Australian Plate with Sunda, the Sunda climate became widely everwet from ~23 Ma onward (fig. 4.6). Subsequently, widespread peats formed across what is now the South China Sea and Gulf of Thailand), indicating an everwet climate there. Parts of the region became periodically more seasonal during the late Miocene and Pliocene epochs, roughly as the Indian monsoon grew in intensity.

Pleistocene climates. The biogeography of the modern flora and fauna can be best understood within the historical context of geological and climatic events during the last 10 million—especially the last 2–3 million—years. The monsoon was wetter 8–10 Ma than at present; it began to weaken 2 Ma, when the advent of the northern ice ages brought dramatic changes to tropical regions. Tibet became continuously white with snow, reflecting incoming solar energy and becoming a zone of constant high pressure. The Indian monsoon probably slackened or even died during the glacial periods, while the East Asian northeastern monsoon may have become wetter. Forests retreated, and deserts likely expanded into South Asia, though conclusive evidence is still lacking. Everwet and seasonally wet evergreen forests may have survived in South Asia only along coasts backed by mountains, where humid onshore breezes would still have been drawn upward and drained of moisture. The drying trend in South Asia was an extreme example of a worldwide pattern during ice ages; the monsoon is currently weaker than it was ~11 Ma. Finally, during the mid-Pleistocene (~900 Ka), the present sequence of ice ages that last ~100 thousand years alternating with warmer periods of ~10 thousand years became established. We currently live near the end of one of these warm periods.

In the Far East, dry seasonal climates and forests spread south at least as far as Kuala Lumpur, where fossil pine pollen has been found dating from glacial maxima. Kuala Lumpur lies in a lowland rain shadow between the Sumatran and peninsular mountains. During the glacial periods, much of Sundaland and New Guinea were still everwet but with reduced or more seasonal rainfall at times of lowest sea levels.

Temperature was reduced on tropical Asian mountains during the Pleistocene epoch, particularly during the last northern ice age (100–112 Ka); surviving glacial moraines descended to lower altitudes, and peat cores preserve pollen from an alpine herbaceous flora, as at 2,000 m on Mount Wilhelm (Papua New Guinea), and 1,500 m in the eastern Himalayas when continuing tectonic uplift is also considered. This implies a descent of the tree line by as much as 1,700 m. The current upper limit of the full lowland rainforest flora, concentrated below 400 m in everwet New Guinea, Sundaland, and the eastern Himalayan foothills, is at ~1,000 m. That suggests reductions of mean annual temperature of less than 6° C. It is therefore unlikely that the overall tree line descended by as much as 1,000 m, or that the altitudinal limits of lowland forest descended more than 400 m in equatorial regions. However, the increased aridity associated with the slackening of the Indian monsoon was likely associated with more drastic temperature depression at the tropical margin in the northeast Himalayan foothills.

John Flenley claimed that upper montane forest (mossy, "elfin" woodland) in the New Guinea massif disappeared during some of these periods, which would have been drier or more widely seasonal at high tropical altitudes.[4] This is consistent with its current impoverishment or absence in seasonal tropical climates (chapter 10). Nevertheless, the persistence on Kinabalu and other major, free-standing equatorial mountains of floristic elements *not* currently occurring in lower montane or subalpine forest types remains to be explained. Surely such forest types are older than the late Pleistocene.

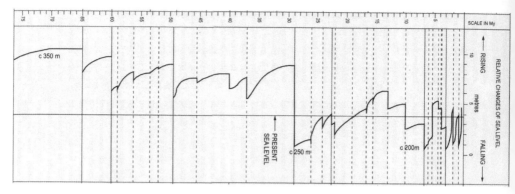

Fig. 4.7 Global cycles of sea level change during the Cenozoic era. A drop of 30 m joined the presently isolated regions of Sundaland. Vertical lines correspond to geological events, often associated with rapid sea level change (I. B., from C. H. Hutchison; see Charles Hutchison, *The Geological Evolution of Southeast Asia*, Oxford Monographs in Geology and Geophysics 13 [Oxford, UK: Clarendon Press, 1989]).

Sea levels and Sundaland. Shallow subsidence of extensive continental surfaces beneath sea level can persist over long periods of time. Though not comparable in duration to those resulting from continental drift, such shallow seas may provide periodic barriers to migration.

Since the onset of the northern ice ages ~2 Ma, global sea-level changes have periodically isolated and consolidated the Sunda uplands (p. 35) as shallow seas alternately invaded and receded from the lowlands between them (fig. 4.7). During the short intervals (~10 thousand years) of high sea levels, such as our own Holocene epoch, sedimentation between these uplands has continued. The uplands themselves consist of previously uplifted rocks, including the sedimentary basins of former seas. Floodplains, and the continental shelf during periods of high sea level or continental down-buckling, receive sediment from surrounding uplands. Deposition under flowing water along rivers, in estuaries, and near the coast was coarser and generally rich in silica sand, irrespective of geology.

Sundaland periodically emerged during the Pleistocene epoch as a continent comparable in area, latitude, and topography to the present northern part of South America. Global sea level was ~250 m below its present level at the end of the Pliocene ~2.8 Ma and may have dropped by up to 120 m below its present level—perhaps only twice—although it has probably fallen more than 100 m more than a dozen times in the last million years, most recently in the last ice age (fig. 4.7). At present, the major epicontinental sea in the Asian tropics is the southern section of the South China Sea, where it overlies Sunda continental rocks. Evidence of a submerged river system, the Proto-Mekong, which drained the region when land was exposed (fig. 4.8), is deduced from subsurface topography. However, global sea levels did not rise more than 5 m during the Pleistocene.

The Proto-Mekong drained a vast area of Sundaland that had been above sea level on the continental shelf during northern ice ages. This land is now below sea level, despite the mass of sediments that have been eroded onto it from wet equatorial Sundaland since the Oligocene epoch. Most were deposited onto the northern continental shelf of Sibumasu and the southern Sunda subplate when this area was covered by shallow seas. Most of these sediments were disgorged from the interior Bornean mountains as the floor of the South China Sea pushed south and Australasia pushed northwest, lifting these mountains.

The everwet Sunda climate started to spread into the extending low-

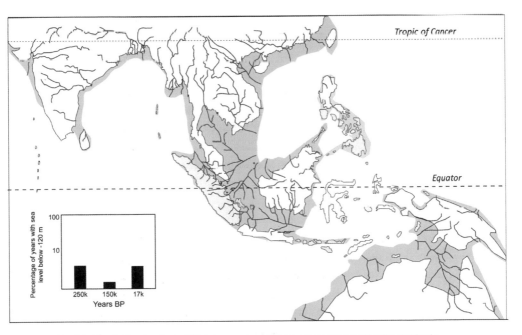

Fig. 4.8 The furthest extent of Pleistocene land linkages in tropical Asia, with riverbeds of most recent exposure illustrated. The bar graph indicates the estimated percentage of time, over the past time period indicated, that sea levels were at or lower than illustrated, or 120 m below present sea level (I. B., after H. Voris; see Harold Voris, "Maps of putative Pleistocene sea levels in Southeast Asia: Shorelines, river systems and time durations," *Journal of Biogeography* 27 [2000]: 1153–67).

lands 23 Ma, during the lower Miocene, as Australia's approach constricted the Indonesian throughflow current and a previously mountainous Sunda was eroded. Complete closure might have occurred during the long periods of low sea level during the Pleistocene, when southern temperate China became drier. The warm waters of the equatorial Proto-Mekong drained the still-wet Sunda core, which is comparable in area to the modern Congo drainage, into the South China Sea, ensuring warm seas and continuing high orographic rainfall in northwest Borneo, and probably also in Peninsular Malaysia east of its Main Range. Wet climate therefore likely continued in the northern Sunda core through periods of moderately low sea levels coinciding with temperate glaciations and South Asian aridity.

During high sea levels, sedimentation continued between the main land masses of Sundaland, whose continental core is Peninsular Malaysia, and Sumatra, Java, Bali, Borneo, Palawan, and the islands strewn among them. Along the former coastlines of Quaternary Sundaland and southwestern New Guinea are numerous raised Pleistocene sand beaches, marking the highest sea levels. Some have also been uplifted tectonically, occasionally by as much as 15 m. The sea level dropped about 2 m at the end of the middle Holocene epoch, ~4,500 BP, exposing shallow sediments and embaying lagoons in dune systems that were down current from river mouths. Eventually, these lagoons filled with fine sediments over which peats developed. These younger plains have left once-coastal dune systems isolated far inland. Raised white sand beaches of pure, fine silica sand are widespread along the coasts of Borneo and behind its coastal peat swamps. They occur along the east coast of Peninsular Malaysia, the western Cambodian coast at the base of the Cardamom Mountains, and the southern coast of Indonesian New Guinea.

Another major epicontinental sea is the Arafura, currently separating New Guinea from northern Australasia, whose bed lies on the Sahul continental shelf. Islands rising from this shelf include the coral Aru archipelago. The Andaman Islands are presently separated from the coastal Arakan hills of Burma by a similar epicontinental sea. In the same way, Palawan—politically part of the Philippine Republic—is geologically linked to Borneo, being attached to the Sunda continental plate (see fig. 3.5).

Thus, the climates of tropical Asia have remained tropical. The patterns of seasonality and rainfall amounts, however, are complex and finely textured, and they have changed considerably over the past 60 million years. Opportunities for migration and survival of the hyperdiverse forests and floras of everwet climates have fluctuated, with long intervening periods too seasonal for survival and with arid conditions leading to major extinctions during the Pleistocene. Those of seasonally dry climates have expanded and contracted simultaneously with climate change during the Cenozoic epoch, periodically isolating and diversifying, as in IndoBurma, during humid periods when dividing barriers of rainforest intervened.

Notes

1. *OTFTA*, sections 1.1.3–1.3.2 and pp. 25—43.

2. Schmidt, F. H., and J. H. A. Ferguson, "Rainfall types based on wet and dry period ratios for Indonesia and Western New Guinea," *Verhandelingen*, no. 42 (Jakarta: Kementerian Perhubungan, Djawatan meteorologi dan geofisika, 1951).

3. Flenley, John R., *The Equatorial Rainforest: A Geological History* (London: Butterworths, 1979).

4. Flenley, *Equatorial Rainforest*. See also Morley, Robert J. "Assembly and division of the South and South-East Asian flora in relation to tectonics and climate change." *Journal of Tropical Ecology* 34 (2018): 209–34.

Descending from a month on Brunei's highest mountain, Pagon Periok, I decided to return by the Temburong river with three of my Iban colleagues. When we reached its headwaters, we found it dangerously rocky, with none of the right trees for a bark canoe. Careering down the rapids on little better than a giant banana skin, we inevitably lost our prahu, clothes, and food over a waterfall. It took us five days to reach a longhouse, but I learned much from my first barefoot hike in the forest: The soil on these high hills, with little litter or surface roots, sticky and moldable, proved on Ladi's recommendation the least unpleasant food substitute—though it did not pass through easily. What a contrast with the gritty soils I had found on the coastal hills, with their carpet of fine roots, tobacco-like raw humus, and deep, leathery litter. Sitting by the torrent in our palm leaf pondok one evening, it dawned on me that these dramatic soil differences might well explain the dramatic changes in tree composition. — P. A.

5

SOILS

Soils are a key feature in tropical Asian environments.[1] Climate and geology interact in the formation of soils and thus collectively influence forests. Soils are produced by the breakdown of organic material, and by the dissemination of minerals, in the form of sand, silt, and clay, from the parent rock minerals beneath—the process of *weathering*. The rate of soil formation increases with higher temperature and greater amounts of moisture (fig. 5.1).

The qualities of soils are determined by the rates of these processes and the qualities of the organic materials available and the mineral composition beneath. The hardness of the minerals determines the sizes of the particles that result from weathering; hard minerals, such as quartzite and feldspar, produce sands; softer minerals produce smaller particles—ultimately the clays that are crucial for determining the physical properties and nutrient storage capacity of soils. Weathering involves physical, chemical, and biological processes; it produces soluble mineral ions that can be taken up by plants.

Biological activity is crucial to the interactions between plants and the soil. Soil animals, particularly earthworms, termites, and ants, move organic materials from the surface deep into the soil. Mycorrhizae (fungi that live within or in close association with roots [p. 269]) move phosphorus and water into the roots, and nitrogen-fixing bacteria in root nodules (primarily in legumes) fix nitrogen gas into forms assimilated by plants. Soils influence the types of vegetation that establish, which affects the quality of litterfall, which in turn affects the quality of soil.

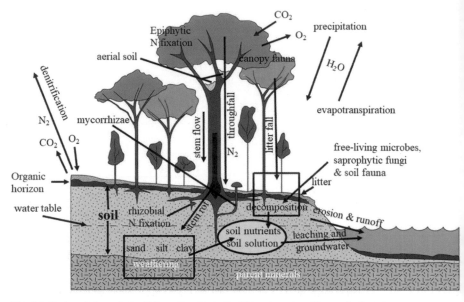

Fig. 5.1 Factors influencing soil formation (including litter decomposition and mineral weathering), highlighting nutrient flows and biological activity (D. L. after I. B.; see Ian Baillie, "Soils of the humid tropics," in *The Tropical Rainforest*, by P. W. Richards, 2nd ed. [Cambridge: Cambridge University Press, 1996], 256–86).

Soil Classification

Attempts at distinguishing among soils began with Russian scientists in the 19th century, were elaborated with the USDA classification in the early and middle 20th century, and have developed further with the FAO-UNESCO system introduced in 1974. All of these systems were designed for agriculture, and they are of limited use for forest soils. First, forest organic matter is rapidly transformed and lost following clearance and cultivation, so they take no account of the crucial role of litterfall and organic matter under forest conditions. Second, they focus on the nutrients available for immediate root uptake and crop productivity, giving less consideration to the nutrient store in the substrate and mineral soil, part of which may become available during the long life of a forest tree. Third, these classifications are based on readily visible variation in the appearance of the profile of the mineral soil, and processes inferred from it, whereas we are also concerned with the appearance of the organic matter and surface horizons. For us, the conventional soil classifications discriminate too finely on mineral soil criteria and differentiated mineral profiles that are poorly correlated with observed forest variation, and they lump changes in those observable organic matter characteristics that are correlated with forest variation.

Ian Baillie has provided a clear review and classification of tropical forest soils, concentrating on the wet tropics.[2] After discussing below the major ways in which all tropical forest soils vary, we focus on soil variation in relation to the climates and surface geology of the Asian tropics.

Tropical Soil Dynamics

The age of a soil is most often expressed in terms of the time since the mineral soil first started to establish by the decomposition of a fresh substrate surface, or to stabilize from debris deposited by a previous physical process. In temperate regions this deposition is often the consequence of the most recent glaciation, so that soils are dated from the end of that event, 10–15 Ka. Such temperate-zone soils are considered young. In the tropics, glaciated soils of this type are confined to the highest mountains and do not occur below ~1,500 m.

The prevalent soils of the Asian tropics are *zonal*, lying over substrates of sedimentary and igneous rocks (p. 48) and primarily influenced by climate. Where erosion is slight, as on the freely draining soils over igneous granites of the Peninsular Malaysian Main Range, the mineral soils can accumulate as deep as 15 m and may be quite ancient (fig. 5.2).

Landscape change and soils. Erosion and other rapid geological processes profoundly affect soil formation, often on quite small, local scales. Young mineral soils in the tropics are mostly caused by landslides on steep surfaces (see fig. 5.2). These are abundant in regions of frequent earthquakes, especially those associated with tectonic uplifting and buckling of sedimentary and other rocks, leading to high, narrow ridges and steep slopes.

Landslides also occur where water percolates through soil and porous substrates onto a sloping, impervious layer beneath. There it may lubricate the interface and allow slippage of the overburden. This is a frequent cause of landslides where uplift has tilted rocks into steeply inclined strata. Impervious rock surfaces may be exposed on the slide itself, while a deep and porous accumulation—a *colluvium*—may be left on the slope below, resulting in two contrasting soil environments for plant establishment. This process generates steep, often narrow ridges held up by less erodible, porous strata such as sandstones. Also, clay-rich and relatively impervious slope soils are

Fig. 5.2 Tropical rainforest udult soils. *Left*, landslip over clay-rich, metamorphic, Precambrian charnokite rock following the exceptional storm of 2003, Sinharaja Research Station, southwest Sri Lanka (U. G.); *right*, zonal clay loam hillslope soil of low nutrient status, over Precambrian metamorphic rock, Kanneliya Forest, Udugama, southwest Sri Lanka (P. A.).

continuously truncated, and thereby rejuvenated, by the runoff from heavy tropical rainstorms. Where they lie over bedrock in such steeply inclined landscapes, both ridge and slope soils are shallow, rarely exceeding a meter in depth, and contrast with the generally deeper, colluvial accumulations below. These landscapes dominate the rugged uplands of inland Borneo, northern Thailand, and northeast Burma, where sedimentary basins on submerged continental surfaces have been uplifted and buckled by tectonic activity.

Where they lie horizontally, porous strata such as sandstones resist surface erosion. Where they overlie impervious, erodible, clay-forming rocks, erosion of those layers and the consequent undercutting of the porous cap leads to breakage along the edges. Such erosion creates a landscape of cuestas or gentle dip slopes following the incline of the porous surface stratum, with steep scarps along the upper edges. The sedimentary landscapes of Borneo and the Himalayan foothills are actively eroded, producing flattened plains, but generally balanced by continuing tectonic uplift.

Soils dry out in climates with long dry seasons. Sudden saturation at the beginning of the wet monsoon can lead to total loss of soils where they overlie inclining rock surfaces. The erosion of ancient, uniformly hard rock eventually generates a rounded topography of extensive bare rock hummocks; viewed from a distance, these give the impression of sleeping elephants (see fig. 3.4). Such *inselbergs* are widespread in regions of ancient continental rocks, such as Sri Lanka and the Eastern Ghats of Peninsular India. Regions of tropical Asia are therefore covered by mineral soils whose prevailing ages vary dramatically. In consequence, soil is generally shallow in steep, erodible, young landscapes and deeper in old landscapes with the associated rounded, "mature" topography.

Litter decomposition. In the prevailing zonal, yellow-red clay and sandy clay loam soils of the lowland wet tropics, rates of litter decomposition generally exceed the rates at which dead organic material accumulates (fig. 5.2): a condition termed *udult*. Leaf and other fine litter often fully decompose in less than one year, or even during a single wet monsoon period in the seasonal climate of Asia. Soil organic matter levels are thus relatively low, and the depths of visible humus discoloration—betraying the presence of particulate organic matter—are generally shallow. Because soil organic matter serves as the major store for nutrients, the lack of it also explains the generally low fertility of tropical lowland udult soils.

In the tropics, fallen debris is initially fragmented primarily by termites and ants, though no leaf-cutting ant species occur in Asia. Their excavations also improve soil texture. Termites have been shown to mitigate drought effects in Borneo's everwet forests.[3] Termites, owing to their ability to decompose plant cell walls through the action of their microbial associates, play a dominant and ubiquitous role in lowland soils.

Decomposition is thereafter carried out by bacteria and fungi (see fig. 5.1). Bacterial decomposition prevails where organic matter pH is relatively high, particularly where exchangeable calcium is a major mediator of soil acidity. Calcium derives from calcium carbonate in limestone; it is also an important constituent of volcanic rocks. Fungal decomposition proceeds at a slower rate than bacterial. Fungal decomposition breaks down cell walls, generates organic acids, and lowers soil pH. Many species of fungi, both within living cells as vesicular arbuscular mycorrhizae—and ectotrophically on roots—form symbiotic relationships with plants (p. 269). These mycorrhizae absorb sugars from the plant host, and the plant absorbs water, phosphorus, and probably some other nutrient ions in solution from the fungus (p. 270).

Soils in which rates of litter decomposition are slower than litterfall and in which particulate organic matter therefore visibly penetrates the upper mineral soil horizon, are termed *humult* (fig. 5.3). Litterfall rates seem not to vary consistently with soil acidity, but rates of decomposition decline with lower pH, lower temperature, and oxygen deficits. Litter fragments accumulate between the intact litter and mineral soil, where surface soils are acid, with a pH lower than ~4.2, as raw humus, or *mor*. Lowland tropical soils

Fig. 5.3 Soils at low elevation in the humid tropics. *Top left*, ultisol in upper dipterocarp forest, Temenggor Forest Reserve, Upper Perak, Peninsular Malaysia; note dense surface root mat and slender descending roots (R. H.), *top right*, Leached humult yellow sandy soil over granite, coastal hills, Kuantan, Peninsular Malaysia (P. A.); *bottom left*, lowland podzol supporting peat forest (*kerangas*), root penetration cut off by cemented humic horizon above them, West Sarawak (P. A.); and *bottom right*, leaf litter with mycorrhizal roots penetrating raw humus over a humic podzol, Bako National Park, Sarawak (P. A.).

bearing raw humus, usually dense with fine tree roots, are confined to climates with less than four dry months. They are found on freely draining, generally sandy mineral soils low in nutrients, especially those on coastal hills, and on high, breezy ridges below the cloud base; these are prone to periodic drought, which itself slows decomposition.

The everwet areas of Borneo have greater areas of raw humus–bearing soils than anywhere else in the tropics. Moisture retention is influenced by slope. On convex slopes and ridges, where water deficits are most frequent and litter decomposition is most retarded, acidity increases in the accumulated surface organic layer, and carbon increases in it relative to nitrogen. Raw humus accumulates most here on flat surfaces where water stands, and the soil beneath is anaerobic.

The ratio of the rate of litterfall to that of its decomposition is also influenced by temperature. Humus accumulates more rapidly at altitude, where temperatures are lower, because decomposition declines with temperature more than litterfall rates do. The ratio comes into balance at ~1,000 m on clay loam soils over basic volcanic rocks in the Javanese mountains. Humus particles can only be drawn down into the mineral soil by organisms. Above 1,000 m termite abundance decreases and earthworm abundance increases. On Kinabalu (p. 201), earthworms ascend to the soil-filled fissures below the summit. Organisms draw humus deep into the soil, and it is rapidly decomposed at lowland temperatures when soils are moist. Deciduous and semi-evergreen forests support large grazing and browsing mammals. Their dung can improve the quality of substrate for earthworms, as has been shown for cows in temperate pastures. In a northern Thai deciduous forest, earthworms produced 133–225 t ha^{-1} of humus annually, equivalent to a layer of soil 9–14 mm deep. Soils approaching these humus characteristics also occur in moist sites on base-rich substrates in the everwet lowlands and in moist swales at higher latitudes (fig. 5.4). Whereas the *climatic* margin of the tropics is defined by incidence of seasonal frost, the *edaphic* margin could be defined by the simultaneous increase of earthworms and decline of termites in zonal soils at the upper limits of the lowland forest (see chapter 10).

Less raw humus accumulates as its depth increases because it all slowly decomposes. At the Lambir ForestGEO site (on Miocene sandstone in northeast Sarawak), Ian Baillie estimated an accumulation rate of ~6 mm per decade. These soils bear depths of raw humus that may reach ~6 cm, but usually average 2–3 cm, implying a turnover rate of 30–90 years. These soils

Fig. 5.4 Soils of mountain forests. *Top left*, upper dipterocarp forest, ancient, deep soil profile over granite, 900 m, Temenggor forest, Upper Perak, Peninsular Malaysia. Peter is pointing to the substrate surface (R. H.); *top right*, yellow-red sandy clay loam, Temenggor Forest, Upper Perak, Peninsular Malaysia. Stuart Davies points to the deep root of *Anisoptera costata* (R. H.). *Bottom right,* humic clay loam lacking surface raw humus, with abundant roots in the mineral horizon, Tia Shola, lower montane forest, Nilgiris, Western Ghats, India, 2,000 m (M. D.); *bottom left*, upper montane forest, shallow acid humult horizon over humus-free mineral soil, Horton Plains, Sri Lanka, 2,100 m (H. H.).

resemble the northern, humus-rich forest soils that Russian scientists called *podzols*. A tropical lowland podzol may bear up to 30 cm of raw humus, with a turnover rate of about 450 years. This time span is crucial to the processes leading to nutrient release. It may seem long, but it is short compared with the time required for a mineral soil profile to reach maturity.

Fertility and leaching. Mineral nutrients are derived from the breakdown of litter, from dust particles in the atmosphere, and from the weathering of minerals in parent rock strata. The latter vary in mineral production. Young igneous rocks are typically rich in the required elements. Sedimentary and metamorphic rocks are less rich, and become even less so from repeated cycles of deposition and erosion. Minerals and clays often lose certain elements through leaching into soil solution and into the groundwater and streams (see fig. 5.1). Nutrients absorbed by roots are at least partially derived from recycled organic material. The following elements are critical, in decreasing order of their needed concentrations: nitrogen (NH_4^+ and NO_3^-), potassium (K^+), phosphorus (PO_4^{-3}), calcium (Ca^{+2}), sulfur (SO_4^{-2}, S^{-2}), magnesium (Mg^{+2}), boron (Bo^{+3}), chlorine (Cl^-), manganese (Mn^+), iron ($Fe^{+?}$), zinc (Zn^{+2}), copper (Cu^{+2}), molybdenum (Mo^{+2}), aluminum (Al^{+3}), sodium (Na^+), and cobalt (Co^{+2}). The first three elements, NPK, are the primary ingredients of fertilizers, in ratios of 2:4:4. The second tier of the next three elements is needed in smaller concentrations. The rest are present in trace amounts and are rarely limiting for plant growth.

The positively charged ions, or cations, are held on negatively charged surfaces in the clay matrix. A mineral's capacity to store these ions is its cation exchange capacity (CEC). Hydrogen ions (H^+) are released from organic acids and weakly acidic compounds reduced from parent minerals. The resulting acidity, measured as pH (on a logarithmic scale), is important in controlling nutrient availability, as hydrogen ions replace other cations on the negative clay charges as conditions favor increasing acidity. Once released into solution, the nutrient cations become leached (mobilized) down the mineral soil profile and lost into the groundwater unless captured by roots. Elements may also be stored in the humus and organic fraction of the soil, with the addition of nitrogen (NO_3^-), which is fixed in legume root nodules and cyanobacteria on canopy leaves, though this is more common in seasonal climates and successional stands than in old growth tropical rainforest.

Nitrate ions diffuse through soil most readily; sodium and calcium, then potassium and magnesium are readily soluble and diffusible, sulfur and phosphorus are less so. Phosphorus, uniquely among the leading nutrients, is generally bound ("occluded") as phosphate anions on positive exchange sites, either on iron (Fe^{+3}) or on aluminum (Al^{+3}) sesquioxides, or as organic phosphates in humus. It is slow to be released in any case but can be absorbed into roots via mycorrhizae. Potassium is similarly sequestered in mica crystals. Calcium may also be immobilized as soil carbonates.

Where leaching of humult soils has been extreme, and the substrate is completely lacking in nutrients, the remaining mineral nutrient source may be organic matter plus the minute input from rainwater. Mineral soils of pure sand, absent clay or sesquioxide molecules, lack exchange sites other than in their organic matter. In low concentrations of nutrients in udult soils, cations are replaced by ions that normally react with Al^{+2} sesquioxides, and with Al^{+3}, which can be an important exchangeable cation in acid udult rainforest soils of seasonally wet climates. This also accumulates those soluble organic acids and phenolic compounds that are responsible for leaching.

Leaching is influenced by several factors. Topography influences leaching by affecting the availability of water. Temperature during rock weathering influences nutrient release rates from mineral soil and from the decomposition of organic matter. Rates are therefore highest in the lowland tropics, and within the tropics they decline in the lower temperatures at increasing altitudes. Water is essential for chemical weathering; too little of it slows activity, and too much fills pores and reduces oxygen diffusion. Rates of litter decomposition, other factors being equal, are highest in equatorial everwet climates. They may be highest during the wet monsoon, but they diminish as soil surface horizons dry out and cool during the dry seasons of higher latitudes, or during droughts in everwet climates.

Diffusible ions, readily released from decomposing tissues in hot climates, are easily leached away during rainfall unless absorbed directly by roots (see fig. 5.1). Soils in everwet climates therefore tend to have few nutrient ions available in solution—especially nitrate and calcium ions. High concentrations of nitrogen or calcium in living leaf tissue relative to their concentrations in the soil indicate that trees reabsorb these elements before leaf abscission, and/or efficiently absorb them immediately following their release as organic matter decomposes. During prolonged dry seasons, the absence of moisture in the soil horizons where feeder roots are concentrated prevents both nutrient uptake and leaching, so dry soils tend to retain exchangeable nutrients. Concentration of nutrients in seasonal forest topsoils is probably mostly biogenic, with some nutrients ascending by capillary ac-

Fig. 5.5 Forest soils of the seasonal tropics. *Top left,* red clay loam soil over basalt trap, supporting short deciduous teak forest, S. N. Rai as scale, Dharwad, Deccan Karnataka (P. A.); *top right,* humus-rich sandy clay soil, semi-evergreen dipterocarp forest, Huai Kha Khaeng (H. H.); *bottom,* ferricrete and edaphically dry soils; *left,* deep profile in the coastal plains of the Carnatic, West India, on a cut of the Konkan railway with deep, indurated ferricrete over pale anoxic profile (D. W.); *right,* ferricrete, mined and cut in its living, cheesy state, heaped to dry and harden into bricks, Malabar Coast, Kerala, India (P. A.).

tion from subsoils during the first few weeks of the dry season, when leaching has ceased. Thus, soils of the seasonal tropics tend to be richer in available nutrients (when moist) than those of everwet regions. In that case they are more fertile for annual crops—as well as for trees—than are soils of the lowland everwet tropics (fig. 5.5).

Soils in many parts of the lowland humid tropics are among the least fertile in the world, but slow nutrient release can produce forest stands of great stature and biomass, if not particularly high growth rates. The turnover rate of forest litter and its fate in the soil are crucial subjects generally ignored by

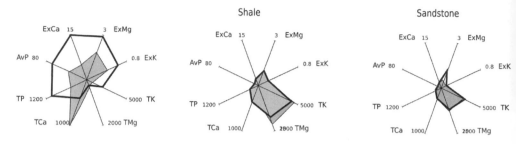

50 ha Seasonal evergreen, HKK

52 ha MDF, Lambir

Shale

Sandstone

Fig. 5.6 Stoichiometric roses, revealing exchangeable nutrient values (Ex) in the top half, and total values (T) in the lower half, on different linear scales. Nutrients include calcium (Ca), magnesium (Mg), phosphorus (P), and Potassium (K). *Left,* semi-evergreen forest, Huai Kha Khaeng, West Thailand; *center,* everwet tropical rainforest, udult clay loam soil, Lambir National Park, Sarawak; *right,* humult yellow sandy soil, Lambir. Nutrient concentrations, particularly exchangeable ones, are higher in seasonal climate soils. In everwet climates, soils—particularly those on sandstones—hold fewer nutrients (I. B.).

commercial crop agronomists, because commercial agriculture produces so little litter—though it is critical for swidden farmers (p. 338).

As soils age over geological time, elements in the soil solution are inevitably diminished. With age, certain constituents, such as siliceous sands and aluminum and iron oxides, may accumulate. One may even speak of the biological life of a soil,[4] which spans millions of years, during which it becomes enriched in sand and devoid of nutrients, with plants recycling minerals from the organic layer at the surface. This cycle, which persists only on gentle topography, can only be broken by erosion from the edge.

Measuring nutrients. Soil fertility is determined by the concentration of nutrients in solution, and tropical soils in general are often low in such nutrients. They may nevertheless contain relatively high insoluble nutrient concentrations (fig. 5.6), either in organic matter or in the mineral soil itself. In the first case, they consist of recycled ions within the forest cycle; in the second, they are a gradually released subsidy from the rock substrate. The second provides a more accurate estimate of the fertility of a soil for long-lived plants. Measures of *total mineral soil* nutrients correlate more closely with tropical forest composition and growth than measurements of any *available* nutrients.

Soils in Tropical Asia

The performance of plants depends in part on the soil in which they grow. The mineral soil's components largely determine soil porosity, which affects its capacity for storing water while allowing oxygen—essential for the re-spiring roots—to penetrate. Moisture availability, in dry months, is also important (table 5.1). All these factors, but especially nutrient availability, influence vegetation litterfall rates, the decomposition of litter (mostly leaves), and the timing of the release and return of nutrients through soil to roots. Thus, the mineral soil influences the structure, phenology, and composition of the forest, while the soil is in turn influenced by the composition and phenology of its forest cover.

Udult clay loams. The soils of the lowlands in the Asian tropics, on granite slopes, shale, and clay substrates, are *friable udult clay, or sandy clay loams* (fig. 5.2). They are relatively low in mineral nutrients, which increase with the length of the dry season (fig. 5.6). Soil patterns are made more complex by erosion and uplift. Even where the geology is uniform, changes in topography due to erosion and drainage affect soil quality on small scales.

Table 5.1 Effects of climate (number of dry months) and geology (from fine sediments to coarse sandstones) on types of soils in tropical Asia

Geology	Climate (number of continuous dry months)				
	0–1.9	2–4.9	5–6.5	6.6–8.5	8.6–11
Least freely draining, most clay rich; base-rich substrates	Friable udult clay loam	Friable udult clay loam	Friable udult clay loam	Friable udult clay loam	Friable udult clay loam
Medium base substrates: dacite, shale	Compact udult clay, silty clay ultisol	Udult clay, silty clay loam	Friable udult silty clay loam	Friable udult clay loam	Friable udult clay loam
Granite, bedded sandstone and shale	Udult/shallowly humult sandy clay ultisol	Udult sandy clay loam	Friable udult sandy clay loam	Friable udult sandy clay loam	Friable udult sandy clay loam
Some granite, sandstone-dominant sediments	Humult yellow sand	Shallowly humult yellow sand	Udult sand	Udult sand	Udult sand
Most freely draining, most siliceous: sandstone, raised beaches	Humic (humult) podzol	Shallowly humult podzol	(Absent)	(Absent)	(Absent)

Soils of base-rich substrates. Volcanic rocks are associated with tectonic movements in tropical Asia, as is evident from the volcanoes through Sumatra and the Indonesian archipelago (the "ring of fire"; see fig. 3.5). These rocks are infused with some silica, but principally include clay-yielding minerals such as andesites. These clays are predominantly long-chain molecules with abundant negative charges, which anchor nutrient cations. The crumbly texture of these soils provides good water penetration and working properties, as well as high cation-exchange capacities. Such regions have traditionally supported productive agriculture and dense habitation.

When magma pours onto the land surface in large, rapid flows, it mixes less with the silica-bearing minerals of surface rocks and is therefore more plastic, so that consecutive eruptions form flat layers. Eruptive basalt plateaus are particularly noteworthy in western India: the Deccan traps (fig. 3.2). The soils on basalt are heavy in montmorillonite, a clay with extremely high cation-exchange capacities. These are among the most fertile agricultural soils in the tropics and support the tallest known tropical trees. The British colonial agronomists called them "black cotton soils," or *regur*.

Humult oligotrophic soils. The term *oligotrophic* is used for soils that are deficient in nutrients, for which rainfall is a significant or the sole external source of nutrients. In the Asian tropics, especially Borneo, seasonal or periodic flooding from nutrient-bearing eutrophic river water restores nutrient loss and reduces acidity resulting from leaching, but anoxic subsoil conditions from waterlogging confine rooting to the region near the surface (fig. 5.4). Beyond the reach of floods, on the extensive plains of everwet regions, anaerobic conditions increase acidity and retard decomposition. Litter accumulates, and tree roots concentrate at the surface because of the anoxic conditions below. A horizon of raw humus, matted by fine feeder roots, develops above frequent flood level. *Peat soil* formation is thus initiated (see fig. 5.3, bottom), and rainwater is left as the sole source of additional nutrients.

In the lowland tropics, peats only form in the absence of a dry season; drought promotes oxygenation of the surface and rapid litter decomposition. The coasts of Borneo and Sumatra, together with more limited areas in Peninsular Malaysia and southwest New Guinea, bear almost 500,000 km² of lowland tropical peat swamp forest. As noted above, peat soil can only accumulate where the soil surface is above the flood limits of the rivers. Although the floodplains may be broader where the rivers emerge from the hills, farther out on the plain the peat gradually invades, so that the floodplains of the slow-moving, snaking rivers seldom exceed 100 m in width. At Marudi (northeast Sarawak), the peat on the broad, flood-free plain has reached a depth of 15 m, implying an accumulation rate of up to 4 mm y^{-1}. Outside Asia, such tropical peat swamps are rare.

On highly weathered sandstone ridges and Pleistocene raised beaches, oligotrophic soils also establish. In these soils, the mineral profile may be sandy or sandy clay loam, yellow to coppery due to thin coatings of iron oxide on the sand grains. Raw humus on such soils does not exceed 6 cm in depth. Alternatively, the mineral soil may be pure white silica sand with no capacity to hold nutrients. Here, rainwater, dust, and leaf litter entering from trees on more fertile soils become the sole nutrient sources; nevertheless, nutrients may accumulate in the raw humus, which can reach depths exceeding 30 cm. Soluble organic compounds—phenols and humic acids— are slowly leached from the raw humus to the base of the white sand, where they may precipitate to form a grey-black humic pan that eventually becomes impervious to water and roots (see fig. 5.3, bottom right). These soils are *tropical humic podzols*. In these soils, organic matter and soil surface horizons vary in pH between ~3.5 and 7, which slows decomposition relative to litterfall and favors fungal over bacterial activity. These soils resemble temperate *podzols* of the Russian soil literature, differing only in the absence of iron oxide in the pan. The forest ecosystem they support is called *kerangas* in Borneo (p. 176), implying land unsuitable for hill rice cultivation.

Substrates Yielding a Diversity of Soils

Particular types of rock produce quite distinctive soils. Four kinds are differentiated below.

Soils on limestone. Limestone is generally free of soil on steep surfaces. In the seasonal tropics, limestone, with its insoluble impurities, weathers to form somewhat brilliant orange-red, clay-rich soils. In the everwet Sunda lands (particularly inner Borneo), on the high karst of New Guinea, and in wet tropical and warm temperate seasonal southeast China and tropical northeast Vietnam, the drought-prone summits support deep blankets of raw humus wherever they escape fire.

Serpentine soils. Serpentine soils are particularly associated with volcanic activity along the edges of continental tectonic plates. They are derived from magnesium and iron-rich rocks, particularly olivine. Such *ultramafic* rocks also are high in nickel, cadmium, and cobalt, which are toxic to most plants at high concentrations. These soils are also very poor in essential nutrients, such as nitrogen and phosphorus. Ultramafic substrates in everwet climates may yield well-structured, base-rich, red-brown udult clay soils. They are usually granular and freely draining; raw humus accumulates, as in the tropical podzols, even though the mineral soil beneath is alkaline. Highly specialized plant communities, with many endemic species, establish on these soils, as on the upper slopes of Mount Kinabalu, Borneo. There, litter accumulation is exacerbated by a floristic change to tree species that bear thick, tough leaves that are rich in lignin. These decompose more slowly and contain fewer nutrients. They also contain more phenolic compounds, which mediate leaching of the phosphorus occluded on sesquioxides in the mineral soil.

Montane soils. With increase in elevation, soil temperatures and termite abundances decline. At intermediate elevations, montane soils characteristic of the tea-growing regions of lower montane and warm temperate climates in East Asia are seen, along with evergreen oak-laurel forests. These soils are well-structured udult loams, but with humus penetration deep into the mineral soil and high cation-exchange capacity, which indicates high fertility (see fig. 5.4). These tropical, mid-mountain soils therefore come to resemble temperate *mull* soils, termed *brown earths*, in which earthworms are responsible for most litter decomposition, burying the humus thereby generated as fine organic matter. Thus, the relatively impervious clay soil becomes crumbly and well aerated, permitting deep rooting.

Initially confined to ridges, *montane acid humult organic soils* may become ubiquitous above 2,000 m, wherever fogginess is not seasonal, and drip from foliage increases leaching rates, although their development is frequently reduced by slippage on steep slopes. On base-rich valley soils, and on mountains where cloud cover and diurnal fog are intermittent during the dry season, the mid-mountain, humus-rich mull soils extend to high altitudes. They become darker and shallower, enriched by humus and densely inhabited by earthworms. They grade into organic muck soils on moist sites. In the subalpine zone these humic muck soils, rich in often giant earthworms, are ubiquitous wherever soil has built up between rocks and on gentle surfaces, mixed with coarse sand and decomposing substrate.

Ferricretes. Ferricrete-bearing soils are widespread in seasonal tropical Asia wherever substrates yield clays rich in iron and aluminum sesquioxides (see fig. 5.5, bottom panels). The only currently active, "cheesy" ferrous and aluminum hydroxide accumulating soils are in the western coastal plains of

India, Burma, East Java, and the Lesser Sunda Islands. These regions have climates with a 5–7 month annual dry season, naturally supporting deciduous or semi-evergreen forest. Sesquioxides tend to concentrate in a single layer in soils of seasonally wet climates, where re-oxidation during dry seasons may lead to precipitation as an impervious, rocklike horizon, often meters thick. The effective depth of soil for root penetration may thus become limited. In more humid seasonal lowland climates, where active sesquioxide accumulation persists, the iron-rich horizon assumes a cheesy texture. Poor aeration at depth during the monsoon partially reduces the yellow-red iron oxides to bluish ferrous hydroxide, as seen in many floodplain soils of ever-wet regions. Exposure to the air through mining or erosion hardens such soil to terracotta, in which form it can be used for house construction. It was originally appropriately named *laterite*, from the classical Greek for "brick," but that term has become so misused that *ferricrete* is recommended instead. Eroding ferricrete pans in everwet forests are evidence of past seasonal rainfall climates.

A Mosaic of Soil Types

All the factors and processes described in this chapter lead to a diversity of soil types at all scales throughout tropical Asia (see table 5.1). Soils of nutrient rich clays extend in a *continuous* tract over lands from South Asia to Australasia. Although nutrient-poor soils widely extend over the metamorphic rocks of seasonal eastern peninsular India, they are restricted to *islands* on lands of varying soil on the metamorphic ranges of everwet southwest Sri Lanka, and throughout the Far East, from extensive plateaus in Indo-Burma, especially southeast Indo-China, to ridge-top archipelagoes along the sedimentary ranges of the everwet Sunda lands and wet seasonal Indo-Burma. Other upland soil types are yet more restricted.

Where regions were saturated with water, large areas of peat swamp soils established. With elevation, humus accumulated, and the montane soils established. For those young mountains established from volcanic activity, the more nutrient-rich, andesitic soils formed, and the specialized serpentine soils were limited to exposures of associated ophiolite rocks. Limestone, as areas of substrate or limited to outcrops, also influenced the types of soils. Local geological history has promoted the formation of quite different soils in relatively small areas, while local differences in topography have influenced the formation of different soils just a few meters apart. These soil distributions contribute to our discussion of tree and forest distributions in later chapters.

Notes

1. This chapter is a condensation of *OTFTA*, pp. 43–59, sections 1.4.1 to 1.4.3.2.

2. Baillie, Ian, "Soils of the humid tropics," in *The Tropical Rainforest*, by P. W. Richards, 2nd ed. (Cambridge: Cambridge University Press, 1996), 256–86.

3. Ashton, Louise. A., et al., "Termites mitigate the effects of drought in tropical rainforest," *Science* 363 (2018): 174–78; Nakagawa, Michiko, Masayuki Ushio, Tomonori Kume, and Tohru Nakashizuka, "Seasonal and long-term patterns in litterfall in a Bornean tropical rainforest," *Ecological Research* 34 (2019): 31–39.

4. Stark, Nellie, "Man, tropical forests, and the biological life of a soil," *Biotropica* 10 (1978): 1–10.

In the summer of 1955, three undergraduate friends and I ascended the lower Amazon in a wood-burning riverboat through the narrow channels that connect the great river with the Rio Para. I marveled at the burgeoning vines; noisy, colorful wildlife; and sumptuous flowers in the riverine forest canopy. It was the vastness of the landscape that impressed, and of the river, which was at times so wide that the other bank was below the horizon, and whose turbid flood rose and fell more than ten meters. Three years later, I entered Borneo dipterocarp forest for the first time, having arrived in Brunei the night before by launch from Labuan Island. The landscape of rolling hills was familiar, friendly, but the forest! Columnar buttressed giants with fissured bark and generous hemispherical crowns rising high above the tangled canopy, providing a fan-vaulted roof over a spacious open understory where walking was a cool delight. The distant calls of wildlife, hardly seen, provided an evocative, echoing, tranquility, the plaintive wail of the gibbons a reassuring background: such a serene contrast to the overbearing, guttural calls of those howler monkeys, their protests still ingrained on my memory. — P. A.

6

PLANTS OF THE
ASIAN TROPICS

Although the overall effect can be overwhelming, it is easy to pick out the different types of plants in a tropical forest, the large and small trees, the palms and woody monocots, the lianas and vines, the epiphytes, and distinctive low plants in the understory. These appear as guilds (plants with similar appearance and function), and their study was set in motion by the 19th-century scientist Alexander von Humboldt and his botanical colleague, Aimé Bonpland, from their voyage of discovery in the American tropics (1799–1804).* Humboldt never traveled to the tropics again, and he spent much of his life writing about the results of their research from this expedition. Humboldt and Bonpland established the principles for the study of the distributions and functions of plants in their *Essay on the Geography of Plants*.[1] In it, Humboldt developed the concept of the *physiognomy of plants*—a term for the analysis of human facial features to predict personality traits! Humboldt argued that the appearance of a plant indicated much about its distribution and function in nature. He also argued that the physiognomy of vegetation, understood as forest or similar plants, has an emotional and esthetic effect on us. Humboldt knew both scientists and artists. His ideas profoundly influenced the botanists who succeeded him, the establishment of the discipline of ecology, and the creative work of artists.

The advent of the theory of natural selection dramatically strengthened

* In this chapter *OTFTA* is cited in small sections throughout, and not as a single large section.

the value of physiognomic thinking. With it, the common forms of plants in distinct plant formations, as in rainforests in different regions of the world, could be understood as the results of evolutionary responses to common physical conditions. Hutchinson's metaphor of the evolutionary theater and the ecological play seemed apt.[2]

Plant Functional Types

The physiognomy of a plant was redefined by succeeding botanists as its life-form and, more recently, as its functional type. The study of the functions of tropical plants was enabled by the establishment of the colonial tropical gardens in the tropics, notably by the Dutch at Buitenzorg (now Bogor) in Java. European scientists studied plants there, contributing to the establishment of an early school of functional ecology: physiological plant anatomy.[3] The Danish botanist Eugen Warming (1841–1924) studied plant life-forms and helped establish the science of ecology, and his student Christen Raunkiaer (1860–1938) developed the first system of describing plant forms, one that is still used today.

Raunkiaer defined plants by their growth strategies and their leaves.[4] The permanent growing parts of the plant define its growth form. Annual plants without such parts are *therophytes*. Those with parts just above the ground are *chamaeophytes*. Shrubs and trees, with erect, permanent aerial growing parts, are *phanerophytes*—all parts of a life-form spectrum. In addition, Raunkiaer stressed the importance of leaf-blade size (width or length of an ellipse, or a disk's diameter; see table 6.1), whether simple or compound: *leptophyll* (≤ 0.25 cm^2), *nanophyll* (≤ 2.25 cm^2), *microphyll* (≤ 45 cm^2), *notophyll* (≤ 67 cm^2), *mesophyll* (≤ 135 cm^2) and *macrophyll* (≥ 405 cm^2). A square inch is 6.5 cm^2, and a typical dipterocarp leaf (fig. 6.7) is a mesophyll. Rankings of leaf sizes help in comparing forest structure (see table 10.1), and forests might be named according to predominant leaf size (e.g., a notophyll forest).

Following this research, Elgene Box of the University of Georgia developed a system of plant functional types (PFTs) based on growth form, leaf type, and leaf appearance.[5] He intended this classification as an aid in the physiological description of vegetation and in vegetation mapping, and it has been widely adopted. Of his 90 PFTs, 42 of them occur in tropical forests, 20 or so have been used in the quantitative modeling of biomes, and 21 of them are tropical trees of different sizes and appearances.

Table 6.1 Leaf sizes, using a modified Raunkiaer ranking, in tree samples of various tropical Asian forests (from tables 4.6 and 4.8 in *OTFTA*). NA = not available.

Forest	Leaf Sizes (% classes)							
	Leptophyll	Nanophyll	Microphyll	Notophyll	Mesophyll	Macrophyll	Megaphyll	Other
Mixed dipterocarp forest, Brunei	0	0	7	20	68	15	0	NA
Peat swamp, Brunei	0	0	12	30	57	1	0	NA
Kerangas, Brunei	4	1	26	37	32	1	0	NA
Lower montane forest, Doi Inthanon	0	0	21	48	31	0	0	0
Lower montane forest, Bhutan	1	0	26	36	25	0	0	17
Hill dipterocarp forest, Kinabalu	1	0	15	34	45	4	0	18
Lower montane forest, Kinabalu	2	2	36	36	23	12	0	7
Upper montane forest, Kinabalu	3	8	57	22	10	0	0	0
Subalpine thicket, Kinabalu	17	18	35	27	2	0	0	0

Trees

In order to elucidate the structures of the tropical forests that are the subject of this book, we need to examine trees for traits that have evolved in different climatic conditions, as well as for the systematic differences that account for the estimated total of 24,819 species in all tropical Asian forests (p. 11). In this book, we consider the outcomes of competition among individual trees, involving their shoots above ground or their roots below, which result in the diverse dynamics and consequent structures of tropical Asian forests. Traits (architectural, anatomical, and physiological) have evolved that provide individual species with advantages in specific environments. The success of these traits frequently involves trade-offs; for example, the advantage gained by early spread of a dense crown that excludes the sun from the leaves of other plants below comes at the cost of water loss and potentially lethal heating. The nature of competition below ground is difficult to study and still poorly understood. We are largely ignorant of the traits involved and their advantages and costs, since the relevant species are inadequately studied, especially in the Asian tropics.

Shade tolerance. Tree heights and foliage layers in a forest determine the amounts and qualities of sunlight penetrating to the forest floor and produce gradients of light at different levels within the forest. Forest structure is revealed by the construction of forest profile diagrams (fig. 6.1) in which the lower levels consist of a few young canopy trees. Most trees do not attain the canopy and grow in more shady conditions.[6] The amount of light available for photosynthesis within the forest is not well documented, but it is reduced exponentially—hence the logarithmic scale in fig. 6.1.[7] Within the canopy there are dramatic reductions in light available for photosynthesis and growth. Light corresponds roughly with the visible region of the electromagnetic spectrum (fig. 6.2), which includes photons in the region of 400–700 nanometers (nm). The units are in $\mu mol\ m^{-1}\ s^{-1}$ within that range, conveniently abbreviated as PAR (photosynthetically active radiation).

Gas exchange. Plants vary in photosynthetic capacity under different light conditions. With the portable equipment available, it is relatively easy to determine a plant's photosynthetic response to changing light conditions by producing a light-response curve (fig. 6.2).[8] Each curve reveals the maximum photosynthesis rate, the light-compensation point (no net CO_2 fixa-

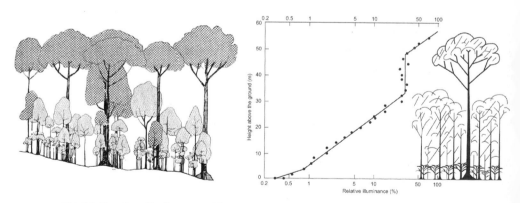

Fig. 6.1 Forest profile diagram and light gradient produced by foliage layers in everwet forests. *Left*, dipterocarp-dominated forest on clay loam, former Bok-Tisam protected forest, Sarawak. Cross-hatched crowns are dipterocarps, lighter crowns are understory species, a late-successional stand (P. A.); *right*, reduction in sunlight passing through forest layers at Pasoh Forest Reserve, Peninsular Malaysia. Levels on the forest floor are less than 3% of the level above the canopy (I. B., after K. Yoda).

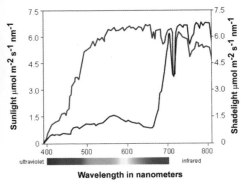

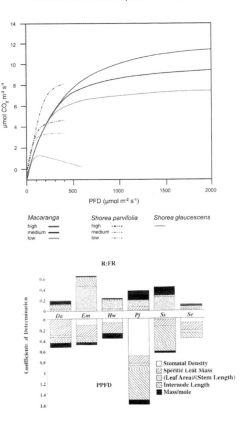

Fig. 6.2 Plasticity of plant responses to shade. *Top left,* spectral distribution of radiation in a gap and secondary forest understory, Peninsular Malaysia (David Lee, *Nature's Palette: The Science of Plant Color* [Chicago: University of Chicago Press, 2007]). PAR is the total of quanta 400–700 nm, or the color bar (synonymous with PFD, photon flux density, and PPFD, photosynthetic photon flux density). R:FR is the quantum ratio 660/730 nm. *Top right,* photosynthetic light-response curves of three trees, early successional *Macaranga beccariana,* late successional *Shorea parvifolia,* and a shade-tolerant climax *Shorea glaucescens.* The first two species were grown in low, medium, and high light environments (I. B.; and see Stuart J. Davies, P. A. Palmiotto, P. S. Ashton, H. S. Lee, and J. V. La Frankie, "Comparative ecology of 11 sympatric species of Macaranga in Borneo: Tree distribution in relation to horizontal and vertical resource heterogeneity," *Journal of Ecology* 86 [1998]: 662–73; A. S. Moad, "Dipterocarp juvenile growth and understory light availability," PhD thesis, Harvard University, 1992). *Bottom right,* responses to low and high PAR versus low and high R:FR, in 6 Malaysian everwet rainforest trees: Da = *Dryobalanops aromatica;* Em = *Endospermum malaccense;* Hw = *Hopea wightiana;* Pj = *Parkia javanica;* Ss = *Shorea singkawang* and Se = *Sindora echinocalyx* (D. L.; see chap. 6, n. 9 in the text).

tion), and the dark respiration rate. The slopes of the rise reveal the quantum efficiency of photosynthesis. The highest maximum photosynthesis rates and light-compensation points here are in *Macaranga,* and the lowest are in *Shorea glaucescens.* In the latter, the significant drop in photosynthesis at higher light levels is probably due to photoinhibition. In both *Macaranga* and *S. parvifolia,* growth conditions strongly influenced photosynthesis. Photosynthesis rates should predict overall growth rates, which integrate the photosynthetic totals under different light conditions over time. Transpiration rates by the same leaves are easily measured. The ratio of molecules of CO_2 fixed per molecules of H_2O transpired is the *water use efficiency*, which is higher in drought-resistant trees. Many other intrinsic plant factors influence photosynthesis, including the stoichiometry of the light-reaction components, the levels of chlorophyll *a* and *b*, leaf anatomy, and leaf angle and distribution.

Plasticity. Trees vary in their capacity to respond to different growing conditions, particularly light. In addition to intensity, or PAR, spectral quality is also important, specifically the reduction of the red (660 nm) to far-red (730 nm) ratio (R:FR), which affects many growth responses via the phytochrome system. In a rainforest understory and gap, PAR and R:FR vary from ~13 μmol m^{-2} s^{-1} and R:FR 0.20 in the understory (~1% of full sun), to 1,528 and 1.34 in full sunlight. In general, light-tolerant and early successional species are more plastic in their responses to shade, with regard to both light intensity and spectral quality. Shade-tolerant understory species and long-lived climax species are less plastic in growth responses (fig. 6.2).[9] The most plastic in light response (especially in PAR) is an early successional tree, *Parkia javanica*, and mature-phase canopy trees are the least plastic. In similar growth experiments with Sri Lankan *Shorea* species, Mark Ashton showed a strong correlation between growth response to shade, as well as allocation of mass to different plant organs, and the distributions of species on ridges and valleys in the Sinharaja forest (p. 172). Plasticity of responses to environmental conditions is evident in many morphological and physiological parameters. Within a species, plasticity can be observed at a single developmental stage (e.g., among seedlings) and in comparing developmental stages, such as saplings versus mature individuals (as in *Endospermum malaccense*).

Leaves. Leaves vary dramatically among tree species in the tropical rainforest (fig. 6.3). Most trees produce leaves that are intermediate in size, elliptical, and often with pronounced, pointed drip tips. The leaves of early successional species, however, can be quite large. The function of leaves of different sizes is partly explained by the boundary layer of still air, which reduces the diffusion of gases and heat. A smaller width (reduced even more by lobing) reduces the boundary layer, which helps remove heat and increase transpiration. Although some leaves are lobed, few leaves in the tropics have teeth. In addition, leaves vary in texture—and quite dramatically in longevity: their lifespan may vary from 2 to 43 months in a single forest.[10]

Fig. 6.3 Leaves. *Left*, Leaf diversity beneath the canopy of everwet forest at Lambir National Park, Sarawak (S. R.); *center*, needle-leaf trees, or *Pinus kesiya*, Mountain Province, Luzon, Philippines (P. A.); *right*, coriaceous leaves of *Leptospermum javanicum*, Genting Highlands, West Malaysia (D. L.).

Although leaves of understory plants may be herbaceous, especially beneath the canopy and among deciduous species, most are leathery (coriaceous). Under more dry conditions, many leaves are extremely tough (sclerophyllous), and leaves often produce pubescence along veins or on undersurfaces, particularly when at or above the canopy.

Since leaves are determinate organs (they reach a final size quickly after initiation), their production allows a plant to most efficiently forage for light. A larger leaf captures light more efficiently, and smaller leaves allow light to penetrate beneath (thereby providing some light for leaves beneath). Really large leaves are more expensive to produce, and are common in hot, wet and sunny tropical climates; very small leaves are more closely coupled to the climate of the surrounding atmosphere, and are common at high tropical elevations (table 10.1).[11]

Leaves, although determinate, vary considerably in size within a species under different growing conditions—often larger when produced in shade. In the Sri Lankan *Shorea* species (p. 172), *S. megistophylla* produced the largest leaves and the greatest response in size to growing conditions. *S. worthingtonii* produced smaller leaves, with the least plasticity in leaf size.

Most trees produce simple leaves, but several (including Fabaceae, Meliaceae, Burseraceae, Sapindaceae, Arialiaceae, Rutaceae, Bombacaceae, Bignoniaceae, and the montane family Cunoniaceae) produce compound leaves. The advantages of compound leaves, in which a large photosynthetic area results from the initiation of a single leaf, compared to that of smaller simple ones, has never been satisfactorily explained. Stephen Jay Gould and Richard Lewontin dubbed them *spandrels of San Marco,* alluding to the ornaments that cap the walls of Venice's cathedral.[12] Such a character appears to be adaptively neutral, escaping competition over geological epochs. But we cannot confirm neutrality, since the possibility remains that an adaptive function might yet be discovered. For comparing leaf size, a single blade unit (i.e., leaflet) can be compared to a simple leaf. We have much to learn before we really understand the functions of leaf sizes and shapes.

Architecture and height. All plants grow by producing leaves on branches, and by producing new branches from axillary buds. The ways in which these buds grow determine the structure of the plant, including trees. The eminent French tropical botanist Francis Hallé observed the different processes for the growth of branches, along with the types of branches composing a tree, and he developed the concept of tree architecture.[13] He wrote, "The concept of architectural modeling is a dynamic one, since it refers to the genetic information which determines the succession of forms of the tree, analogous to the blueprint which is the plan of a machine." He based his classification, in which he named different models after scientists who had studied the relevant plants, on the types of branches produced during tree growth (fig. 6.4).

Every branch has an apical meristem, which produces leaves and elongates to allow the leaves to be positioned along its length. Additional (axillary) meristems establish from the base of each leaf. The concept of individuality for a plant is different than for most animals, and a plant is sometimes considered as a "metapopulation" of these semi-autonomous units, the shoot apical meristems and underlying stems. Shoots vary in a variety of ways. First, they differ in the positions of leaves. Shoots may produce leaves around the stem, seen as a spiral, whether in an alternate or opposite position (*orthotropic*), or they may produce leaves in a flat plane (*plagiotropic*). In some cases, the shoot extends very slowly at first and more rapidly later. Alternatively, the shoot may extend rapidly at first and very slowly later. The slow growth then crowds the leaves at the tip of the shoot. Finally, some shoots may only grow very slowly, showing little branch extension but producing many leaves together. Some shoots grow continually, and then apical meristems some distance below the tip may produce secondary shoots. If the apical meristem is not damaged, the shoot can grow indefinitely, perhaps even for centuries (*monopodial*), and lateral branches are produced beneath the tip. However, the apical meristem can become reproductive, producing

Fig. 6.4 Tree architecture. *Top left*, model of Corner; *left center*, tree fern (*Cyathea latebrosa*), Rimba Ilmu, Kuala Lumpur, Malaysia; *right center*, model of Aubréville; *right*, pagoda tree (*Terminalia catappa*), Miami, FL. *Bottom left*, model of Troll; *center*, poinciana (*Delonix regia*), Miami street; *right*, kempas (*Koompassia malaccensis*), below Frazer's Hill, West Malaysia (D. L., and drawings after F. H.).

a flower (or inflorescence—many flowers) and fruit. This stops the future growth of the shoot tip, but additional growth results from the activity of one or more axillary meristems just beneath the tip (*sympodial*). Thus, the positions of flower production can dramatically alter the growth of a plant. From his study of the branching of seedlings and saplings, primarily in tropical trees, Hallé discovered 24 architectural models.

In the architectural model of Corner (fig. 6.4), the single trunk is produced from a single meristem at the tip, which continues to grow for the life of the tree. The leaves are produced as a rosette at the tip, and flowering does not disrupt the activity of this meristem. The most obvious example is the palm (or Box's palmiform tuft trees and treelets), but a few broad-leaved tropical trees also grow by this model. In the model of Troll, the single plagiotropic shoot grows up and then bends over to grow laterally. Additional height is produced by a succession of plagiotropic shoots, and a thick trunk is ultimately produced from the secondary thickening of the branches. This model usually produces a low and spreading tree, such as the ubiquitous poinciana tree (*Delonix regia*), but extremely lofty trees, such as *kempas*

(*Koompassia excelsa*), also grow by this model. Both of these are members of the Fabaceae, where this model is universal among its shrubs and trees. The archaic pantropical tree family Myristicaceae also has only one model (Massart's), with single leader and plagiotropic branches forming a spire like a fir tree. They share other restricting characteristics and have survived at least 40 million years, yet they are confined to the rainforest subcanopy, where they are still abundant. Other families manifest diverse architecture and habit. Five models have been observed among the dipterocarps (p. 104), rarely the models of Corner and Troll. Most commonly, family members grow by a monopodial trunk that produces plagiotropic lateral branches, continuously (Roux) or rhythmically in tiers (Massart). The striking growth pattern of the pagoda tree (*Terminalia catappa*) results from the production of two branch types (model of Aubréville). The initial shoot grows orthotropically. At intervals it produces rosettes of lateral branches that extend and then cease growing to produce bunches of leaves at the tips; further lateral growth continues by relays of sympodial branching. This pattern is often referred to as terminalia branching, as it is common in this pantropical genus. Although this pagoda branching appears to provide exposure to sunlight on different planes, this same model, with different branch angles, can produce quite a different crown, such as that of the gutta percha tree (*Palaquium gutta*). Several other models can produce a pagoda-like crown. The rainforest has harbored a diversity of successful tree architectures over geological time; extraordinarily, however, the functional differences among architectural models are difficult to discern, and those spandrels must again be recalled. Corner had also observed in trees that thicker branches (more pachycaulous) bear larger and more elaborate appendages, such as the fruits of a durian tree.[14]

In tropical forests, trees are adapted to grow to certain heights (fig. 6.1), such as extremely tall emergents above the canopy (e.g., *kempas*). Others grow to attain the canopy, forming a mostly continuous layer with other species. Then subcanopy trees grow to lower heights (these are described as functional types by Box), and very short trees and shrubs occupy the low understory.

Sean Thomas has shown that the general mechanism by which species attain a maximum height is *asymptotic growth*.[15] In older trees, the units of growth increment become miniaturized, including leaf size, and rate of increase is steadily decreased toward a maximum height. Once that is achieved, periodic die-back of the crown often occurs following a drought. Maximum height varies within species, becoming shortest in drought-prone habitats. We used to think that the highest rainforest trees were around 70 m tall, but high-quality remote sensing of Sabah forest has revealed individuals within the yellow merantis (*Shorea* sect. *Richetioides*) of more than 90 m. The tallest one was discovered by Greg Asner (Carnegie Institution Airborne Observatory LiDAR) at 94.1 m.[16]

Roots. Like shoots, roots grow asymptotically, but their extension, intermittent or continuous, and in great contrast to that of shoots, is nonmodular. A descending tap root arises first from the seed; then, on the axis of the seedling stem, lateral roots arise from it adventitiously, concentrating at depths where nutrients and water are available. Mycorrhizae (p. 269) are associated with roots in most tropical trees.

Growth and lifespan. Trees grow at different rates.[17] Although photosynthetic light response is a primary influence, many other traits are just as important: relative allocation of energy to producing leaves versus roots, stems, or reproduction; efficiency of leaf display; and leaf anatomy. Early successional trees grow rapidly, and mature-phase trees grow more slowly. Among the latter, understory trees adapted to shade grow slowly, and light-hardwood species generally grow more rapidly than heavy-hardwood species. (This does not apply so well among deciduous tropical trees, teak, and some rosewoods [*Dalbergia*], which grow surprisingly fast.) Thus, wood density is generally inversely correlated with tree growth rate. Also, large trees have low productivity rates in relationship to biomass.

High wood density is the result of vessel elements with very small diameters and thick, lignified walls, as well as greater density of highly lignified fibers. Such wood is less efficient in transporting water under negative tension toward the crown, but the small diameters reduce the risk of air embolisms (or cavitation). Dense hardwood trees grow on ridgetops and in freely draining soils, where drought could increase the risk of cavitation. The greater carbon investment in such wood reduces overall growth rates, as measured by trunk diameter or height increase. Such high-density hardwoods are economically valuable but often endangered from over-exploitation. Large trees are disproportionately important in tropical forests, predictors of total above ground biomass and basal area of the surrounding forest.[18]

Risks of mortality are present at all stages of tree growth: lack of light and herbivory in seedlings, branch-fall in saplings, liana and epiphyte loads, disease, storm damage (wind and lightning), and extreme drought in mature canopy trees.[19] Since trees in the everwet tropics do not produce reliable annual growth rings, it is difficult to assess their ages and life spans. Stem-diameter increments measured over a decade or more can be extrapolated at considerable risk in accuracy. Tree life span typically ranges from 200 to 400 years, but the slow-growing, heavy hardwoods may reach ages 2–3 times greater. Large individuals of ironwood (*Eusideroxylon zwageri*) may reach 1,000–1,500 years, and large *Shorea* species (e.g., *Shorea superba*) are similarly dense and large. Radiocarbon dating is not very accurate at these age ranges.

In contrast, the early successional species may live 10–20 years. In tropical deciduous forests, more species do produce growth rings, but they are not well studied. The largest specimens of teak (*Tectona grandis*) in India and Thailand may reach ~1,500 years in age.

Phenology. Trees vary greatly in the frequency and duration of the production of expendable organs: leaves, flowers, and fruits. We are largely ignorant of this "behavior" in most tropical trees. The classical study of Lord Gathorne Medway in Malaysia documented the diversity of patterns within an everwet tropical forest (fig. 6.5).[20] Three species were evergreen, with periodic production of new leaves. Two of the species were briefly deciduous, every year and once in five years. One species produced fruits (fruits within syconia—figs) frequently each year and asynchronously among individuals. Individuals of another species, *Shorea dasyphylla*, flowered and fruited mas-

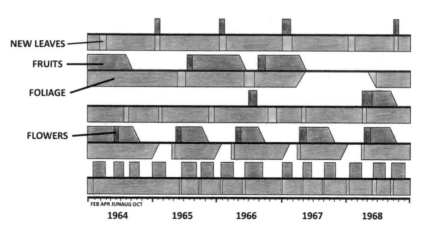

NEW LEAVES

FRUITS

FOLIAGE

FLOWERS

FEB APR JUN AUG OCT   1964   1965   1966   1967   1968

Fig. 6.5 Phenology diagrams of five everwet forest species in Gombak, West Malaysia. From top to bottom: *Cynometra malaccensis* (Fabaceae), *Garcinia* sp. (Clusiaceae), *Shorea dasyphylla* (Dipterocarpaceae), *Erythroxylum cuneatum* (Erythroxylaceae) and *Ficus sumatrana* (Moraceae; D. L.; data from L. Medway, "The phenology of a tropical rainforest in Malaysia," adapted by D. Lee in *Nature's Fabric* [see chap. 6, n. 1).

sively only once in the five years, an example of *mast flowering and fruiting,* an important phenomenon in the Asian tropics (p. 255).

In seasonally dry climates, virtually all species drop their leaves after the end of the rainy season and regain them before or immediately after the arrival of next year's rains. Box referred to these trees as "raingreen" in contrast to evergreen in everwet forests and "summergreen" in temperate deciduous forests. Such raingreen species include those that are shortly deciduous, but many remain evergreen, dropping leaves and simultaneously flushing new ones at the same time during the dry season. Phenology is an important and evolved response to trees' survival in different climates.

Seeds. Tropical trees vary in the size and viability of seeds (fig. 6.6).[21]

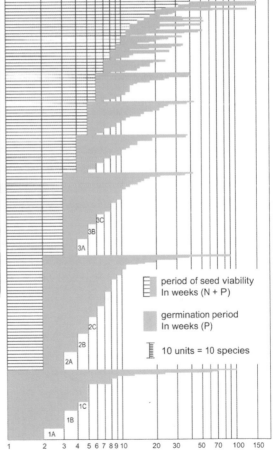

period of seed viability
In weeks (N + P)

germination period
In weeks (P)

10 units = 10 species

Fig. 6.6 Seeds and seedlings. *Left,* seedlings of *Shorea laxa* and saplings from an earlier fruiting event, Lambir (S. R.); *right,* frequency of peninsular Malaysian tree species according to seed viability (potential dormancy: N + P) and time to germinate (I. B., after Francis Ng, "Germination ecology of Malaysian woody plants," *Malayan Forester* 43 [1980]: 406–37).

Large seeds provide a greater store of nutrients for germination and early growth into seedlings, including descent of a taproot, while small seeds provide less storage, and propagules require optimal conditions for survival from the get-go. A general problem with tropical seeds is their lack of dormancy, although the length of time from dispersal to germination varies. Since such seeds cannot be stored for later use by reducing moisture content and lowering temperature, such seeds are said to be *recalcitrant*. This is particularly true for the understory and canopy trees of mature, everwet forests, whereas in deciduous and semi-evergreen species of the seasonal tropical forests, seeds may remain dormant for months until the rains arrive. In early successional plants, such as *Macaranga*, seeds may remain dormant in the soil for years; germination may be triggered by light penetration and temperature after disturbance. In a few late successional leguminous trees and *Anisophyllea* (Rhizophoraceae), seeds remain dormant until their shells are bitten off or rot (p. 258). Fruits and flowers are discussed in separate chapters (11 and 12).

Such plant functional types, including leaf sizes, can be used to model forest dynamics and productivity.[22] All the traits just discussed may vary within and among tree families. Although the concept of functional types may be generally useful in studying forest structure and mapping biomes, there is considerable variation within just a single functional type.

Diversity and dipterocarps. The largest difference between the tropical forests of Africa, the Neotropics, and Asia is a single, ecologically dominant family in the Asian tropics: the dipterocarps, or trees of the family Dipterocarpaceae (fig. 6.7; table 6.2).[23] This family consists of trees, and all 475 or so species within the subfamily Dipterocarpoideae are distributed in tropical Asia. These comprise the great majority of all emergent canopy trees, but many species are of the main canopy or subcanopy. In South America, there are 2 species within other subfamilies, and there are 29 species across Africa and Madagascar. The evergreen and some deciduous forests in tropical Asia feature dipterocarps in the genera *Shorea* (196 species), *Hopea* (104 species), *Dipterocarpus* (70 species), and *Vatica* (65 species). This is the single most important family of tropical hardwoods; many of the species are ecologically well known, and timber properties are well documented. This family is so important in the region that we call the prevailing evergreen, lowland forests where several of their emergent species dominate, *mixed dipterocarp forests* (MDF). The flowers are bisexual and vary considerably in size. The fruits are mostly winged, a nut with one seed carried by

Table 6.2 The taxonomic hierarchy in the classification of a single Asian tropical forest species, *Shorea curtisii*, seraya. Family, scientific, and common names are emphasized in this book. The general name at any level is *taxon* (pl., *taxa*), and clades are intermediate levels of classification without formal names.

Taxon	Level	Common names
Plantae	kingdom	all plants
Spermatophytes	division	seed plants
Angiospermae	class	flowering plants
Eudicots	clade	higher dicots
Rosidae	clade	rosids
Eurosids II	clade	malvids
Malvales	order	
Dipterocarpaceae	family	dipterocarps
Dipterocarpoidae	subfamily	Asian dipterocarps
Shorea	genus	
Mutica	section	
Shorea curtisii Dyer ex King	species	seraya (common Malay name)

two or more wings derived from the flower sepals. Members produce resins, perhaps conferring some herbivore resistance and providing chemical signals to mycorrhizae. We offer much more discussion of this family in later chapters.

Another family of importance in the region is the Fagaceae, with ~25 species of *Quercus*, 2 of *Trigonobalanus*, <120 species of *Castanopsis*, and ~300 species in *Lithocarpus*, all but a few of which are tropical.[24] These are important in lowland and montane forests. The family Moraceae is well represented in Asian forests, and vastly richer in species than elsewhere. There is a much greater diversity in the Moraceae, both in figs as trees and climbers, and in the main canopy tree, *Artocarpus*. Because of their abundance and frequent fruiting, they are extremely important food sources.[25] Within the Myrtaceae, whose understory trees are common throughout the tropics, is a single very large genus in tropical Asia: *Syzygium*, with some 600 species.[26] In tropical America there is a parallel genus with a similar diversity: *Eugenia*. *Garcinia* (Clusiaceae) and *Diospyros* (Ebenaeae) are also common understory tree genera.

Fig. 6.7 The Dipterocarpaceae. *Left*, flower and foliage of *Vateria acuminata*, Tropical Botanical Garden and Research Institute, Kerala, India (D. L.); *center*, *Dipterocarpus obtusifolius* (J. G.); *right*, fruiting crown of *Shorea amplexicaulis*, Gunung Palung National Park, West Kalimantan (© Tim Laman).

Palms and Other Woody Dicots

Palms are iconic tropical plants, and they are relatively abundant throughout the tropics. In Asia, they are particularly important in the understory. Remarkable in Asian palm diversity are the rattan climbers, particularly in the genus *Calamus* (~400 species; fig. 6.8).[27] With spines on their flexible stems and spiny hooks on their fronds, these plants climb large trees and stretch from crown to crown, reaching lengths of 200 meters and more.

Screwpines are woody monocots in the old-world tropical family, Pandanaceae (fig. 6.8). Almost all species (~1,000) are in a single genus, *Pandanus*. They occur in a variety of habitats, from forest understory to forest margins. In contrast to palms, they produce aerial branches. An ecologically equivalent family, the Cyclanthaceae, occurs in the Neotropics.

As PFTs, bamboos are called *arborescent grasses*—not exactly trees and unbranching, but tree-like in their size and structural influence. Although bamboos occur in the Neotropics and Africa, they are particularly important and species-rich in Asia (fig. 6.8). Their economic importance rivals that of the rattans.

Understory Plants

Most of the families important in understory shade are herbaceous: Araceae, Begoniaceae (which are particularly diverse in tropical Asia), Gesner-

Fig. 6.8 Palms and woody monocots. *Left*, a rattan palm rises through a narrow gap, Lambir National Park, Sarawak (S. D.); *center*, bamboo (*Bambusa bambos* thicket), Huai Kha Khaeng, Thailand (H. H., P. A. for scale); *right*, female plant of *Pandanus tectorius*, Fairchild Tropical Botanic Garden, Miami (D. L.).

iaceae, Rubiaceae, Melastomaceae, and Acanthaceae, are found throughout the tropics. There are some differences among genera and PFTs. In Asia, rhizomatous monocots in the order Zingiberales are important components, spreading laterally (fig. 6.9). In the Neotropics the most important genera are *Heliconia* (Heliconiaceae) and *Calathea* (Marantaceae). In Africa, members of the Marantaceae are important. In the Asian tropics, bananas (Musaceae) are common, particularly in regenerating forest. A striking number of herbs, more than in other tropical regions, produce blue, iridescent leaves and variegated color patterns.

Lianas and Vines

Climbing plants, both herbaceous climbers and woody vines, or lianas, are an important component of tropical forests—the wetter, the more. There are some differences in climbers in the Asian tropics (fig. 6.10). First, are the contributions of woody monocots, the rattans and climbing pandans. Second is the much-reduced diversity and abundance of fleshy, especially aroid, climbers. In the New World tropics, *Philodendron* (~500 species) and *Monstera* (50 species) dominate, but these are absent from the Asian tropics. The smaller aroid climber *Scindapsus* is present, and *Rhaphidophora* (~100 species) is a good substitute. All these are root climbers, and most

Fig. 6.9 Forest herbs. *Left*, wild bananas in a forest gap, Salawin National Park, Thailand (H. H.); *center*, rare *Begonia raja*, Rimba Ilmu, Universiti Malaya, Malaysia (D. L.); *right*, understory ginger, *Etlingera* sp., Ulu Gombak Forest Reserve, West Malaysia (D. L.).

Fig. 6.10 Vines and lianas. *Left*, *Piper* sp., Ulu Langat, West Malaysia (D. L.); *center*, aroid vine *Rhaphidophora megaphylla*, Menglun Forest Reserve, south Yunnan, China (D. L.); Liana, cf. *Bauhinia* sp., Menglun Forest Reserve (D. L.).

grow as hemi-epiphytes; they start out from the ground, grow up into trees, and eventually sever their ground connections. Some figs grow into stranglers, especially on nitrogen-rich soils. As throughout the tropics, spectacular lianas, with twisted and often flattened stems grow into the canopy. *Bauhinia* is an important liana genus in Asia, as it is elsewhere in the tropics.

Fig. 6.11 Epiphytes, from lowland forests. *Left*, orchid *Grammatophyllum speciosum* (© Tim Laman); *center*, fern epiphytes in upper canopy of Bupang Forest Reserve, Xishuangbanna, Yunnan, China (D. L.); and *right*, phytotelmata-producing epiphytes, *Drynaria* and *Asplenium*, Ulu Kenaboi, Negri Sembilan, Malaysia (D. L.).

Asian lianas are light-tolerant, and their abundance indicates past canopy damage.

Epiphytes

These plants grow on branches in tropical forests—the wetter, the more abundant. The diversity and abundance of epiphytes are reduced in the Asian tropics, compared to the Neotropics. Two important New World epiphytic families, the Cactaceae and the Bromeliaceae, are completely absent from Asia. Orchids are important, but the species-rich genera are different (fig. 6.11). In the Neotropics, the important genera are *Pleurothallus* and close cousins, with ~4,000 species. In Asia, *Bulbophyllum* (~4,000 species) and *Dendrobium* (~1,000 species) are abundant. Epiphytic ferns may be more important in the Asian tropics, and several genera produce nest-leaves (phytotelmata) that collect litterfall and form habitats for other organisms, such as the birds-nest fern (*Aspleniun nidus*), *Drynaria* sp. and *Platycerium* sp. In the Apocynaceae, *Hoya* and *Dischidia* are common, and often associated with ants. *Rhododendron* (Ericaceae) of the mountain forests is replaced in the Neotropics by *Cavendishia*.

A feature of many epiphytes throughout the tropics is their modified photosynthetic uptake of CO_2, crassulacean acid metabolism (CAM). These plants produce succulent leaves in which they store CO_2 at night as 4-carbon organic acids, when water loss via transpiration is minimal. During the day, with stomata shut, they decarboxylate those acids and release the CO_2 for carbon uptake through the Calvin cycle.

Animals and Plants

Similar stories can be written about animals, about functional types and convergent evolution.[28] Animal families differ among the tropical regions. The vertebrates of Asian tropical forests are far less diverse than those of the Neotropics (p. 224). The New World tropics have the extraordinarily species-diverse hummingbirds (Trochilidae) as important pollinators of many plants. These are absent from the old-world tropics, but in Southeast Asia sunbirds (Nectarinidaea) and, further east, honey eaters (Melaphagidae) perform similar functions. Tropical forests feature large-billed frugivores (and seed dispersers). In the Neotropics, toucans (Ramphastidae, 40 species) are highly visible. In the Asian tropics, the hornbills (Bucerotidae, 53 species) serve a similar ecological function. Parrots occur in all regions, but the genera vary. Bats are important pollinators and fruit dispersers throughout the tropics but are less diverse in the Asian tropics. There, members of the Pteropididae (an old-world and primarily tropical family) feed on nectar, pollen, and fruits. Thus attracted, they pollinate flowers and disperse seeds (p. 243). Their most spectacular members are the flying foxes. In the African and Asian tropics, apes are consumers and dispersers in their forests, although there are only four species overall. Elephants are also potent ecological forces in both areas. Vertebrates are concentrated in Asian rainforests wherever giant figs, free standing and hemiparasitic, are abundant. Animals are important in the pollination and dispersal of trees, as discussed in chapters 11 and 12.

Notes

1. Lee, David, *Nature's Fabric: Leaves in Science and Culture* (Chicago: University of Chicago Press, 2017), chapters 1 and 4; Humboldt, Alexander von, and Aimé Bonpland, *Essay on the Geography of Plants* (Chicago: University of Chicago Press, 2013); 1805 edition, trans. Stephen T. Jackson.

2. Hutchinson, G. Evelyn, *The Evolutionary Theater and the Ecological Play* (New Haven, CT: Yale University Press, 1965).

3. Lee, *Nature's Fabric*, chapter 4.

4. Wikimedia: https://en.wikipedia.org/wiki/Christen_C._Raunkiær

5. Box, Elgene O., *Macroclimate and Plant Forms: An Introduction to Productive Modeling in Phytogeography*, Trends for Vegetation Science, vol. 1 (The Hague: Dr. W. Junk, 1981); Box, Elgene O., "Global and local climatic relations of the forests of East and Southeast Asia," in *Vegetation Science in Forestry, Handbook of Vegetation Science*, ed. Elgene Box et al., 12/1 (Dordrecht, NL: Kluwer, 1995), 23–55.

6. *OTFTA*, 63–68.

7. *OTFTA*, 95–96.

8. *OTFTA*, 71–72.

9. Lee, David W., K. Baskaran, M. Mansor, H. Mohamad, and S. K. Yap, "Light intensity and spectral quality effects on Asian tropical rainforest tree seedling development," *Ecology* 77 (1996): 568–80.

10. Russo, Sabrina E., and Kaoru Kitajima, "The ecophysiology of leaf lifespan in tropical forests: Adaptive and plastic responses to environmental heterogeneity," in *Tropical Tree Physiology*, ed. G. Goldstein and L. S. Santiago (New York: Springer Nature, 2016), 357–83.

11. Wright, Ian J., et al., "Global climatic drivers of leaf size," *Science* 357 (2017): 917–21.

12. Gould, Stephen J., and Richard Lewontin, "The Spandrels of San Marco. The Panglossian paradigm: A critique of the adaptationist paradigm," *Proceedings of the Royal Society B* 205 (1979): 581–98.

13. *OTFTA*, 68–70; Hallé, Francis, Roelof A. A. Oldeman, and P. Barry Tomlinson, *Tropical Trees and Forests: An Architectural Analysis* (New York: Springer, 1978). This is the most important book on tropical tree architecture, but the basic idea is Hallé's; Lee, *Nature's Fabric*, chap. 7.

14. Lauri, Pierre Eric, "Corner's rules as a framework for plant morphology, architecture and functioning—issues and steps forward," *New Phytologist* 220 (2019): 1679–84.

15. *OTFTA*, 103.

16. Klein, Alice, "Tallest known tropical tree discovered in Malaysia's lost world," *New Scientist*, 8 June 2016; Mascaro, Joseph, Gregory P. Asner, Stuart Davies, Alex Dehgan, and Sassan Saatchi, "These are the days of lasers in the jungle," *Carbon Balance and Management* 9 (2014): 7; Disney, Mathias, "Terrestrial LiDAR: A three-dimensional revolution in how we look at trees," *New Phytologist* 220 (2019): doi: 10.1111/nph.15517.

17. *OTFTA*, 92–94; Kohyama, Takashi S., Matthew D. Potts, Tetsuo I. Kohyama, Kaoru Niiyama, Tze Leong Yao, Stuart J. Davies and Douglas Sheil, "Trade-off between standing biomass and productivity in species-rich tropical forest: Evidence, explanations and implications," *Journal of Ecology* 108 (2020): 2571–83.

18. Bastin, Jean-François, et al. (98 coauthors), "Pan-tropical prediction of forest structure from the largest trees," *Global Ecology & Biogeography* 27 (2018): 1366–83; Lutz, James A., et al., "Global importance of large-diameter trees," *Global Ecology & Biogeography* 27 (2018): 849–64.

19. *OTFTA*, 92–93; McDowell, Nate G., "Deriving pattern from complexity in the processes underlying tropical forest drought impacts," *New Phytologist* 219 (2018): 841–44; McDowell, Nate G., et al., "Drivers and mechanisms of tree mortality in moist tropical forests," *New Phytologist* 219 (2018): 851–69.

20. *OTFTA*, 319–28; Medway, Lord, "The phenology of a tropical rainforest in Malaysia," *Biological Journal of the Linnean Society* 4 (1972): 117–46.

21. *OTFTA*, 79–81.

22. Rüger, Nadja, et al., "Demographic trade-offs predict tropical forest dynamics,"

Science 368 (2020): 165–68; Li, Yaoqi, et al., "Leaf size of woody dicots predicts ecosystem primary productivity," *Ecology Letters* 23 (2020): 1003–13.

23. *OTFTA*, 323–46; Ghazoul, Jaboury, *Dipterocarp Biology, Ecology and Conservation* (Oxford, UK: Oxford University Press, 2016).

24. *OTFTA*, 247–48, 254–56.

25. *OTFTA*, 355–60.

26. *OTFTA*, 397–99.

27. *OTFTA*, 78.

28. *OTFTA*, chap. 5; Corlett, Richard T., and Richard Primack, *Tropical Rainforests: An Ecological and Biogeographical Comparison*, 2nd ed. (London: John Wiley, 2011); Fleming, Ted H., and W. John Kress, *The Ornaments of Life: Coevolution and Conservation in the Tropics* (Chicago: University of Chicago Press, 2013).

Back in 1964, I clambered up through the hill padi *of Rumah Jelian and through the little trial plantation of oil palm . . . then through a wall of secondary growth into the grandest, darkest, yet most open and vault-like forest I had seen during five years in Borneo. The sweet, fermenting aroma coming up from the soil reminded me of autumnal temperate forests of elm and ash. Even the call of the argus pheasant echoed as if in a great building.* — P. A.

7

LOWLAND EVERWET FORESTS
Structure and Dynamics

This was the great dipterocarp forest dominated by *paji* (*Dryobalanops lanceolata*), a camphor tree on the lower slopes of Bukit Mersing in Sarawak, an isolated Tertiary basalt ridge 1,000 m high, whose chocolate-brown, spongy clay soils allow deep rooting and resistance to drought. Such cathedral like forests have always been rare in Borneo, though they are widespread on the volcanic soils of Sumatra, the Philippines in the lee of typhoons, and (perhaps centuries ago) in West Java. Enough lowland Sunda forests remain (though logged over outside of parks) to reveal that the "pile" of the landscape surface—that is, the forest canopy structure—is quite diverse. It varies with topography and is particularly obvious along the defiles of the narrow, steep-sided, parallel ridges that are such a feature of upland Borneo (p. 44). The narrower the ridge, the smaller and denser is its pile of the forest carpet.[1]

We owe a similar description of Borneo to Alfred Russel Wallace:[2]

The observer new to the scene would perhaps first be struck by the varied yet symmetrical trunks, which rise with perfect straightness to a great height without branching, and which, being placed at a considerable distance apart, give an impression similar to that of columns of some enormous building. Overhead, at a height, perhaps, of a hundred and fifty feet, is an almost unbroken canopy of foliage formed by the meeting together of these great trees and their interlacing branches. . . . The great trees we have been hitherto describing form, however, but a portion of the forest. Beneath their lofty canopy there often exists a second forest of moderate-sized trees, whose crowns, perhaps forty or fifty feet high, do not touch the lowermost branches of those above them. . . . Yet beneath this second set of medium-sized forest trees there is often a third undergrowth of small trees, from six to ten feet high, of dwarf palms, of tree ferns, and of gigantic herbaceous forms. Yet lower, on the surface of the ground itself, we find much variety.

The Mixed Dipterocarp Forest

Forest structure (fig. 7.1; table 7.1) is determined by several factors. The structure we see has a history about which we can infer little in the absence of records from plots or tagged trees, except in strongly seasonal areas, where annual growth rings allow age estimates. Forests without large-scale destruction consist of individual patches of canopy species. Each such patch

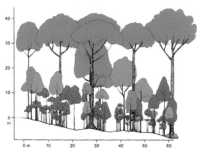

Fig. 7.1 Profiles of mixed dipterocarp forest. (MDF). *Top left*, forest margin of MDF from roadside, an almost continuous emergent canopy on the skyline. Danum, Sabah (P. A.). Profile diagrams within MDF on different soils, 8 × 64 m; *A*, Bako National Park, shallow, leached humic yellow sands, stand recovering from major canopy mortality; *B*, Andulau Forest Reserve, Brunei, deep, humic yellow sandy loams, intermediate canopy disturbance and recovery; *C*, the former Bok-Tisam protected forest (now oil palm), udult yellow clay loams, mature phase with an almost continuous, emergent, light hardwood dipterocarp canopy. Tree types are indicated in colors: blue = emergent; green = flowering in the main canopy; pink = flowering in the subcanopy. Juveniles of all occur beneath the canopy (P. A.; see Peter Ashton, "Ecological studies in the mixed dipterocarp forests of Brunei State," *Oxford Forestry Memoirs* 25 [1964]; Peter Ashton and Pamela Hall, "Comparison of structure among mixed dipterocarp forests of northwestern Borneo," *Journal of Ecology* 80 [1992]: 459–81).

Table 7.1 Species diversity at three sites with structures analogous to those illustrated in fig. 7.1 and enhanced by differences in canopy diffuseness: greatest at Bako, where the emergent stratum merges with the main canopy and is dominated by notophyll and microphyll leaf size classes; intermediate at Lambir, where notophylls dominate; and least at Bukit Mersing, where the dense, emergent canopy is mesophyll. Species diversity of functional types is the percentage of total forest tree diversity.

Locality	Bako N. P.	Lambir N. P.	Bukit Mersing, Anap
Site characteristics	Drought-prone coastal slope; interfertile, freely draining sandy soil	Moderately drought-prone, moderately infertile, sandyloam soil; ridge and gentle slope	Mesic lower slopes; deep, fertile basaltic clay loam soil
Number of species	223	321	143
Pioneer spp. (%)	18 (8)	30 (9)	15 (10.5)
Emergent spp. (%)	10 (4.5)	40 (12.5)	8 (5.5)
Main canopy spp. (%)	111 (50)	147 (46)	43 (30)
Subcanopy spp. (%)	84 (38)	104 (32)	77 (54)

Source: OTFTA, p. 467, fig. 7.5b.

is the outcome of an opening in the canopy long ago, which permitted a group of juveniles to survive and grow. Foresters call these patches of trees with a shared history *stands*. The stands in a forest are at various stages of recovery back into the canopy. Trees reach a maximum height and then persist, in some species for more than a century, while continuing to reproduce and to grow in girth. Therefore, most stands in old-growth forest have attained full height and a semi-stable, vertical structure. This stage, approaching equilibrium, is called the *mature phase*. It generally occupies at least 80% of the area of a little-disturbed forest, and therefore characterizes it. Mature phase structure varies with habitat. Structure partly determines economic values and management options; understanding the processes that produce forest structure is therefore an important goal of forest science.

Forest ecologists have concentrated on *primary* forests: those in which disturbances of the canopy and regeneration can be traced to natural causes. Their influence on forest structure and dynamics is profound. They can be studied by documenting a forest sample, waiting for a natural catastrophic event, and then recording the consequences. The study of *secondary* forests, those that are regenerating after human activity (from burning and cultivation to selective logging), can only be fully rewarding with careful documentation of the forest *prior* to the disturbance, which is often absent from silvicultural research in forests subject to logging.

Vertical Stratification

Wallace and others claimed that the mixed dipterocarp forests (MDF) of the everwet Far East are vertically structured into three or more distinct tree strata of the main canopy, with emergent crowns a stratum above, and sub-canopy trees below both (fig. 7.1). Species diversity at three sites with structures analogous to those illustrated is enhanced by differences in canopy diffuseness (table 7.1): it is greatest at Bako, where the emergent stratum is merged with the main canopy and dominated by notophyll and microphyll leaf size classes (see table 6.1); intermediate at Lambir, where notophylls dominate; and least at Bukit Mersing, where the dense emergent canopy is mesophyll. These lowland MDFs include the tallest stands in the tropics (p. 101). Products of a humid and windless climate, they set the standards against which to measure other regional forest types. Even in mature stands, however, the strata are rarely distinct, mainly because of juveniles growing among and between the strata below those that they will occupy at maturity.

The upper stratum of *emergent* canopy crowns is the most distinct, since their branches rise above the main canopy foliage (fig. 7.2). In everwet tropical Asia, the emergent stratum of mature phase stands is overwhelmingly of dipterocarps. These forests are called *mixed dipterocarp* (p. 104) because of the many dipterocarp species; the term differentiates them from swamp and deciduous forests in which a single dipterocarp species is dominant. Several other Malvalian species grow in the emergent canopy, along with scattered, emergent leguminous trees—including *Koompassia*, which is among the tallest in stature in an MDF.

Some pioneer and early successional species also become emergent, growing on fertile clay soils and in periodically flooded valley bottoms, such as *Albizia* and *Parkia* (Leguminosae-Mimusoideae,) and *Alstonia* and *Dyera* (Apocynaceae).

This emergent canopy is limited to windless and everwet sites. Wind and rainfall seasonality reduce canopy height and contract the emergent canopy into the main canopy.

Emergents growing on each topographical element (plain, slope or ridge) trend toward a uniform height (fig. 7.2). In contrast, main canopy crowns vary widely in height, and subcanopy individuals yet more so.

Ecologists have long discussed whether this stratification actually exists and, if so, whether it is imposed by the physical environment of trees within a forest stand irrespective of species, or whether the species themselves are genetically constrained to certain statures at maturity.

Fig. 7.2 Emergents and canopy of MDF. *Top left*, Pasoh Forest Reserve, Peninsular Malaysia (D. L.); *top center*, successional climax species *Shorea parvifolia* emerging above main canopy, Danum, Sabah (P. A.); *top right*, height class histogram (in feet) for two 200 × 25 ft transects, Kuala Belalong, Brunei (I. B., after P. A.; see chap. 7, n. 8). *Bottom left*, canopy and emergents, Pasoh (D. L.); *bottom right*, break in main canopy foliage beneath dense MDF canopy, Gunung Palung National Park, West Kalimantan (© Tim Laman).

Stratification by maximum tree height. Tree stratification in tropical rain-forest could be caused by an aggregation of the maximum height of tree species at maturity into vertical guilds, groups of species sharing a common ecological attribute. If such aggregation exists, the reasons for it are unclear. Woody plants lack genetic control of a specific height at maturity. Instead, their highest buds continue to expand, but eventually their shoot extension is reduced by stress and halted by apparent damage. Even so, each species reaches a distinct maximum height. The cause among emergents, not diag-nosed in rainforest trees, may be the inability to lift water during periodic drought, because maximum heights vary with topography and soil. Their heights are asymptotic (p. 100). Trees also grow taller at different rates. Vertical stratification by maximum tree height alone is therefore improbable, despite the apparent uniformity of emergent canopy stature.

Stratification by branch architecture. Trees grow by genetically determined patterns of branching and growth (by the architectural models of Hallé, p. 98). In Asian forests, the predominantly Malvalian species (such as dipterocarps) generally start with an orthotropic leader (p. 124). They often grow to canopy height by lofty extensions of that leader followed by lateral extension of a dense layer of leaves on near-horizontal, plagiotropic branches. Once they break into the full sun, their branches become orthotropic, ascending with spirally arranged leaves equivalent to the leader, which eventually becomes indistinguishable from the others. The crown then becomes dome-shaped, the leaves smaller and bunched. Inflorescences are in part terminal, forcing the twigs to branch more.

The canopies of African and Neotropical forests, in contrast, are comprised substantially of leguminous trees, conforming to Troll's model, with sequential replacement of axes. It has proved immensely adaptable. In the rainforest understory, the shoot may remain vertical through much of its growth and extension, so that the next "relay" can grow vertically above it. Leguminous giants of Far Eastern forests, such as *Sindora* and *Koompassia*, develop among the most columnar boles in these forests.

A rich diversity of architectural models occurs beneath the canopy. Optimal placement of flowers and fruits appears to take precedence over growth in height; light-capture in this challenging light climate is achieved by a variety of branching and juvenile leaf forms. Among pioneers of mesic sites, leaves quickly form a dense umbrella above the trunk. This too can be achieved by several models, but the most common are Troll's and others that form dense horizontal or drooping twigs, and those that produce orthotropic branches bearing dense rosettes of large leaves.

These models are often phylogenetically conservative. They are generally constant among related species, and frequently within genera and even families. Trees must then adapt within the possibilities of a set of genetic constraints imposed by evolutionary history. Many basic models, such as Troll's, are sufficiently versatile to allow species sharing the same model to develop into different roles in mature forest structure. Also, juveniles of species of any model must pass through lower strata on their way up to their full height at maturity. Because there are always more juveniles than mature trees in any stable population, juveniles inevitably obscure any stratification by architecture that may exist among mature individuals. As trees pass through life's vicissitudes, including herbivory, insufficient light, and drought (even in everwet regions), all follow by erratic recovery through reiteration (p. 133). This, too, obscures the overall model of the individual, which cannot then contribute to the pattern of forest structure.

Stratification through gross crown architecture and presentation of flowers. The climax species of equatorial lowland forests fall into three guilds distinguished by their crowns and flowering characteristics at maturity. These guilds form three height "strata" at maturity, giving a characteristic form to mature stands, to the extent that they result in stratification of the forest; this is due to the characteristics of individual species.

Emergent canopy flowering guild. The uppermost, emergent stratum is composed of a distinct group of species, the main orthotropic branches of whose crowns at maturity arise once the leader ascends into full sun above the main canopy. Broad, hemispherical crowns form, with leaves often in a single layer concentrated at the twig endings. These species present their flowering twigs in full sun. Among dipterocarps and other Malvales, heavy flowering in closed forest is confined to those individuals that have attained mature, dome-shaped crowns. The emergent dipterocarps of MDF flower intensely within a well-defined season, at several-year intervals. Most other emergent taxa do likewise, albeit less intensively (p. 233).

Main canopy flowering guild. Species in this category also flower only along crown branches in prolonged sunlight. Their persisting mature branches arise beneath the main canopy itself, and their crown branch form differs from that in immature individuals only in its degree of inclination. Flowering is less seasonally concentrated and less confined to certain years. In Far Eastern MDFs, the main canopy is the stratum richest in families, genera, and species (see chapter 14). Species of the main stratum vary more in maximum height than emergents, and their leaves are less concentrated in a single layer—a more mixed and less well-defined profile.

Subcanopy flowering guild. This guild comprises species that reproduce in the patchily sunny conditions beneath the canopy of the forest. They flower synchronously within populations, often over long periods and varying less in intensity between years. The majority apparently require *some* direct sunlight to reproduce. They are diverse both in maximum height and in architecture, though they change little architecturally during ontogeny. In the everwet lowlands of Asia, the tree flora of the subcanopy is dominated by juveniles of canopy and emergent species, especially dipterocarps. Subcanopy trees thus rarely form a distinct layer in the forest profile. The stratum can, however, be distinguished by its distinct reproductive guild, even though this is diffuse owing to the interspecific diversity of heights at reproductive maturity. The components include genera and families of monopodial habit (p. 120), with a persisting vertical leader and plagiotropic, mostly horizontal branches bearing distichous leaves. Many stemless palms occur in the subcanopy (*Licuala* and other species), sometimes abundantly or dominantly, along with slender understory palms of *Pinanga* and other genera.

In the understory, juveniles of climax canopy species have more slender trunks than do subcanopy species of the same size. Understory species tend to have broader crowns than canopy saplings, and a higher proportion bear plagiotropic branches with small leaves. Leaf size correlates with tree architecture: species with extending plagiotropic branches bear smaller leaves than those mostly canopy juveniles, which extend tall leading shoots with short lateral branches. Thus, height growth (enabling potential future access to much-increased sunlight) trades off with leaf extension (capturing the energy of fleeting sun-flecks).

Some species in all three vertical guilds are cauliflorous; they develop flowers directly from their trunks. Most cauliflorous climax trees belong to the subcanopy guild, including Myristicaceae and Annonaceae, and to understory species in other families. Yet there are also some cauliflorous main canopy species, such as the *cempadak* (*Artocarpus integer*) and wild durians (e.g., *Durio malaccensis*, *D. wyatt-smithii*), which flower from the crown branches downward and are pollinated by fruit bats (p. 241), visually attracted at dusk.

An overriding question. Tree architecture, reproductive structures, and breeding systems are each highly diverse in every vertical guild of tropical rainforests. This diversity must reflect different solutions among different groups, which are all evolutionarily successful. Does that imply that species are, in effect, ecologically complementary, or do they occupy different ecological niches in *other* respects (p. 122)? The function of a tree's different parts can only be understood in the context of the whole plant, and the whole plant must be understood in relation to its habitat, the forest.

Leaf layers as a source of stratification. Observant visitors may be amazed by the diversity of leaf size and shape (or *geometry*) in the "jungle" along the roadside. Upon first sight of mature, old-growth rainforest, they may also notice the unevenness of the canopy, causing it to resemble piled morning clouds. Canopy leaves, especially if small, are bunched toward the ends of twigs, arching over the domed crowns.

All tropical tree species are identical in the biochemistry of photosynthesis (p. 123). A leaf of even the fastest-growing tree can use no more than a third of the incoming energy from direct sunlight for photosynthesis. The rest is either converted to heat, reflected, or transmitted. Heating of leaves above 35° C inhibits photosynthesis, and another 10° may damage leaf tissue. Leaf temperature is mostly reduced through transpiration. Even in everwet climates, multiple-day droughts occur once a year or so, much longer during ENSO years (p. 11). Consequently, canopy leaves are well-adapted to conserve water. Leaves are insulated from the atmosphere by the boundary layer, which also resists the diffusion of water and carbon dioxide (p. 96). A canopy tree in direct sunlight therefore achieves a compromise between avoiding overheating, water loss, and out-shading its competitors with a tight canopy of overlapping leaves.

Temperature moderation restricts leaf size, shape, inclination, and density. In the more variable diffuse light and high humidity of the subcanopy, constraints on leaf size and shape should diminish. There, the relative cost of constructing leaves and petioles of different form and size versus the more durable twigs that bear them becomes the primary constraint. In the calm air of everwet equatorial Asia, the tension between these conflicting needs may play the major part in resolving the differences in canopy structure and subcanopy density that characterize forests in different habitats.

In lowland rainforest subcanopy (fig. 7.3), sun-flecks appear and disappear with the passage of the sun. Leaf shadows become more diffuse with distance below the canopy, since the sun provides light from a disc rather than a point (penumbral radiation). Conditions immediately beneath the canopy foliage are predominantly shady, with insufficient light for net photosynthesis during most of the day, though with occasional flashes of direct sunlight (sun-flecks). Further below the canopy, the light becomes more uniform, with enough diffuse light for some net photosynthesis among shade-tolerant species. Wind hardly penetrates, and atmospheric humidity remains high. The advantages of different leaf sizes and inclinations are largely lost. This stratification of leaves, in layers that decline in concentration from the canopy to the understory, is regulated by the stand, not the individual species. Thus, leaf and crown characteristics lead to canopy stratification.

Evergreen broadleaf forest emergent crowns, such as those of MDF, bear their leaves in shallow, dense layers. Once they have attained the canopy, individuals in a still climate therefore do not generally continue to carry leafy branches deep within the canopy. Subcanopy tree crowns seldom reach as

Fig. 7.3 Subcanopy in the MDF. *Top left*, subcanopy palms and trees, Lambir National Park, Sarawak, all on humult sandy soils (© Christian Zeigler); *top center*, hemispherical photograph from ground toward canopy, Lambir (S. R.); *top right*, dense understory, Lambir (S. D.). *Bottom left*, diffuse MDF understory beneath dense canopy of giant yellow and red meranti, Gunung Palung National Park, West Kalimantan (© Tim Laman); *bottom center*, leaf diversity in Bukit Lagong Forest Reserve understory, Peninsular Malaysia (D. L.); *bottom right*, ground herbs in dense understory shade, Bukit Lagong (D. L.).

high as the canopy foliage overhead, except where there is a gap between emergent crowns. this intense stratification of foliage near the canopy gives the visual appearance of distinct stratification within mature lowland mixed rainforest.

Subcanopy leaves. These tend to display horizontally at maturity and separate loosely and haphazardly in the vertical dimension to capture light. Many species (both juveniles of canopy trees and understory species) do bear deep crowns in which the leaves are layered on spreading, plagiotropic branches. However, leaves are not stratified beneath the canopy of mixed species stands in the mature phase. Leaves on plagiotropic branches tend to be small, with short petioles supported by the permanent branch structure, while leaves of the orthotropic leading shoots are larger, with longer petioles providing support and presentation to light.

Shade-adapted species, and shade leaves of canopy species' seedlings reduce leaf mass and synthesize less chlorophyll per unit leaf area; they nevertheless photosynthesize at the same rate as leaves of sun-adapted species. Many understory ground herbs bear leaves with a layer of purplish anthocyanin pigment within their lower surfaces, which may enhance photosynthetic efficiency. Leaves of tropical Asian evergreen trees are rarely lobed. If they are, large leaves are seen on juveniles or understory individuals as well as on successional *Artocarpus* and pioneers (fig. 7.3).

Leaves are vertically stratified in mature individuals of climax species in the mature phase of evergreen forests, primarily owing to the height stratification of their species into emergent canopy, main canopy and subcanopy guilds. These guilds arise because zones beneath which light is at first insufficient for photosynthesis, below and between which the main canopy foliage does the same to a lesser extent, and each is associated with the floral presentation of their associated species guilds. Leaf stratification is often obscured by individual age and performance; but leaf stratification becomes most distinct in mature stands, and it is consistent throughout communities within a uniform habitat.

Crown shyness. In some canopies leaves on twigs sharing the same branch are tightly packed, but there are sharp-edged open channels between branches, and the longer the branch, the greater the width of the spaces (fig. 7.4). The canopies that have this pattern are those of young and maturing emergent trees whose crowns are still dense and undamaged, yet whose mature orthotropic branches have developed. Crown shyness is a misnomer, since the pattern is due to wind-shearing of extending shoots while they are still fleshy and brittle. It results in some increase in both sun-flecks and diffuse light in the subcanopy. It is confined to evergreen trees with small to mesophyll leaves in a single, dense outer layer, such as dipterocarps and durians, in climates that are breezy during the season of leaf flush.

The MDF Cycle: The Life of a Stand

A stand's composition and performance are determined by its constituent species (and their ecophysiologies).[3] It is important to distinguish the process of stand reestablishment from the attributes of the species that participate in and define that process. Although reestablishment is fundamentally a continuous process, the initial phase of stand reestablishment immediately following canopy gap formation is the *gap phase*. *Succession* then follows, continuing until canopy individuals have established full height, and the

Fig. 7.4 Crown shyness in a diffuse, emergent canopy of *Shorea albida*. Ulu Mendalam peat swamp forest, Brunei (A. C.).

vertical structure of the mature stand is reestablished; this then is the *mature phase*. The participating species are recognized as *pioneer* or *climax* species (table 7.2). Both participate at different stages of stand development, or succession. The term *climax* is fraught with difficulty, formerly used for a hypothetical, stable asymptote that plant communities might reach under a specific climate. In our contemporary concept of the mature phase, individual tree mortality continues owing to competition and external catastrophe. Climax *species* as defined here differ greatly from one another in their response to light, but all differ from the pioneers in their requirements for establishment.

The regeneration niche: Pioneer (and climax) species. Seeds of rainforest trees vary in size and dormancy (table 7.2). Following germination, seedlings lose mass, depending more on seed reserves than on photosynthesis. This period may last for months in shade-tolerant rainforest plants, the large seeds of which are mostly *hypogeal* in germination (p. 104), and seedlings grow slowly. Both size and dormancy constrain seedling germination and

Table 7.2 Characteristics of pioneer and climax woody plants

Character	Pioneer	Climax
Longevity	Short to medium	Medium to high
Ontogenetic change	Low	High
Reproduction:		
—age at first flowering	Early	Late
—flowering frequency	High	Low
Seeds:		
—size	Usually small, epigeal, produced copiously ± continuously from early in life	Larger, mostly hypogeal, fewer, produced annually or less frequently, starting as full height is approached
—quantity	Large	Small
—dispersal	By wind or animals, often far Continuous or regularly intermittent	Diverse, including by gravity, mostly short Often unpredictable
—dormancy	Mostly capable	Lacking dormancy or with delayed germination
—soil seed bank	Mostly abundant	Seldom abundant
Germination	Mostly reliant on the influence of direct sunlight, either on light or temperature	Involuntary
Seedlings	Require bare soil for successful establishment Few successfully establish on raw humus in absence of fire Cannot survive prolonged shade	Establish on a variety of soil surfaces, depending on capacity to extend tap roots through unfavorable surface materials Variably survive shade, as a "seedling bank"
Photosynthesis/respiration:		
—light compensation point	Variable, relatively high	Relatively low
—light saturation point	High (above 400 μmol m^{-2} s^{-1}	Relatively low
—optimum light conditions	Full sun	Variable
—max. photosynthetic rate	High (<15 μmol m^{-2} s^{-1})	Relatively high (3–6 μmol m^{-2} s^{-1})
—unit leaf rate in shade	Low or negative	Relatively high
—night respiration	Often high (0.4–1.2 μmol CO_2 m^{-2} s^{-1})	Often low (>0.2 μmol CO_2 m^{-2} s^{-1})
—growth rate	Relatively fast	Relatively slow
Height growth rate	High	Variably lower

(*continued*)

Table 7.2 continued

Character	Pioneer	Climax
Branching	Strongly orthotropic or strongly plagiotropic	Various
Growth periodicity	Continuous or frequent intermittent	Various
Leaf life	Short	Longer
Canopy structure	Low LAI, canopy often monolayer, adapted to maximize shade creation within the limits of prevailing soil water economy	Variable but higher LAI, adapted first to withstand all but the most extreme water stress
Herbivory	High	Lower, with more protection
Root/shoot ratio	Relatively low	Relatively high
Wood	Light, with living wood parenchyma and often no heart; resisting rotting until the tree dies, when it is rapid	Heavier, with heartwood, which is dead and protected by tannins and other chemicals which eventually degrade, exposing the heart to rotting
Population structure	Single age cohorts	Multiple age cohorts
Longevity	Short	Longer

Source: Tim C. Whitmore, *Tropical Rainforests of the Far East,* 2nd ed. (Oxford, UK: Clarendon Press, 1984); Francis S. P. Ng, "Tropical sapwood trees," in *Colloque International sur L'Arbre* (Montpellier: Naturalia Monspeliensis, 1986).

establishment. Large seeds, usually associated with climax tree species, have a higher proportion of mass in cotyledon reserves for survival and growth. However, seed carbohydrate reserves alone would be unlikely to help a seedling survive without the ability to achieve net photosynthetic gain in low light. Seedling growth rates are negatively correlated with dry mass of seed reserves, but positively with leaf area (fig. 7.5).

Species whose seeds germinate in response to canopy gaps are *pioneers.* Their two diagnostic characteristics are (1) seed dormancy, with light-induced germination and (2) establishment therefore limited to gaps. All pioneers require high light intensities to survive, grow fast in response to them, and are relatively short-lived. Growth rates and survival among forest species increase along a continuum of overlapping photosynthetic responses to increasing light, extending from the slowest growing, most shade-tolerant hardwood climax species to the fastest growing pioneers (see fig. 6.2). Some climax species demand high light intensities to survive, and

Fig. 7.5 Tree growth in MDF. *Left*, relative growth rates in saplings of three *Shorea* species in relation to estimated sunfleck light availability, light-demanding, mid-successional *S. parvifolia,* and two later successional species (I. B., after A. S. Moad, [cited in fig. 6.2 caption]); *center*, layered crown of a pioneer, *Alstonia scholaris*, that persists in succession to reach the emergent canopy (P. A.); *right*, the opening orthotropic branches of a maturing macaranga, left, beside a young *Shorea argentifolia* crown with branches beginning to change from plagiotropic to orthotropic, Kuala Belalong, Brunei (P. A.).

they grow at rates comparable to the fastest-growing pioneers in sunlight. Therefore, although most pioneers grow in response to full sun more than do climax species, their requirements for establishment distinguish pioneers from climax species.

Pioneers often germinate in response to the increase in the red/far-red ratio of the light spectrum that accompanies the change from diffuse canopy to direct sunlight (p. 129); others respond to increased daily temperature range in gaps where sunlight reaches the soil surface, and particularly to the difference between day and night temperatures. Their generally small seeds are mostly *epigeal* (p. 104), their true leaves open rapidly, and seedlings grow quickly. A few pioneers in Asia become emergent—notably *Alstonia*— possessing tiny seeds, wind-borne by tufts of long silky hairs. Pioneers are adapted to the gap habitat in other ways (table 7.2). They begin reproducing when still small, and only in full sunlight. Dormant seeds of pioneer species germinate wherever sunlight penetrates to unlittered forest floor. A few pioneer trees are large-seeded, though, including the candlenut (*Aleurites moluccana*) and successional climax *Artocarpus.*

Small-seeded pioneers only establish consistently within forests on bare mineral soil, and soil nitrogen accelerates seedling growth. Small-seeded subcanopy species also establish only on surfaces free of litter, such as slopes, periodically flood-washed ground, and places where animals have disturbed the soil.

Those species whose seeds do not germinate in response to light vary widely in their durations of seed dormancy. Some climax species delay germination and are not truly dormant. Emergent legume species and dipterocarps mast-fruit synchronously at intervals of several years (p. 232). Delayed germination of some leguminous species leads to continuous but sparse production of seedlings in the following years, with survivorship comparable to that achieved by dipterocarps lacking seed dormancy. Leaves of leguminous trees have relatively high nitrogen concentrations and are therefore preferred by browsing animals over dipterocarp and other seedlings, which are relatively low in nitrogen, resinous, rich in phenols, and suffer little herbivory.

Gap phase and succession. In general, we are short of long-term studies of these two stages—studies that involve tagging plants over decades and passing on the baton to succeeding generations. Our knowledge base is therefore weak.

Gap origins. Gaps have different causes (fig. 7.6). The smallest are caused by the death of a single standing tree, from lightning strike, drought, or disease. Such mortality minimally disturbs the forest floor. Climax juveniles can survive and establish, and the litter layer beneath remains unbroken, depriving seeds of pioneer species access to the direct sunlight necessary for germination. Larger gaps may be formed, in order of magnitude, by (1) local windthrow, (2) logging, (3) landslides and floodwater, (4) downdrafts and cyclonic storms and (5) volcanic lava and ash. In these cases, the soil surface becomes patchily or wholly disturbed, and climax regeneration is reduced or absent. Such gaps can destabilize adjacent canopies, producing more falls and increasing gap size over time. Always, the tallest adjacent survivors influence the light climate below and, consequently, the fate of shorter individuals. The size, nature, and origin of a gap thus strongly determine the course of tree succession within it.

Early gap phase dynamics. When a canopy gap is formed by the death of a canopy tree, a patch of sunlight penetrates to the subcanopy for part of the day (fig. 7.7); the extent of penetration, as determined by the height of adjacent forest and the diameter of the gap, mediates the effectiveness of the gap in spurring regeneration. Every gap includes a diversity of light regimes, and the surviving adjacent canopy is also affected.

With enough light, dormant seeds may germinate. Pioneer species—

Fig. 7.6 Gap formation in MDF. *Left*, a fresh landslip in Lambir National Park, Sarawak (P. A.); *center*, small gap left from a leafless crown of a single dead tree, Temenggor Forest Reserve, Perak, Peninsular Malaysia (R. H.); *right*, single dead emergent and small gap, Lambir, Sarawak (C. Z.).

more fecund, fruiting more frequently, and in many cases with more effective seed dispersal mechanisms—at first are more competitive than climax species, because the climax species' seeds lack dormancy and reproduce less often. Eventually they re-invade, and their more shade-tolerant seedlings ascend through the senescing pioneer canopy.

Tropical rainforest dynamics differ from those of other forests, because the great majority of species are climax and lack seed dormancy—germinating over ensuing weeks or months. Only removal of regeneration at gap formation, as in landslides, gives the advantage to those few species whose dormant seeds remain, or that fruit frequently and disperse effectively.

The leaves of pioneer species grow quickly and change little to attain maximum size (fig. 7.8); seedling leaves resemble their mature crown leaves. In most, they are presented in a single, densely overlapping layer. They develop secondary branching earlier, and their foliage has limited plasticity, being intolerant of shade.

The atmosphere in rainforest canopy gaps is generally still. The leaves of pioneer crowns, tessellated like slates on a roof, are immersed in a thick layer of humid air. Leaves may overheat in direct sunlight, compensated for by transpirational cooling, by wilting and hanging down, or by leaf movements.

Many pioneer tree species produce orthotropic branches, establishing

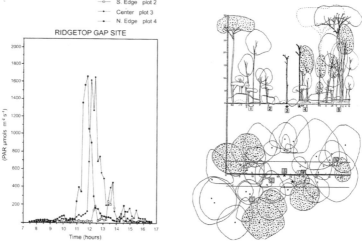

Fig. 7.7 Gaps and light environments. *Top,* small gap in forest transect, Pasoh Forest, Peninsular Malaysia (P. A.; see Peter Ashton, "Crown characteristics of tropical trees," in *Tropical Trees as Living Systems,* ed. P. Barry Tomlinson and Martin H. Zimmerman [Cambridge: Cambridge University Press, 1978], 591–615.); *bottom right,* transect through a single tree gap on ridge, Sinharaja, Sri Lanka (M. A.); *bottom left,* light intensities of transects through and on edge of the same gap (M. A.; see Mark Ashton, "Some measurements of the microclimate within a Sri Lankan tropical rainforest," *Agricultural and Forest Meteorology* 59 [1992]: 217–35).

Fig. 7.8 Early successional plants. *Top left,* moisture and light promote the growth of *Macaranga hypoleuca*, with dense crowns, yet small, palmate leaves adapted to periodic water stress, Danum, Sabah (© Chien Lee); *top right, Macaranga siamensis* shoot with leaves overlapping like slates on a roof, which bear a high heat load and transpirational cost (S. D.); *bottom left,* diffuse crown and plagiotropic branches of the light-demanding successional climax species *Shorea trapezifolia* sapling (M. A.); *bottom center,* wild banana, *Musa violescens,* Peninsular Malaysia (D. L.); *bottom right,* ginger, *Zingiber* cf. *spectabile,* Ulu Langat, Peninsular Malaysia (D. L.).

dome-shaped crowns early on, and some also ascend into or above the canopy. Tall pioneers, such as *Alstonia*, initially extend tiers of densely set leaves on plagiotropic branches between lofty extensions of the leader, thereby out-shading competition (fig. 7.5). On reaching the forest canopy, they develop a domed crown, usually by reiteration (p. 121), with smaller leaves more diffusely arranged.

Trade-offs between Rapid Growth, Competitive Exclusion, and Adaptation to Drought

Pioneers grow most rapidly in large gaps because their leaves attain high maximum rates of photosynthesis in response to direct light (fig. 6.2). In areas of relatively high (but still diffuse) light intensities below the canopy, their daily gross photosynthesis soon falls beneath respiration, and the pioneers die. Climax species, including those of late succession, have lower maximum rates of photosynthesis in full sun, but higher rates in shade than pioneers. In moderate shade, slow growing, shade-tolerant, heavy hardwoods can usually sustain net photosynthesis (fig. 6.2).

Pioneers and vines invest more in shoots than roots compared to climax species, and in wide diameter vessels that permit rapid movement of water up the stem, enabling rapid growth. These, and their dense, shallow crowns, allow them to out-shade individuals of lower stature, but expose them to high risk of vessel cavitation (embolism) during drought. Pioneers are at risk for mortality from drought, predation, and mechanical damage to their weak, low-density wood. The pioneer's high-risk life history strategy is mitigated by high fecundity and seed dormancy.

Early succession. The new seedlings, together with established juveniles able to respond to the new conditions by increasing their growth rates, form the regenerating stand. Many canopy openings produce little disturbance or mortality near the ground. These initiate a truncated stand succession in which climax species' juveniles survive, and pioneers may play little or no part. Full sun over long periods is harmful to the growth and survival of shade-tolerant climax species' seedlings. Established seedlings of such climax species nevertheless compete in the mostly diffuse light environment toward the gap edge and in the adjacent subcanopy with increased light. The advantage always rests with those juveniles that survive the gap event and are tall enough to create a shadier light environment that shorter adjacent competitors and seedlings must endure. During the first five years, successfully established pioneers otherwise become dominant. Early on, little vegetation survives beneath the thin but dense layer of leaves across their crowns, except for the subsequent establishment of a few scattered, seedlings of shade-tolerant climax subcanopy and canopy species. As they grow, pioneer crowns become more diffuse as the increasing weight of leaves bows their branches out. Then the climax saplings eventually emerge between the pioneers.

Gaps with little direct light promote the growth of species with high survivorship in low understory light regimes, such as *balau* or *selangan batu* (*Shorea* sect. *Shorea*). These may have already reached pole size in an earlier gap phase. The calm and equitable micro-environments of the lowland tropical rainforest produce the tall, columnar boles and symmetrical domed crowns typical of these forests.

Primary forests are therefore comprised of a mosaic of stands whose origins are diverse. In the Asian tropics, most gaps are caused by catastrophic, climate-related events at the local or landscape scale, and they tend to be synchronized. The clumping so characteristic of emergent crowns, especially dipterocarps, may thus reflect shared beginnings as a stand in a gap.

Deflected succession. Vines, like pioneers, grow quickly in sunlight and moisture into dense, thin crowns, supported by trees and by vessels of exceptional diameter and length. Such vines may form blanketing canopies over competing trees. The gap, or early successional phases, may thereby persist for decades, eliminating regeneration of all but a few of the less light-demanding climax species beneath. *Dicranopteris linearis,* a climbing and creeping rhizomatous fern, commonly invades landslides and degraded soils following cultivation. Tree regeneration under its continuous canopy is inhibited and, once the fern has invaded, a gap can only be gradually reduced by shading from the edge.

Later succession. The canopy of the regenerating stand increases in height as the stand grows, and the crowns within it do not conform to the stratification of the mature-phase stands nearby. Until a stand approaches maturity, and a few individuals begin to dominate, the tree crowns within the stand continue to lack any vertical stratification other than an increasingly dense upper layer of foliage in their canopy. The canopy structure of the stand therefore continues to obscure the overall stratification of the forest profile.

Climax species of mid-succession. MDF differs in mid-succession from many other rainforest types because the subcanopy and understory are dominated by juveniles of canopy trees, mostly emergent dipterocarps. Within the decade following gap formation, the new canopy, of moderate-size and with no vine cover, becomes dominated by species that do not flower until their crowns have emerged at least partially into full sun. These are mostly faster-growing climax species.

Light-demanding, successional climax canopy species maintain their access to light through a single leader until full height is approached. Crown depth declines once the full, domed emergent crown is formed. Many late successional climax individuals have orthotropic branches, or these develop as their crowns emerge above the canopy. Their leaves vary in size, and they generally hang or are upwardly inclined, reducing transpiration without affecting photosynthesis, and allowing light to penetrate when competitors have been overtopped. Within a given size class, maximum growth rates are proportional to light interception and to wood density. Many slower-growing climax species are inhibited by full sunlight.

The faster-growing climax species that first overtop the pioneers are themselves relatively shade-intolerant. They may achieve net photosynthesis and growth rates comparable to those of pioneers. Since their growth rate is inversely correlated with wood density, these species include the light hardwood mahoganies and merantis that comprised most internationally traded wood in the period of 1970–1995. Up to 70% of the MDF canopy on moist fertile sites may be composed of these timber species. They flower en masse at around five-year intervals; afterward, the ground is dense with their seedlings (p. 103). Although they suffer high mortality in the shade, enough seedlings survive the five years between fruiting seasons to insure propagation.

A suite of tall pioneer species may persist within stands into late succession. These include *Octomeles, Alstonia*, and *Dyera*. Their wind-borne and partially dormant seeds germinate in response to light. Some early successional climax species, too, are persistent. These include species with large seeds, dispersed by birds (*Sterculia foetida*), mammals (*Artocarpus, Terminalia*), or water (*Terminalia phellocarpa*). These are "pagoda trees," conforming to several different architectural models (p. 120) but in each case gaining height rapidly by periodic extensions of a vertical leader, while plagiotropic branches extend horizontally in false whorls, their dense foliage layers out-shading competitors. These species can gain height faster than adjacent climax individuals. Furthermore, like pioneers, they grow most rapidly on well-watered fertile soils, including floodplains. *Octomeles* and *Alstonia* commonly persist in groves along riverbanks, where their columnar juvenile and maturing crowns continue to receive lateral sunlight. A mature *Alstonia* crown may eventually exclude all other canopy species beneath it.

Climax species of late succession and the mature phase. In Sundaland, the slow-growing, heavy, hardwood canopy species include not only the emergent *balau* and *cengai* (*Shorea* sect. *Shorea* and *Neobalanocarpus*), but also heavy red merantis, such as *S. curtisii*, also *sepetir, merbau*, and *tapang* (*Sindora, Intsia*, and *Koompassia* of the Fabaceae). Seedlings of these species survive many years beneath the mature-phase canopy. They may grow into the canopy continuously over the lifetime of the gap, or intermittently following small openings.

The mature phase. Stand maturity may be reached when the upper canopy species have reached full height, though they continue to grow in diameter. Their crowns then broaden to their full dimensions, while their leaves become concentrated in one shallow layer at the domed crown surface. In calm climates, this phase may occupy more than 80% of the canopy, but the proportion varies both over time and with topography and soil. The main canopy between emergents may itself be mature and relatively stable, or it may contain emergent species in late succession.

The ever-changing canopy. MDF has a unique canopy structure. It includes the tallest forests in the tropics, with the sole exception of stands of the conifer *Araucaria* in New Guinea. In MDF, the density of emergents is unparalleled, owing to the abundance of dipterocarps. Additionally, except in narrow valleys humid with the spray of whitewater rivers, MDF canopies are only sparsely adorned with epiphytes. Asian forests generally lack tank epiphytes; the giant *Platycerium* ferns and giant orchids, such as *Grammato- phyllum speciosum*, are rare, except at the forest edge (p. 109). The emergent climax species eventually outgrow the main canopy species, which survive between their crowns.

The leaves, or leaflets, of rainforest canopy trees vary in size with habitat. They are mostly elliptic or lanceolate, leathery, hairless, dark, and often shiny above but dull and sometimes hairy or waxy beneath. Petioles are short and stout, and they do not articulate, holding the blades in rigid positions on the twig. The leaves do possess stomatal control, at least following first expansion, but they transpire relatively freely; water loss is restricted by leaf shape, orientation, and arrangement within the crown.

Droughts may be the nemesis of emergent pioneers in the wet tropics, but the final stature and dominance of emergent dipterocarps and legumes may be explained by their greater access to soil water during dry periods, thanks to their extensive ectotrophic mycorrhizal mycelia (p. 269). Few climax species can repair mature crowns by epicormic branching. Trees with such reduced crowns do not recover maximum growth rates. Competition among crowns thereby accentuates the stratification of the leafy layers.

The emergent canopy structure of lowland evergreen tropical rainforest, composed of isolated individual and clumped giants, reflects the vestiges of stands arising from gaps of different sizes. After reaching the canopy stratum, these individuals continue to die from drought and/or disease, or from lightning. This background pattern is overlain by periodic catastrophic events: major droughts, landslides, or volcanic eruptions. The pattern and density of the emergent canopy, and therefore the extent of the main canopy, may reflect the history of canopy mortality, as well as the scale and frequency with which canopy gaps are formed.[4] They are also specific to a habitat. Again, continuous, long-term observations are important.

Mortality. All tree communities consist of individuals whose numbers decline as stands grow. Even among the fastest-growing individuals, only a very few ever reach reproductive maturity; competition for light and space kills most of them early. Thus, the great majority of individuals in a stand are growing at less than their maximum rates—indeed, many are not growing at all, but are trapped in a slow decline toward death.

Trees of the past, present, and future. Oldeman introduced this concept to identify stages through static observation.[5] He defined them respectively as those individuals that are in decline through competition or disease; those that have reached full height and reproductive maturity; and those in active height growth. Decline through senescence per se appears relatively rare in Asian rainforests, since it is generally preempted by natural catastrophe. Trees at the onset of flowering are characteristic of his "trees of the present," with the ready identification of canopy dominants in successional stands. Identifying the "trees of the future" among juveniles of shade-tolerant species is more challenging, and it requires prior knowledge of their physiology. Mortality at first remains constant as tree size increases, but eventually increases with it. Trees generally don't live for long, perhaps 200–300 years, but there are exceptions (p. 102).

Structural and Physiognomic Variation of MDF within Everwet Landscapes and Regions

The *uniform* everwet and windless climate of this region makes it possible to study the influence of other factors on forest structure, dynamics, and composition, free of any regional variations in climate. In these tropical lowlands, rainforests vary considerably in these characteristics, more than has been observed anywhere else within a uniform climate. In fact, the variation is arguably as great as between other evergreen broadleaved forests of *differing* climates!

Geology, topography, and soils. Within an effectively uniform rainfall regime, this variation is particularly marked on the steep ridges and siliceous rocks so widespread in sedimentary Borneo, on granitic Peninsular Malaysia, and on the ancient metamorphic mountains of southwest Sri Lanka in the southwest monsoon climate. Variability is enhanced by the emergent canopy dominance of dipterocarps in most forests in the Asian tropical lowlands. Their stature and density vary in a pattern consistent with changes in the physical landscape.

The lowland mixed dipterocarp forests. Wyatt-Smith[6] recognized a zonal MDF in everwet West Malesia growing on udult yellow-red soils of moderate clay content: the *red meranti-keruing forest*, as well as other forest types on other soils and geology (p. 170). In everwet southwest Sri Lanka, where MDF also dominates the lowlands, the forest is shorter, mostly 30–40 m tall, with a few emergent species or individuals reaching 50 m in sheltered coves. Shorter emergents and consequently smoother canopy may be attributed to strong breezes throughout the year and gales during the southwest monsoon.

Variation in species growth. Forests vary within the landscape because species with different growth patterns and mortality prevail in each.[7] Climax species are the principal determinants of stand dynamics. Variables such as physiology; maximum growth potential; leaf longevity, number, and turnover rate; crown structure; and relative allocation of biomass to shoot and root *together* determine ecological distribution, including successional role. The size and shape of gaps and the stature of the surrounding canopy provide a diversity of light conditions and soil moisture levels in which regenerating seedlings compete, setting the stage for niche differentiation (p. 307). Even species of the pioneer genus *Macaranga* that were sharing a single gap at Lambir (Sarawak) had distinct light responses, minimum stature at first flowering, phenology, and fecundity.[8]

Heavier hardwood species. These are distinctive in their tolerance of shade and drought. High wood density reflects anatomical features that reduce cavitation and increase transport (p. 101). As might be predicted, heavier hardwood species are thus relatively more abundant in Sundaland and in southwest Sri Lanka on narrow ridges, on freely draining sandy soils, in the drier climates of coastal forests, and in *kerangas* (below, p. 176)—all habitats prone to periodic severe soil-water stress. Other lowland everwet forests occur on very different soil substrates.

Variation in canopy disturbance. Emergents are the principal mediators of canopy structure, and therefore of the light climate beneath. On steep slopes with landslides and frequent tree falls, fewer large emergents are seen in MDF, and they are often grouped on more stable surfaces. In contrast, on the udult clay loam soils of mesic undulating land, a continuous emergent canopy, broken only by lightning strikes and narrow slices caused by local windthrows, may prevail. This outshades and reduces the density and diversity of main and subcanopy guilds and may reflect late secondary succession following canopy catastrophe at differing scales. Even where past catastrophes are no longer discernible after attainment of full height, succession persists as the accumulation of biomass among the largest size classes, by diameter if not height growth.

Cyclonic storms affect the forests of seasonal regions most (p. 147), but they also influence the everwet northeastern Philippines as well as Taiwan. The Philippine forests (except those in Mindanao) lie within the *typhoon* belt. They receive storms that increase in frequency and intensity toward the north, and endure persistent northeast trade winds from November to March. Each such storm can defoliate the canopy and create massive windthrow gaps, creating subcanopy light and soil-surface conditions favorable to survival and regeneration of light hardwood seedlings and subcanopy juveniles.

Other forests. MDF occupies the yellow-red soils that dominate the lowland landscapes of everwet regions. It is replaced in extreme habitats: on dry skeletal soils of rocky bluffs and limestone, the freely draining podzol and ultramafic soils (p. 45), by other forests which differ floristically (chapter 9) but are similar in structure. On the wettest soils they are replaced by peat swamp forests. The forests on freely draining white sand podzols, known in Sunda lands as *kerangas*, and on peats share a surface of acid raw humus which is exceptionally low in nutrients, derived solely from rainfall and the vegetation itself (p. 77).

These soils represent the extremes of edaphic drought and waterlogging, and make informative contrasts. In both forests, the number of vertical strata is reduced to two. But whereas it is the emergent stratum that is lost in *kerangas*, it is the main canopy in the tallest forests of the most constantly waterlogged peripheral peat swamps. This appears to be due to differences in their canopy catastrophe regimes and water economies, and in canopy foliage. *Kerangas* includes the shortest of lowland forests, sometimes less than 10 m, whereas the great alan (*Shorea albida*) stands bordering the Brunei peat domes can exceed 70 m. Emergent dipterocarps are absent or restricted to isolated clumps in *kerangas*, and they probably suffer high mortality in exceptional droughts. The peat swamp surface is dry except following storms and consists of a platform of interwoven surface roots perched above the water table, on which raw humus accumulates, supported by coarse branching roots that descend into the mire beneath. Contiguous emergent alan crowns extend over many tens of kilometers, resembling a plantation but not of the same age; they may never experience drought. The tallest trees, toward the swamp periphery, reproduce, but few of their seedlings survive. *Kerangas* trees produce small leaves (table 6.1), and ecologists have argued whether this is due to the influence of nutrient restrictions or to drought. Though they share the same nutrient availability, *kerangas* species mostly have high wood density—implying slow growth—whereas alan is a light red successional meranti until it reaches the canopy, when its growth slows. *Kerangas* lacks pioneers: its dense rooting resists windthrow, and its understory is seldom affected by canopy gaps, whereas peat swamp forests are prone to windthrow, and their peripheral forests include some pioneer species, including tall *Alstonia pneumaphora* and *Dyera polyphylla*. Their dominant canopy foliage characteristics also contrast. *Kerangas* canopy consists of microphylls, narrow, upwardly oriented and diffuse, yielding light shade with a dense subcanopy almost reaching the main canopy above. Swamp forest canopy consists of mesophylls (table 6.1), close to horizontal in alan (and other peat swamp dipterocarp species), yielding subcanopy shade in which the main canopy stratum is absent, and the subcanopy is confined far below. These differences mirror those along the soil-moisture gradient within the diversity of MDF: the leaf-size spectra of alan peat swamp forest and MDF on mesic clay loams hardly differ, although their soil nutrient levels vary greatly.

Equilibrium

Over the last 10,000 years, trees in the calm everwet tropics have lived in a stable climate, but in great diversity of topography and in soils differing in nutrient availability and water stress. They have faced the comparative costs of excluding competitors by out-shading versus the costs of withstanding occasional extremes of water stress. Within the demands of each forest community and its habitat, each species has evolved its own solutions to these dilemmas, reaching a given height and specific light environment that are optimal for it; then, the additional cost of reproduction must be borne.

Rainforests are in dynamic equilibrium. At a larger spatial scale, none of the 50 ha ForestGEO tree demography plots (p. 429) that have been remeasured have maintained a constant total basal area. Overall, they have gained, implying that the whole forest is in late succession, although some patches within them have lost mass, and others remain close to equilibrium.

Careful analysis of biomass increase in individual stands suggests a prosaic explanation for this gain: few forests ever actually reach climax equilibrium. The gradual rise of the main canopy during forest succession is accompanied by a change in the proportion of above-ground biomass from the smaller to larger trees. This generally occurs at landscape as well as stand scale. The frequency and intensity of canopy catastrophe are mostly due to storms and droughts at landscape or larger scales. Natural catastrophes are generally frequent enough at the landscape scale to set forests back before equilibrium is reached. We need to test this hypothesis by monitoring forest change over decades.

Notes

1. *OTFTA*, chap. 2, 63–121.

2. Wallace, Alfred. R., *Natural Selection and Tropical Nature: Essays on Descriptive and Theoretical Biology* (London: Macmillan, 1891).

3. Turner, Ian M., *The Ecology of Trees in the Tropical Rainforest* (Cambridge: Cambridge University Press, 2001).

4. Vincent, John B., Benjamin L. Turner, Clant Alok, Vojtech Novotny, George D. Weiblen, and Timothy J. S. Whitfeld, "Tropical forest dynamics in unstable terrain: A case study from New Guinea," *Journal of Tropical Ecology* 34 (2018): 157–75; Arellano, Gabriel, Nagore G. Medina, Sylvester Tan, Mohizah Mohamad, and Stuart J. Davies, "Crown damage and the mortality of tropical trees," *New Phytologist* 221 (2019): 169–79.

5. Hallé, Francis, Roelof A. A. Oldeman, and P. Barry Tomlinson, *Tropical Trees and Forests: An Architectural Analysis* (New York: Springer, 1978).

6. Wyatt-Smith, J., *Ecological Studies on Malayan Forests* (Kuala Lumpur: Malayan Forest Department Research Pamphlet 52, 1966).

7. Ashton, Mark S., "Seedling growth of co-occurring *Shorea* species in the simulated light environments of a rainforest," *Forest Ecology and Management* 72 (1995): 1–12; Ashton, Mark S., C. V. S. Gunatilleke, and I. A. U. N. Gunatilleke, "Seedling survival and growth of four *Shorea* species in a Sri Lankan rainforest, *Journal of Tropical Ecology* 11 (1995): 263–79; Ashton, Mark S., R. R. Hooper, B. Singhakumara, and S. Ediriweera, "Regeneration recruitment and survival in an Asian tropical rainforest: Implications for sustainable management," *Ecosphere* 9 (2018): e02098; Sukri, R. S., R. A. Wahab, K. A. Salim, and David Burslem, "Habitat association and community structure of dipterocarps in response to environment and soil conditions in Brunei Darassalam, Northwest Borneo," *Biotropica* 44 (2012): 595–605.

8. Davies, Stuart J., "Tree mortality and growth in 11 sympatric *Macaranga* species," *Ecology* 82 (2001): 920–32.

We arrived in Thane District, northeast of Mumbai, in September of 1984. I soon found some protected forest at the foot of Mandagni Mountain (fig. 4.2), a low peak showing the layers of the basalt Deccan trap in relief. The last rains fell on October 8, and did not begin again until June 7: a dry season of about 8 months. My first visits in the forest reminded me of previous ones in Malaysian MDF, the understory green with a rich ground flora of gingers and aroids, and woody vines ascending into a thick canopy, with trees 25 m high. During the following months, the streams and soil dried out, the understory herbs vanished, and the crowns of most trees lost their leaves. Surprisingly, a few were evergreen near ephemeral streams, and the other trees recovered their foliage in the month prior to the return of the rains. The changes to the forest were dramatic, from luxuriance to bare survival.[1] — D. L.

8

FORESTS OF THE
SEASONAL TROPICS

The structure and dynamics of seasonal forest are a departure from everwet.[2] The above example of Indian tall deciduous forest (p. 23) is midway in the continuum from the minimum of seasonality of one dry month per year to the extreme of more than ten dry months. The stature of primary seasonal forests declines with increasing length of the dry season, and along a soil gradient from deep clay mesic loams to shallow, drought-prone soils, irrespective of mineralogy or fertility. The tallest trees in semi-evergreen forest can reach 50 m, although the canopy generally stands at about 40 m. Tall deciduous forest can reach 35 m but is more generally 20–30 m, while short deciduous forest is 10–20 m, and as low as 5 m. In semi-arid regions, thorn woodlands rarely exceed 10 m (table 2.1). The mass flowering and mast fruiting so characteristic of forests in everwet regions is replaced by annual reproduction (p. 233). Aside from seasonality, fire is an important factor in these forests. Human disturbance from fire and grazing have become more important in the past two centuries.

Seasonal Evergreen Dipterocarp Forests

When the Golden Blowpipe Express passed into Thailand, Peter observed southern seasonal evergreen dipterocarp forest (fig. 2.4), which extends north into Thailand and Burma. These forests establish in moist sites of drier climates and in climates with an annual dry season sufficiently short that fire does not penetrate primary stands—up to about four months. They were originally found wherever native, wild durian species occurred and *Durio zibethinus* is cultivated. But annual rainfall varies greatly, from ~2,000 mm to more than 5,000 mm where the summer monsoon confronts the coastal mountains along parts of the northern Western Ghats and Burma's Arakan (Rakhine) coast.

Canopy. These forests rarely attain the stature of MDF, although scattered emergents may reach 50 m in sheltered coves. At 35–40 m, emergent canopy crowns are less clearly raised above the main canopy than in MDF (table 2.1). Another difference from MDF is the greater frequency of deciduous canopy species. Pioneers include *Tetrameles nudiflora, Alstonia scholaris,* and *Bombax* species, while *Lagerstroemia* and *Dalbergia* species are present in later succession. The proportion of deciduous species in the canopy depends on the frequency of canopy and soil disturbance.

Near the northern edge of the tropics, from northeast India into China, a distinct, northern evergreen forest grows, with emergent dipterocarps. These forests exist because of dew formation during the first three of the four dry winter months or, toward the east, because they receive winter rain off the China Sea from the northeast monsoon. Such forests vary considerably in structure. In coastal areas subject to typhoons, they may be no taller than 10 m, with a dense even canopy. Coastal forests of northern Vietnam and Hainan rarely exceed 30 m in height; emergent dipterocarps and other species hardly emerge above the main canopy. On sheltered, south-facing slopes at Mengla in Xishuangbanna (southernmost Yunnan) and eastward, patches of triple-stratum evergreen forests grow, in which the dipterocarp *Parashorea chinensis* reaches true emergent status at 50 m (fig. 2.3). Structurally similar forests have been recorded in Arunachal and the northeast Burma valleys. Although some deciduous pioneers and several successional species occur, the canopy is overwhelmingly evergreen, and fire does not penetrate primary stands.

Understory. The understory of seasonal evergreen forests is generally less dense in juveniles of canopy species. Cohorts of juvenile dipterocarps are sparse, while subcanopy tree species' juveniles are more abundant. Dipterocarps are often less abundant in the canopy, but they flower more frequently: at least one species flowers each year in the more strongly seasonal regions. An annual dry season is associated with more consistent fruit production and a higher density of browsing animals than in everwet climates.

Dynamics. These remain poorly understood in seasonal evergreen forests. Growth rates are not yet available for stand-level comparisons, but there are few light hardwood species in mature phase stands—and probably lower species growth rates than in MDF. In many respects, these forests more closely resemble the evergreen rainforests of Africa and the Amazon Basin, which share a similarly seasonal climate, than they do the forests of everwet Asia.

Semi-Evergreen Forests

Semi-evergreen forests are a mix of barely emerging canopy dipterocarp species and an evergreen mature phase, with predominantly deciduous pioneer and successional canopy species (figs. 2.4, 8.1). Cyclones (fig. 4.4) affect

Fig. 8.1 Seasonal forests and disturbance. *Top left*, monsoon wind-pruned canopy of the Pilarkan sacred forest, Western Ghats, India; southern seasonal evergreen dipterocarp forest, with semi-evergreen forest below and mixed deciduous forest in the distance (D. W.); *top right*, *Wyall* savanna in Mudumalai Wildlife Sanctuary, India, with sedges and *Imperata* (H. H.); *bottom left*, narrow crown shapes from decades of firewood cutting in tall deciduous forest, Maharasthra, India (D. L.); *bottom center*, the wood gatherer may walk several kilometers into the forest to collect firewood (D. L.). *Bottom right*, anthropogenic palmyra palm savanna, Phuket, southern Thailand; fire has suppressed the woody vegetation and promoted the growth of grasses, except for the tall, fire-resistant palms. Vegetated islands off the coast indicate what could grow on the mainland if protected from fire (D. L.).

these forests. Often, tall bamboos establish early in windthrows, retarding and perhaps deflecting succession (p. 135). Coastal forests of the southeast Indo-Burma coast and the western slopes of the Western Ghats have a continuous, even, emergent canopy (fig. 2.7). Crowns are pruned to bend away from the prevailing wind.

Although semi-evergreen and tall deciduous forests occupy similar rainfall regimes, tall deciduous forests extend into drier regions, especially along river banks. These differ in their proportion of evergreen trees, including dipterocarps, and successional deciduous species. Teak is among the few deciduous species that require fire for their regeneration, but all need freedom from evergreen competition (therefore an open understory) in early regeneration. Occasional catastrophic windthrows insure an abundance of deciduous climax successional species in the regenerated canopy (fig. 8.1). However, fire may be necessary in disturbances of this scale to suppress fire-sensitive vines and trees. Moderately small gaps, caused by lightning and small windthrows, provide too little light for the reliable seed germination of dormant pioneer and deciduous successional species, but enough for survival and growth of the deciduous, late successional quality hardwood species that do not require light for germination and early establishment.

Semi-evergreen forests have survived better on freely draining sandy soils in part because the dense, tall bamboo brakes—highly combustible after mass fruiting and mortality—are confined to moist sites among them or to clay, thereby restricting crown fires to those sites. The semi-evergreen forests on the Deccan Plateau are likewise restricted to freely draining, coarse, yet well-watered sites. Some semi-evergreen forests also survive in hilly regions, where clay loams are the preferred lands of swiddening minorities. In the plains, they have also been centers of civilization for more than 15 centuries (p. 344).

With increasing seasonality, more deciduous successional species grow, while true pioneers decline; some pioneers establish following fire and are long-lived. This increase is associated with more frequent and widespread natural catastrophes, especially extensive cyclonic storm damage and increasing frequency of ground fires, which destroy evergreen regeneration.

Semi-evergreen and tall deciduous forests subject to occasional ground fires may have an understory rich in broad-leaved monocotyledons, bamboos, and gingers, and a well-developed understory of deciduous shrubs and canopy saplings of both evergreen and deciduous species. Once they attain a certain diameter, they withstand ground fires.

In summary, while semi-evergreen forest may persist in the absence of fire, where storm damage is sufficiently drastic and frequent to ensure opportunity for regeneration of their deciduous component, tall deciduous forests in Asia depend on fire for their existence. This is in apparent contrast with Africa, where fire is not considered essential. The numerous and diverse browsing animals there may kill evergreen species during the dry season.

Growth rates. Maximum growth rates in these forests, at least of evergreen species, are similar to those in MDF. The much higher soil-nutrient levels and longer growing seasons contribute to growth of semi-evergreen forests. Among the fastest-growing species are many figs, both free-standing and hemiepiphytic. The Asian seasonal dipterocarp and semi-evergreen forests share a similar climate (including windstorms) and many figs (and the associated arboreal vertebrates) with the well-studied forest at Barro Colorado Island (BCI) in Panama.

The semi-evergreen/deciduous forest mosaic. The Huai Kha Khaeng and Tung-Yai Naresuan Wildlife Sanctuaries in Thailand are examples of how a history of fire and windstorms has produced a mosaic of semi-evergreen and tall deciduous forest. This is the largest continuous area of protected forest in Southeast Asia (fig. 8.2).[3] Forest altitudes range from 500 m to 1,500 m, with a diversity of vertebrates expected in this regional climate of 5–6 dry months. The area is rimmed on the west by low mountains bearing lower montane oak-laurel forest (p. 202) and blocks of semi-evergreen forest, and on the east and north by low uplands. Unexpectedly, the central alluvial lowlands and the low hills around the stream-heads are covered by tall deciduous forest, except where semi-evergreen forest penetrates along riparian fringes. Their soils are predominantly clay loams. Curiously, where the tall deciduous forest grows on the clay loams of the alluvial plain, the semi-evergreen forest is confined to freely draining, greyish sandy loams which, although of moderately high nutrient status, are drought-prone. The semi-evergreen forests may have survived because their soils were avoided by farmers.

Bamboo brakes occur only in patches through the tall deciduous forests, and they may hold the key to the forest differentiation. Two clumping genera predominate: *Dendrocalamus* and *Bambusa*. *Bambusa bambos* forms dense brakes and reproduces over whole landscapes, at long intervals. Reproduction and death throughout the catchment lead to periodic crown infernos, which may explain the distribution of the deciduous forest and the

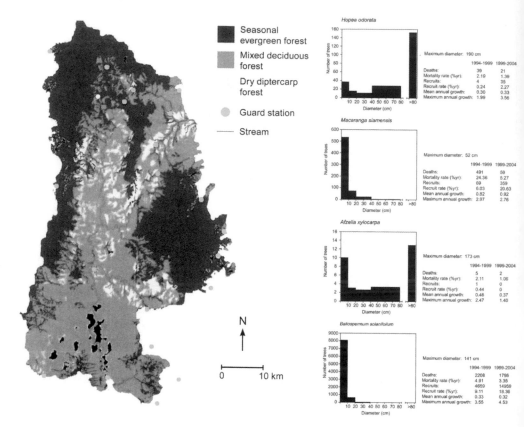

Fig. 8.2 Huai Kha Khaeng World Heritage forest and wildlife forest. *Left,* map of preserve, showing distribution of different forest types. Note the penetration of semi-evergreen forest along water courses within the tall deciduous forests, and the concentration of deciduous, dry dipterocarp forest within the latter (see chap. 8, n. 33). *Right,* population structure of four tree species within the 50 ha ForestGEO plot (both, I. B. after Bunyavejchewin; see S. Bunyavejchewin, J. V. La Frankie, P. J. Baker, S. J. Davies, and P. S. Ashton, *Forest Trees of Huai Kha Khaeng Wildlife Sanctuary, Thailand: Data from the 50-Hectare Forest Dynamics Plot* [Bangkok: National Parks, Wildlife and Plant Conservation Department, 2011]; L. Co et al., "Palanan forest dynamics plots, Philippines," in *Tropical Forest Diversity and Dynamism: Findings from a Large-Scale Plot Network*, ed. E. C. Losos and E. G. Leigh Jr. [Chicago; University of Chicago Press, 2004], 574–84).

rarity there of evergreen canopy species. *Dendrocalamus strictus,* which is scattered in semi-evergreen and tall deciduous forest on upland sandy soils, reproduces erratically and with less synchrony. Consequently, the semi-evergreen forest has received only ground fires every 3–10 years in recent decades.

At Huai Kha Khaeng, evergreen species dominate the canopy on sandy soils, except in patches where deciduous species common to moist deciduous forest predominate. Twenty or so abundant canopy species are distributed in a distinct population structure at scales from patch to landscape. Larger diameter trees outnumber smaller ones. They include both evergreen (such as the light hardwood dipterocarp *Hopea odorata*) and deciduous species (fig. 8.2). Deciduous species probably benefited from catastrophic canopy openings, but they were part of a cohort whose canopy repressed further regeneration of their own species. In contrast to MDF and seasonal evergreen forests, deciduous species dominate succession, though they vary greatly in the duration of their leaflessness and are mainly overtaken by evergreen species in mature stands (fig. 8.3).

Many canopy trees have misshapen boles, evidence of recovery from past wind damage (fig. 8.3). Deciduous pioneer and successional species with annual rings allow estimates of the age of evergreen species. This forest consists of a patchwork of stands, varying from few to hundreds of hectares, arising from catastrophes at different times and scales. About 150 years ago, a landscape-scale windthrow, apparently without ensuing fire, supported the establishment and growth of evergreen forest species, notably *Hopea odorata*, which regenerated during years without ground fires, cyclones, and some shorter-than-average dry seasons. Following later windthrow events, smaller canopy openings appeared in which deciduous species, but not evergreen *H. odorata* and its associates, successfully regenerated. These burned, accounting for the absence of evergreen competition, which is necessary for the success of deciduous successional species. Thus, the combination of windstorm damage, variations in wet season length, and fire helps to explain the forest mosaic seen at Huai Kha Khaeng and in other locations in tropical Asia.

Semi-evergreen notophyll forest (fig. 2.6). In the extreme south of India, and in northern and eastern Sri Lanka, a drought-tolerant, semi-evergreen forest persists, although mostly destroyed by grazing. In contrast to the larger megaphyll and macrophyll leaves frequent in deciduous forests (p. 92), the small leaves of the dominant evergreen species are notophylls,[4] though deciduous forest species are frequent in succession. The evergreen habit is surprising in such a xeric environment.

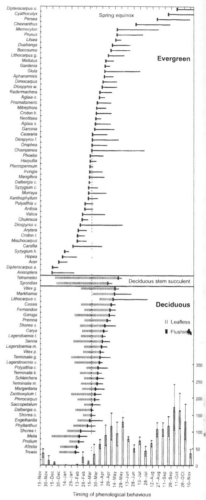

Fig. 8.3 Huai Kha Wildlife Sanctuary, 50 ha ForestGEO plot, Thailand. *Left*, stem irregularities used for inferring age from the pattern of branching in *Hopea odorata* once the timing of canopy catastrophes had been inferred from other evidence. Vertical scale is in meters; the straight-stemmed large tree is typical of canopy individuals (I. B., after P. Baker; see P. J. Baker, S. Bunyavejchewin, C. D. Oliver, and P. S. Ashton, "Disturbance history and historical stand dynamics of a seasonal tropical forest in western Thailand," *Ecological Monographs* 75 [2005]: 317–43). *Right*, leafing phenology of 85 deciduous and evergreen tree species over 4 years. Error bars are the standard deviation of inter-annual variation in the mean flushing date. The dry season begins in November and ends in late April. Leaf fall is gradual in the forest, and late for many species (I. B., after I. Williams; see I. J. Williams, S. Bunyavejchewin, and P. J. Baker, "Deciduousness in a seasonal tropical forest in western Thailand: Interannual and interspecific variation in timing, duration and environmental cues," *Oecologia* 55 [2008]: 571–82).

Deciduous Forests

In India, the seasonal evergreen and semi-evergreen forests of the Western Ghats form ecological islands in a sea of deciduous forests and woodlands that otherwise occupy the lowlands of the entire peninsula. These tall (moist) and short (dry) forests (p. 184) form a continuum. Anthropogenic fire and cattle browsing can degrade the former into a woodland more resembling the latter in structure. This most widespread of Indian forests is broadly distinguishable into two subtypes.

Tall deciduous forests. These forests differ from semi-evergreen forests primarily in their stature, and also in the dominance of deciduous species in the canopy (table 2.1), and the frequency and intensity of fire. In tall deciduous forest, fire has eliminated evergreen species from the canopy, though juveniles and some subcanopy species may persist in the understory. The interrelationship between the two forest types is therefore dynamic and shifting. Tall deciduous forests may only exist in nature where hot fires are frequent enough to suppress or eliminate evergreen tree regeneration. These conditions have been maintained by our ancestors for 2 million years, and much later intensified by domestic cattle and the annual burning that encourages the young grasses on which they depend. Periodic hot fires favor bamboo, which in turn leads to further hot fires. The conditions under which deciduous trees can successfully compete with bamboo are poorly understood. Where deciduous species manage to close the canopy, bamboo density declines. Thereafter, ground fires must be sufficiently frequent to eliminate early evergreen invasion, which patchily persists along water courses (fig. 8.2), but they must periodically abate to allow deciduous seedlings to grow roots and shoots to fire resistant size. This can only occur where browsing and trampling are light. Such conditions have become rare in South Asia.

The deciduous species that regenerate in gaps caused by major windthrows and occasional ground fires in semi-evergreen forests, and which comprise the entire mature canopy of deciduous forests, are light-demanders. However, in Indian tall deciduous forests, the often-dominant teak and sal are only deciduous for 1–2 months, February–April (see also fig. 8.3). In effect, the gradient in light response, the main difference among species of evergreen forest, is replaced by gradients primarily related to drought tolerance, resistance to fire, and browsing.

Short deciduous forests. These differ from tall deciduous forests not only in height (table 2.1), but also in not reverting to semi-evergreen forest if fire is excluded. Short deciduous forests lack the juvenile evergreens that persist in the tall deciduous understory by stump-sprouting or spreading from refuges in nearby moist, fire-free sites. Tall deciduous forests regenerate periodically, predominantly by seed in large-scale cohorts, while short deciduous forests regenerate more patchily, by coppice, and less from seed (at least now). Short deciduous forests and thorn woodlands bear a sparse, grassy understory, fueling less intense and frequent fires, though human-caused fires in tall deciduous forests obscure that difference.

In India, semi-evergreen and tall deciduous forests have dominated regions with up to 6 dry months; short deciduous forest, from 6–9 dry months; and thorn woodland, from 9–10 dry months (table 2.1). Increases of domestic cattle can reduce a once-productive field layer to short sparse grassland and prevent tree regeneration by coppice, thereby impoverishing the tree flora. This promotes the expansion of drier forests into moister climates. *Lantana camara* (see fig 8.7) is now pervasive in the widespread transition between tall and short deciduous forests of Peninsular India.

All tree species of short deciduous forest produce dormant seeds. Seed and seedling survival are influenced by the intensity of drought and by the variable duration of the dry season. Large seeds may be more resistant to drought than small ones, although seeds in short forests are generally small. Seedling survival in short deciduous forest is very low because of drought, fire, and especially browsing—at present overwhelmingly by domestic cattle and goats. On the other hand, frequent fire, by accelerating nutrient release from litter, can also increase growth rates for the surviving trees in short deciduous sal forest. Light-demanding and shade-tolerant juveniles are not distinguished in these often open-canopied woodlands.

A tall/short deciduous forest mosaic. Ongoing research at Mudumalai Wildlife Sanctuary and Tiger Reserve (Tamil Nadu, India) documents the balance between tall and short deciduous teak forest (fig. 8.4). It is dominated by teak; many individuals are 150–190 years old, and the oldest is 270. Suppression of fire at the sanctuary has promoted the shift from grass to *Lantana*, which has suppressed teak regeneration. Currently, *Lantana* is in decline, and the native, shrubby understory trees *Helicteres isora* and *Kydia calycina,* though favored by elephants, are increasing, as is spiny shrub *Xeramphis spinosa*. These species are typical of short deciduous forests in the absence of frequent fire. Nevertheless, rainfall at Mudumulai is similar in amount and seasonality to that in semi-evergreen forest at Huai Kha Khaeng, though the latter does receive more pre-monsoon, hot dry season showers. Evergreen *Mangifera indica* occurs in the Mudumalai swales, and *Syzygium cumini* on the uplands, while *Lagerstroemia* and *Terminalia,* important genera of tall deciduous forests, are abundant in higher rainfall regions of Mudumalai. On the northwest margin of the Mudumalai reserve, in a rainfall regime of approximately 5 dry months and 1,600 mm/y, a semi-evergreen forest patch survives, sharply separated from the deciduous forest by a perennial stream. The canopy palm *Caryota urens* is present. Only long-term monitoring of this forest mosaic and fire-exclusion experimentation can resolve the status of its deciduous forests. Mudumalai Tiger Reserve is

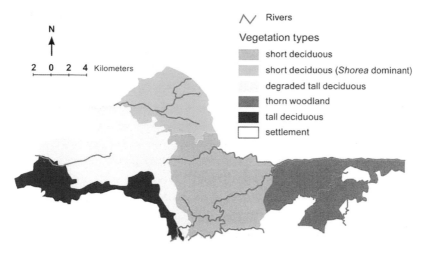

N

↑

2 0 2 4 Kilometers

∿ Rivers

Vegetation types

short deciduous

short deciduous (*Shorea* dominant)

degraded tall deciduous

thorn woodland

tall deciduous

settlement

Fig. 8.4 Vegetation map of Mudumalai Wildlife Sanctuary, Tamil Nadu, India (I. B., after H. Suresh; see H. S. Suresh, H. S. Dattaraja, N. Mondal, and R. Sukumar, "Seasonally dry tropical forests in Southern India: An analysis of floristic composition, structure and dynamics," in *The Ecology and Conservation of Seasonally Dry Forest in Asia*, ed. W. J. McShea, S. J. Davies, and N. Bhumpakphan [Washington, DC: Smithsonian Institution Scholarly Press, 2011], 37–58).

at 860–910 m, with a band of semi-evergreen forest extending further along its western margin, somewhat higher up and against the Nilgiri slope. It includes lower montane forest species (p. 202).

Distinguishing between moist and dry deciduous forests is difficult because millennia of anthropogenic fire and recent grazing by domestic cattle have eliminated the evergreen element on clay soils; its usual persistence near streams betrays its likely wider historic distribution. In Thailand and elsewhere in Indo-Burma, much of the deciduous forest, described as "dry deciduous forest" by local foresters, is tall deciduous forest that has lost its evergreen element through increased fire—consistent with its tall stature (usually 30 m but sometimes reaching 35 m). The understory of bamboos, tree regeneration, patchiness, and frequent absence of grass, absence of *Lantana*, and local abundance of broad-leaved monocotyledons, including gingers, all place it there. These features of Indian tall deciduous forests abate toward the upper slopes of ridges, where height declines to 15–20 m, bamboos become sparse, and monocotyledonous herbs are rare. However, even there, a grassy field layer is generally patchy and thin or absent, while the tree flora is an impoverished subset of tall deciduous forest; Instead, the frequent appearance of deciduous dipterocarps presages an ecotone to that forest type (p. 147).

Complicating the relationship between deciduousness and the physical environment is the higher proportion of deciduous species on clay loams rather than siliceous soils, seen throughout Asian forests. Within a climate of extreme seasonality, teak, which dominates clay loams, is leafless for a longer period than sal, which dominates siliceous soils (p. 202).

All Asian tropical deciduous forests recall temperate deciduous broad-leaved forests in the shortness of the trunks of canopy individuals relative to the depth of their crowns; the drier the habitat, the more extreme is this relation. Columnar trunks become rare in shorter forests; misshapen trunks may be a consequence of crown fires, browsing by large mammals, or early flowering of the leading shoot among individuals of species with terminal inflorescences, including teak.

Leaf size and shape. An extraordinary feature of the tall and short deciduous forests, as well as deciduous dipterocarp forests, is that the leaves of canopy dominants (teak, sal, or deciduous dipterocarp species, often only shortly deciduous) are macrophyll or large mesophyll (p. 92), are leathery, and are dispersed sparsely in crowns. Other tree species of deciduous dipterocarp forest develop large leaves. The proportion of pinnately compound leaves with microphyll or smaller leaflets does increase in short deciduous forest and thorn woodlands of drier climates, as in other parts of the tropics. Nearly all leaf and leaflet blades are entire. Petioles are short, and leaves are held at various angles and dispersed within their crowns.

Growth rates and productivity. A dry season reduces time for growth. Evergreen species are adapted to annual soil water deficits; species with narrow water-conducting elements (and therefore heavy timber and slower growth rates) are favored (p. 139).

Canopy tree species in the semi-evergreen forests at Huai Kha Khaeng, with 1,500 mm of rain restricted to seven months, have unexpectedly high growth rates, almost identical to those at Lambir in everwet Borneo. The soils in these seasonal regions, even on unpromising substrates, are much more fertile than in everwet regions (fig. 5.6), while the longer periods of cloudy weather in the everwet tropics may reduce tree photosynthesis and growth. The annual number of days of full sun may also contribute to the formidable growth rates achieved by some evergreen tree species in the seasonal tropics;[5] this is also true in West Africa and at Barro Colorado Island in Panama, under a three-month dry season (fig. 8.5).

Stem Density (per ha)

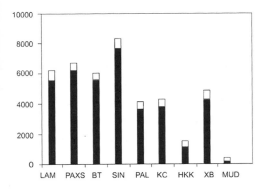

Basal Area (m²/ha)

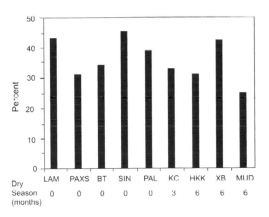

Fig. 8.5 Mean basal area and stem density in ForestGEO sites in tropical Asia; black part of each bar is for trees more than 1 cm and less than 10 cm DBH (diameter at breast height). Forest acronyms: LAM = Lambir, Sarawak; PAXS = Pasoh, Peninsular Malaysia; BT = Bukit Timah, Singapore; SIN = Sinharaja, Sri Lanka; PAL, Palanan, Philippines; KC = Khao Ban That, Thailand; HKK = Huai Kha Khaeng, Thailand; XB = Xishuangbanna, Yunnan, China; MUD = Mudumalai, Tamil Nadu. Dry season in months is given for each site (see citation in fig. 8.4 caption). The effective length of dry season is shorter if fog drip is considered for HKK (at 600 m), which should be 5 months, and for Xishuangbanna, which effectively has no dry season (I. B., after Bunyavejchewin; see citation in fig. 8.2 caption).

Most species flush their leaves as atmospheric humidity increases in the hot dry season, beginning in early April, and are at their physiological peak during the southwest monsoon, when skies are overcast for days, and the atmosphere is saturated. The climate in these respects resembles that of the evergreen forest understory. Higher diffuse light intensity during the wet monsoon, which is windy, may account for high net rates of photosynthesis. The wide spacing of canopy trees in shorter deciduous forests indicates relatively low above-ground competition but implies intense root competition for water. The ensuing dry season is cool at first, reducing evaporation.

Species of later succession and the mature phase. Climax canopy species of the mature phase should not match the extraordinary growth rates among pioneer species (and canopy figs). Most of the deciduous species, including sal, teak, and most legumes, although they adopt a successional role in semi-evergreen forests, are heavy hardwoods. Nevertheless, teak in managed tall deciduous forest may reach 50 cm DBH (diameter at breast height) in ~80 years, a mean diameter growth rate of 6.3 mm/y. Sal reaches 50 cm DBH in 70 years, or 7 mm/y. Such rates compare favorably with growth rates of MDF light hardwood dipterocarps in plantation or managed regenerating stands. Surprisingly, sal forests are distributed on siliceous substrates, and their grey-brown soils are less rich in nutrients than the calcium-rich clay loams to which teak forests are restricted in nature. Water availability rather than nutrient concentration likely restricts maximum growth rates on the higher-nutrient soils of the strongly seasonal tropics.

The growing season of deciduous forests is extended beyond the onset of the wet monsoon by the precocious leafing of most species during the hot dry season, which starts in March at lower latitudes. These early shoots are boosted by reabsorption of nutrients from senescing leaves of the previous season: 20–100% of phosphorus may be retained. Biomass has been calculated for a mature, managed tall deciduous sal forest in the Bhabar Terai, with a remarkable estimated total biomass of 710 t ha^{-1}. This is higher than the estimate for the 50 m tall MDF at Pasoh Forest Reserve (p. 119). Net primary production, without the inclusion of litter, was 18.6 t ha^{-1} y^{-1}, of which the dominant sal comprised 12.8 t ha^{-1} y^{-1}, less than at Pasoh.

The regeneration niche. Seeds germinate and seedlings establish in seasonal evergreen and deciduous forests during the wet monsoon. As the wet monsoon shortens, seedlings establish earlier, and their leading root extends deeply in their first wet season. Variability in the annual rainfall pattern lends selective advantage to seed dormancy and to responsiveness in germination to increased humidity. For stumps, dormancy in sprouting allows reserves to be built up in the roots so that a high root/shoot biomass ratio is achieved early, permitting maximum access to water and growth of a leading shoot that can survive in a relatively fire-free year. Although fire kills first-year seedlings of most species, the dormant seeds of many species are fire resistant. Fire thus increases the germination rate in teak. It reduces the thick layer of large fallen leaves of teak and sal, which resist seedling root penetration. It also exposes a friable surface rich in ash and favorable to seedling establishment. Nevertheless, fire prevention is necessary for a few years initially if rootstocks, and coppice shoots in later years, are to survive. Fire frequency throughout fire-prone forests in the Asian tropics has increased so much that seedling establishment has become rare even in many semi-evergreen forests. Regeneration is thus increasingly dependent on stump sprouts, which are also increasingly repressed.

The length and variability of the dry season influence all aspects of the forests of the seasonal tropics. In the occasional, supra-annual lengthy droughts associated with anomalies in the ENSO (p. 11), significant mortality occurs in both the upper canopy and lower understory; without fire, however, no permanent influence is apparent in primary forest, other than a reduction in the abundance of certain successional species on drier sites.

Bamboo brakes are patchy but can be extensive in some regions, especially northeast India and Indo-Burma. Brakes occur mostly on clay soils or in moist valleys. Bamboos are light-demanders and do not survive under established canopies. Bamboo seeds are dormant, germinating in response to light. Their seeds are dispersed by rodents and birds, and they are highly invasive. Bamboo brakes are a major impediment to forest regeneration. Regeneration is most successful during years immediately following mass bamboo fruiting, death, and fire, which opens the canopy. Seedlings of more shade-tolerant tree species survive beneath the bamboo canopy, hardly growing in height. Bamboo thereby favors survival of those shade-tolerant, slow-growing, heavy hardwood tree species, mostly deciduous, that invade after fire.

Vines. Vines are more abundant in the semi-evergreen and tall deciduous forests than elsewhere in the Asian tropics. Large vines are particularly abundant in windfall gaps on clay and sandy clay soils and readily form the dense monolayer leaf blanket that is also seen in gaps on mesic sites in the everwet tropics. Many vines of the seasonal tropics are deciduous. They are less abundant, and shorter and slimmer, in short deciduous forest, and are absent from thorn forest.

Browsing and grazing. The biomass of browsing and grazing vertebrates was formerly vastly greater in the seasonal than everwet tropics, because of the high seasonal production of accessible biomass in the herbaceous field layer and as coppice. Although the species diversity of ungulates in Asia is not great overall (but see p. 256), the impact of elephants and gaur (*Bos gaurus*) is strong. Grassland, whether induced by flooding, frost, or fire, attracts grazing and browsing herds. Bamboo attracts elephants, but some bamboos are poisonous. Their rhizomes resist fire, but shoots and seedlings are browsed by cattle and deer. Browsing intensity is greatest where food and shelter are most abundant, in the semi-evergreen/tall deciduous forest mosaic, where large browsing mammals congregate as the dry season progresses. Browsing mammals have a variety of food preferences. Many vines, especially *Merremia* and other fast-growing species, are favored by elephants, and these also are attracted to forest disturbance by the fleshy growth of successional species. Elephants are fond of *Artocarpus* in evergreen forests, and understory *Kydia calycina*, *Helicteres isora*, and young bamboo shoots in deciduous forest. Deer and langur monkeys consume sal regeneration. Firewood collection also takes its toll on trees (fig. 8.3; see also chapter 17).

Tall deciduous forests establish in southwest China, along the tributaries to the Mekong and other rivers. The understories of these forests are particularly rich in evergreen species, likely due to the lack of cattle grazing: dairy consumption is low because many of the inhabitants are lactose intolerant.[6]

Browsing intensity fluctuates in the deciduous forests with the seasons and from year to year. Although the best-managed parks in seasonal Asia still support close to natural populations of wildlife, regional patterns of forest vegetation are no longer influenced by wildlife; instead, they are primarily influenced by cattle. The relative impacts of browsing versus fire in restricting juvenile survival and growth have not been assessed.

Drought mortality. Increasing length of the dry season is accompanied by its greater variability and occasionally by extreme droughts. Species vary in mortality rates, implying that these occasional events play a crucial role in determining species composition. Such catastrophes, in effect, turn back the forest to early successional stages. Studies of tree mortality in catchments can reveal the effects of the depth of aquifers on mortality of different trees, as has been shown at Mulehole, near Bandipur National Park, India.[7] More research like this recent study is needed to understand tree demography in the seasonal tropics.

Semi-Arid Vegetation

In semi-arid climates with more than 8 dry months, forest in India is increasingly replaced by *thorn woodland* with a deciduous open canopy reaching to 5–10 m, which nowadays is severely overgrazed. The vegetation is too sparse for fires to spread; regeneration dynamics are dominated by grazing and browsing, especially by cattle. Saplings may escape the browsing when they establish within the pervasive thorny shrubs.

A *savanna* is woodland with grassy field layer, in which the trees are isolated or in scattered clumps. Such woodlands are now widespread in continental Asia (figs. 2.6, 8.1). Excepting pine forests (p. 216) and some deciduous dipterocarp forests (p. 148), tropical Asian savannas appear to be caused by human activity: overgrazing by domestic cattle, cutting fuelwood, and burning to regenerate grass. Anoxic, still flood waters, which prevent tree regeneration, may produce savanna in the floodplains of some of the great Asian Rivers, such as the Brahmaputra.

The Influence of Fire

Fire, and consequently grass and ungulates (including domestic cattle), are widespread in the seasonal lowland forests of continental Asia (fig. 8.6). Fire starts naturally by lightning strikes when rain is not falling. Additionally, our progenitors in Asia, as in Africa, burned forest to raise game for almost 2 million years, and for at least 4,000 years to encourage grass for cattle. The history of fire as an ecological factor has been dramatically different from its history in the Neotropics. In Asia, only one hominin—us—has been actively burning, and only for the last 20,000 years, and in Australia, for 50,000.

Fires start in and are fueled by dry combustible material; they are consequently rare in everwet regions. Fires are infrequent at the beginning of the dry season, but litter and ground vegetation become increasingly fire-prone as the season progresses. Rainfall seasonality influences fire incidence and intensity, as well as affecting forest structure, phenology, and growth. Importantly, increasing seasonality increases variability in the length of the wet monsoon, the amount of precipitation, and in its timing (p. 148). As regional aridity increases, annual biomass production declines, and thereby the frequency and intensity of fire falls. From Myanmar eastward, shower frequency increases during the hot, dry season, starting in April, when trees begin to flush before the onset of the monsoon; thus, natural fire frequency declines. As the tropical margin is approached, the cool dry season and dewfall extend closer to the beginning of the monsoon, and forests again become resistant to burning. Fire-maintained *Imperata* grasslands are, of course, ubiquitous in wet regions, originating from degraded swidden agriculture (p. 338; fig. 8.7).

Types of fire. Fires differ in intensity and impact. *Ground fires* are of low intensity, occurring where combustible material is sparse, consisting mostly of leaf and twig litter or short grass. Fire only affects the field layer, occasionally entering the hollow trunks of canopy trees (fig. 8.6), especially those already wounded by crown fires. Temperatures may reach 400° C at 50 cm above ground.

Coarse litter of dead bamboo and tall grasses, fallen branches and other

Fig. 8.6 Fire in seasonal forests. *Top left*, hot crown fire in short deciduous forest, 2/19, Bandipur National Park, Karnataka (Wikimedia: Naveen Kadalaveni); *top right*, ground fire burns toward, but is excluded by protected southern seasonal evergreen forest in Kanyakumari, South India, 1/89 (D. L.); *bottom left*, ground fire in tall deciduous forest, Mae Hon Son, Thailand (H. H.); *bottom center*, burnt tree trunk of *Hopea odorata*, caused by ground fire in semi-evergreen dipterocarp forest, Huai Kha Khaeng ForestGEO plot, Thailand (H. H.); *bottom right*, teak with epicormic shoots, two years after hot fire, 2010, Bandipur National Park, India (H. H.).

material, and combustible subcanopy vegetation (shrubs and bamboo) feed *hot fires*. These can ascend to the canopy, reaching 700°–800° C at 50 cm, and often killing mature, fire-sensitive trees (fig. 8.6). Fire-resistant species, all deciduous (but for eucalypts), including teak and sal, recover by epicormic branching, but their crooked trunks and branches recall their tribulations. Fire exclusion leads to accumulation of fuel and increases the risk of crown fires.

Fig. 8.7 The aftermath of fire. *Top left*, short deciduous forest following a hot fire, Bandipur National Park, India (R. S.); *top right*, same forest type three years after fire, with *lantana* and returning trees shooting epicormically (H. H.); *bottom left*, lantana (*Lantana camara*; D. L.); *bottom center*, lantana habit, India (S. D.); *bottom right*, stand of dwarf bamboo (with trees emerging from clumps), *Vietnamosasa ciliata*, in fire-degraded deciduous forest, Khorat Plateau, Thailand (S. Dr.).

Various fuels feed *hot fires*. One is combustible vegetation in the understory. Bamboos—especially those many genera that reproduce and die synchronously over whole landscapes and even regions (p. 148)—are associated with devastating fires. Bamboo germinates and establishes once rains arrive, and the opening of the bamboo canopy after flowering provides the sole opportunity for tree regeneration. Trees with seed dormancy and fire resistance, shade-tolerant seedlings and, especially, fire-resistant rootstocks have the competitive edge and survive sporadically thereafter. The Neotropical shrub *Lantana camara* has, during 150 years, invaded deciduous forests of the drier parts of South Asia to become the dominant understory shrub (fig. 8.7). As it matures, it becomes flammable. Its canopy is neither so dense nor so tall (less than 2 m) as to exclude all other species. However, the fires that feed on it can reach the canopy of short forests. Tall grass in dense swards (notably *Imperata,* which can grow to 1 m), feeds fires of intermediate intensity, killing evergreen understory species but not reaching the canopy. The greater fire frequency associated with such grasses destroys all seedlings.

Fire frequency. Infrequent fires may favor invasion of bamboo. Vast regions of continental Southeast Asia are nowadays dominated by forests with a bamboo understory, and bamboos also invade landslides and heavily logged forests. Frequent fire, on the other hand, leads to the death of bamboo and its replacement by grassland. Frequent fire dramatically changes tree species composition, phenology, and forest structure. Thus, fire can transform multilayered evergreen forest into deciduous forest with a single-layered canopy and an understory of grass, bamboo, or (in moister sites) woody or herbaceous dicots.

Fires have increased, particularly over the last century, as human populations have grown and pressure on forest resources has increased. In Indo-Burma, ground fires from burning rice stubble spread for tens of kilometers into parks and sanctuaries. Many forests now receive almost annual ground fires. Fires are set in India to encourage grasses for the ubiquitous cattle, which also severely reduce regeneration by their browsing. There, the interval between fires in deciduous forests has decreased dramatically over the last century.

Interactions between fire and forest. The impact of fires on forests, although of central importance to forest management in the seasonal tropics, still awaits careful systematic observation and experimentation. Evergreen tree species in continental Asia generally do not survive persistent fire as juveniles, although *Syzygium cumini, S. nervosum,* and *Mangifera indica* are more resistant than most and occur in forests that are otherwise deciduous. Seedlings and young saplings of evergreen species are particularly sensitive. Larger individuals of many species have thin bark, and ground fires can wound the trunks. Nevertheless, many species recover by coppice shoots if fires are infrequent or not too intense. Widespread semi-evergreen forest pioneers are particularly sensitive, confined to moist valley sites.

Rootstocks of many seedlings of deciduous forest species resist ground fires after the wet season and after germination (fig. 8.7). Tree seedlings are often absent in short deciduous forests. The importance of burning, browsing, or drought in killing trees varies with general forest conditions. Deciduous forests include a guild of fire-tolerant deciduous species which, as first-year seedlings, may be killed by fire but shortly thereafter can survive as rootstocks, resprouting from the ground each year until there is either a run of years without fire, or the rootstock has grown sufficiently to support a shoot that can grow tall enough to escape a ground fire in one season. Fire and browsing kill above-ground parts of young trees, but the root may continue to grow slowly, forming a large woody "lignotuber." Each year, it has more resources for a reiterating leader to grow taller in the following dry season; eventually, following several years without fire, the sapling develops bark sufficiently thick to insulate living tissues from fire. Species with such bark are overwhelmingly deciduous. Fire resistance is therefore the primary motor driving the distinctive composition and phenology of tropical deciduous forest.

Because the density and nature of the understory affects the amount of combustible material, the most intense crown fires are in moist tall deciduous and the taller margins of dry short deciduous forests. Fires rarely invade thorn woodland owing to lack of fuel. Increased fire frequency is currently reducing fuel accumulation and therefore crown fires. This will favor species of the shorter communities. Ground fires are beneficial because they burn the leaf litter accumulated over the dry season, which can become matted during the rains and deter seedling emergence. Fire is therefore central to the ecology of seasonal forests.

Notes

1. Lee, David W., "Canopy dynamics and light climates in a tropical moist deciduous forest," *Journal of Tropical Ecology* 5 (1989): 65–79.

2. *OTFTA*, 139–52, 202–29.

3. Bunyavejchewin, S., J. V. La Frankie, and S. J. Davies, "Seasonally dry forests in continental southwest Asia: Structure, composition and dynamics," in *The Ecology and Conservation of Seasonally Dry Forest in Asia*, ed. W. J. McShea, S. J. Davies, and N. Bhumpakphan (Washington, DC: Smithsonian Institution Scholarly Press, 2011), 9–35.

4. Chitra-Tarak, Rutala, et al., "The roots of the drought: Hydrology and water uptake strategies mediate forest-wide demographic response to precipitation," *Journal of Ecology* 106 (2018): 1495–1507.

5. Ashton, Peter S., and H. Zhou, "Tall tropical deciduous forests in SW China," *Plant Diversity* 43 (2021): in press.

6. Ashton and Zhou, "Tall tropical deciduous forests."

7. Dong, S. X., et al., "Variability in solar radiation and temperature explains observed patterns and trends in tree growth rates across four tropical forests," *Proceedings of the Royal Society B* 279 (2012): 3923–31.

I went to Brunei as forest botanist almost one century after Odoardo Beccari,[1] but having first familiarized myself with the more recent work of Paul Richards.[2] He knew that the apparently chaotic jumble of species in biodiverse rainforests varied across the landscape. It took six months to reliably identify the perhaps 200 commoner species in Brunei, taught to me by my Iban Dayak colleagues. I began to keep notes about the species, especially the dipterocarps, that we encountered along the forest tracks as we collected specimens for a forest flora. I was soon struck by the very different tree flora of the MDF in the Belait coastal hills compared with that of the steep inland ridges draining into the whitewater Temburong River. The Belait hills were soft young Mio-Pliocene sandstones; the Temburong were older, harder, lower Miocene shales. The soils were distinctly different, deep and sandy versus shallow and clay. These floristic and landscape differences precisely followed the boundaries on the geological map. Later, when exploring other regions in Asia, I saw that the sandy habitats form islands within a regional matrix of clay soils. — P. A.

9

TREE SPECIES COMPOSITION IN TROPICAL LOWLAND FORESTS

The Bornean Flora assumes multiform aspects, and its components are of
a varied nature according to the localities, the elevation, and the physical
conditions of the soil. Thus, distinct areas of varied extent can be recognized,
some much restricted, on which a special vegetation grows, different from that of
adjoining lands.

Odoardo Beccari, Wanderings in the Great Forests of Borneo

Tree species composition varies across diverse landscapes of lowland tropical Asia. Patterns of variation can be discerned at both local and regional
scales. Identifying these patterns and their possible causes is one approach
to understanding the general question of biological diversity in tropical
Asian forests; other approaches are addressed in chapters 13 and 14.

The Forest Formations

Regional-scale variations in forest structure and phenology in the major
forest formations are correlated with patterns of rainfall seasonality (chapter 8), but there are floristic characteristics too, relating both to climate and
geographic history, especially during and following the Pleistocene epoch.
The decline in species richness of all forest formations, and within them
most floristic communities, is notable from equatorial Sundaland (p. 45)
and seasonal, easternmost Indo-Burma westward, with a major decline
among deciduous forests across the evergreen forest barrier of Chatterjee's
Partition at the India-Burma frontier (p. 290), and across the deciduous

forests of the Ganges-Brahmaputra Plain for evergreen and semi-evergreen forests.

The MDF and related forest types are highly diverse at all taxonomic levels, have emergent canopy dominance of Dipterocarpaceae, and contain series of congeneric species within communities (table 7.2). Diversity peaks only in Sundaland (chapter 14), with nearly 400 dipterocarp species, whereas there are but 25 dipterocarp species overall in East Malesia, and ~50 dipterocarp species in Sri Lanka, of which all but one are endemic. The Philippine tree flora has high species endemism as well (p. 297).

The seasonal evergreen dipterocarp forests differ similarly, with highest endemism in Peninsular India. The northern block is also distinguished by high endemism and a suite of lower montane species that descend to the lowlands in its dew-rich climate. The South Indian, southern seasonal evergreen dipterocarp forest, with the widespread and locally abundant emergent *D. indicus* and several *Hopea*, but no *Shorea*, is less diverse in dipterocarps, and includes the deciduous species *Lagerstroemia lanceolata*.

Semi-evergreen and deciduous forests, as well as thorn woodland, harbor many deciduous species that are rare or absent in evergreen forests. Richness at all taxonomic levels declines as the dry season increases in length. *Terminalia* and *Lagerstroemia*, tall deciduous canopy trees that are well represented in semi-evergreen and tall deciduous forests, hardly appear in short deciduous forest, where *Anogiessus*, *Cochlospermum*, and *Acacia* and other legumes of thorn woodland become common. Deciduous forests of Indo-Burma are richer than those of Peninsular India, with high regional and local endemism (chapter 13).

Floristic Communities within the Formations

The rich species diversity in MDF and other forests of everwet climates is the most studied. Rainfall seasonality influences species diversity in other forests. Several other physical factors also interact powerfully with climate to produce the current patterns of floristic distribution, particularly geology and soils (and consequently nutrients), and topography and drainage (and water availability on a local scale). It is surprising that so little attention was paid by early foresters (of temperate origin) to the relationships between forests and soils in much of tropical Asia, particularly in the lately forested terrains of the everwet Far East.

The everwet tropics. From an ecological perspective, the everwet climate is effectively uniform over vast areas of tropical Asia because precipitation, although variable, is surplus to evapotranspiration except during unpredictable droughts (p. 51). There are major forest types circumscribed by other variables, such as drainage and soils. Those types influence, for instance, the total diversity of the Kinabalu region of North Borneo (fig. 9.1).

Within this zone, soils are seldom dry enough for nutrients in solution to remain unattached or unleached, let alone to be drawn upward and concentrated by capillary movements of solutes, as may happen during annual dry seasons. Instead, relentless leaching enhances the differences between mineral soil types. Nevertheless, at time scales comparable to the life span of a tree, nutrients may become available to its roots (p. 81). The forests may therefore be exceptionally variable (table 9.1). In Brunei, the total species richness among these different everwet forests is ~3,000 species, which is perhaps ~12% of tree species for all of tropical Asia, with 38% of them endemic to Borneo. More than climate is necessary to explain these landscape patterns of diversity, and forest and soil variation are interdependent.

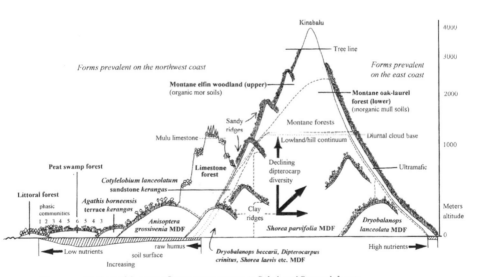

Diagrammatic chart of the major floristic associations in Sabah and Sarawak forests
MDF stands for Mixed Dipterocarp Forest

Fig. 9.1 Schematic diagram of the major floristic associations in the forests of northern Borneo (P. A.; see Peter Ashton, "Towards a regional forest classification in the humid topics of Asia," in *Kluwer Handbooks of Vegetation Science*, ed. E. O. Box, et al. [Dordrecht: Kluwer, 1995], 453–64). The *Anisoptera grossivenia* MDF is on humic yellow sandy loams, the *Shorea parvifolia* MDF on silty clay loams, the *Dryobalanops lanceolata* MDF on friable clay loam, while the *Dryobalanaops beccarii* MDF is on sandy clay loams. Other forest types that occur on different geological substrates add to the floristic diversity of the region.

Table 9.1 Approximate number of species in different forest types in Brunei, Northwest Borneo. The figures in total exceed the number of species known from Brunei because many occur in more than one forest type.

Forest type	Number of tree species	Borneo endemism (%)
Sea shore	25	0
Mangrove	33	0
Back-mangrove, brackish river banks	50	5
Plains rivers, alluvial banks	200	15
Rocky, whitewater river banks	200	35
Floodplains (seasonally swamped)	600	25
Peat swamp forest	200	25
Kerangas	300	50
MDF, sandy, sandy clay soils	1,200	50
MDF, clay soils	900	35
Upper dipterocarp forest	600	40
Lower montane forest	500	45
Upper montane forest	100	50

Source: Peter S. Ashton, *A Field Guide to the Forest Trees of Brunei Darussalam* (Bandar Seri Begawan: University of Brunei Darussalam, 2003).

Variation in diversity with topography and soil. Wyatt-Smith recognized a zonal MDF in everwet West Malesia growing on yellow-red soils of moderate clay content: the *red meranti-keruing forest*.[3] The upper stratum is dominated by light hardwood *Shorea* of the light red and yellow meranti field groups and species of *Dipterocarpus*, with scattered white meranti, all of which are moderate light-demanders as juveniles. He also recognized that some inland MDFs also carry dense populations of single, emergent, shade-tolerant heavy hardwood species, including *Neobalanocarpus* and *Shorea* sect. *Shorea* (balau, selangan batu), which there dominate the light hardwood dipterocarps. Symington recognized that forests of the coastal hills of Peninsular Malaysia were floristically distinct, dominated by slower-growing dipterocarp species of denser wood.[4] Most of this *coastal hill dipterocarp forest* was dominated by balau but occurred in areas also dominated by slow-growing, dark red meranti (*S. curtisii* sect. *Mutica*). Similar increases in heavier hardwood species were seen on high inland granite ridges, where a *hill dipterocarp forest* association was recognized. These forests share sites prone to soil water stress.

Forest variation with topography. Comparing plots in a Brunei forest showed that species associations were related to topography. In forests on ridges versus riverbanks, quite different tree associations were observed (fig. 9.2). At Sinharaja Forest, in Sri Lanka, two species of ironwood (*Mesua*) were plotted on a forest ridge with spurs and gullies. They were largely segregated by habitat, with gullies having more moisture than spurs (fig. 9.2); topography mediates water stress.

Species responses. Correlating species distributions with variation in topography within a large plot, as in the ForestGEO system, and among many small plots, is a powerful technique for determining species and community relationships to the physical environment. It is also important to directly

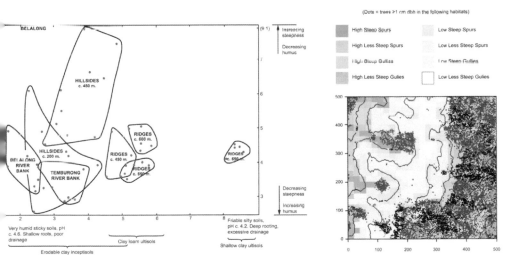

Fig. 9.2 Tree distributions and topography. *Left*, Bray and Curtis ordinations of 50 0.4 ha plots, on clay loams at Kuala Belalong Forest Reserve, Brunei. Dots are species assemblages plotted on two gradients, steepness and humus, and on soils from deep moist to shallow friable loam (P. A.; see Peter Ashton, "Ecological Studies in the Mixed Dipterocarp Forests of Brunei State," *Oxford Forestry Memoirs* 25 [1964]). *Right*, distributions of co-occurring sister taxa in *Mesua* (Clusiaceae); *M. ferrea* (red dots) is a main canopy species on the spurs, and *M. thwaitesii* is a subcanopy species in the intervening gullies. Color blocks indicate steepness and spurs or gullies in the 25 ha ForestGEO plot in Sinharaja Forest, Sri Lanka (M. A.; see C. V. S. Gunatilleke and I. A. U. N. Gunatilleke, *Ecology of Sinharaja Rainforest and the Forest Dynamics Plot* [Columbo: WHT, 2004]).

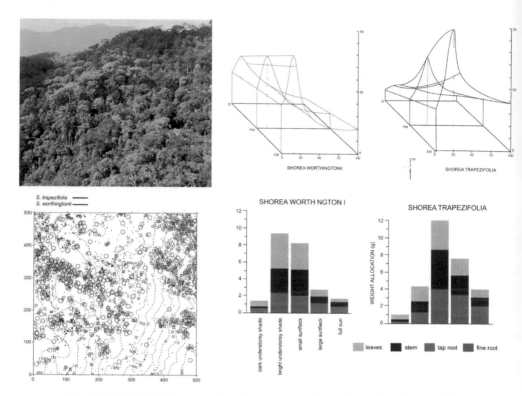

Fig. 9.3 Distribution and growth of two *Shorea* species, *S. worthingtonii* and *S. trapezifolia* at Sinharaja Forest, Sri Lanka. *Top left*, aerial view of the forest setting (M. A.); *top right*, height growth of seedlings after 800 days (vertical bar) in response to soil moisture; in the left line (D = drought, PM = periodically moist, EM = continuously wet) and in the bottom line (light, daily PAR as percentage of full sunlight; M. A.; see Mark Ashton, *Seedling Response of* Shorea *Species Across Moisture and Light Regimes in a Sri Lankan Rainforest* [New Haven, CT: PhD thesis, Yale University, 1990]). *Bottom left*, distribution of these two species in the Sinharaja ForestGEO plot (Gunatilleke and Gunatilleke, *Ecology of Sinharaja Rainforest*). *Bottom right*, dry mass increase and allocation among seedlings of these two species grown in differing amounts of light (M. A.; see Mark Ashton, "Seedling growth of co-occurring *Shorea* species in the simulated light environments of a rainforest," *Forest Ecology and Management* 72 [1995]: 1–12.).

study the growth and physiology of related species growing in these environments. A model system for study is the *Shorea* species (section *Doona*) in the ForestGEO plot of Sinharaja forest in southwest Sri Lanka, studied by Mark Ashton and colleagues (p. 171; fig. 9.3; cited in n. 6 below). There, species distributions varied in relation to topography and physiology. The landscape is of steep-sided hills up to 500 m high, with narrow ridges supported by harder more siliceous Archaean metamorphic rocks, and narrow valleys bearing colluvium. Soils are well-drained clay loams. Convex surfaces of upper slopes and ridges are more likely to dry out than concave valley slopes. Raw humus is patchy and thin on ridges, absent elsewhere. Canopy gaps along ridges are small, predominantly due to the mortality of individual standing emergents. Valley gaps, in contrast, are larger, more variable in area, and more sun-exposed, mostly caused by windthrows or occasional slips. Large gaps tend to retain more soil water toward their centers. The canopy is most diffuse on ridges and upper slopes, and more lateral light reaches the subcanopy. The topographic gradients in these Sri Lankan forests are seen throughout hills in the Far East, and the results are generally applicable there as well.

Four species have overlapping ranges from ridge (a heavy hardwood, *Shorea worthingtonii*) to valley (a light hardwood, *Shorea trapezifolia*). Observations included planted seedlings across gaps on ridge, on slope, and in valley, which have been monitored for more than a decade, and potted seedlings in shade houses whose light regimes simulated those of the three gaps.

The mean area of individual leaves of each species consistently declined from valley to ridge. Photosynthetic compensation points were at a PAR of 300 for *S. trapezifolia* and ~100 for *S. worthingtonii* (p. 000). Nevertheless, seedlings of the ridge species *S. worthingtonii* were still surviving in the understory 10 years after planting, while 99% of those of *S. trapezifolia*, a light-demander of large gaps, had died within two years. The topographic distributions were explained by differences in their combined responses to light and soil moisture.

Survivorship in Sinharaja was higher in the shade-tolerant ridge species *Shorea worthingtonii* than in *S. trapezifolia*. The heavy hardwood *S. worthingtonii* has less leaf plasticity, survives long periods in understory shade, attains low maximum growth rates in sun, and its leaves are photo-inhibited. It is best adapted to the periodic soil-water deficits on the ridges, while attaining maximum photosynthetic rates in the moderate light of the diffuse canopy there, with single tree gaps and exposure to lateral light. Research on photosynthesis in seedlings of other dipterocarps in Asia has also revealed physiological differences with some correlation to ecology.

Ashton and colleagues later studied seedling recruitment, growth, and mortality at Sinharaja for a decade. They studied thirty species in total, seventeen of which germinated during the study period.[5] Seven species of seedlings were differentiated between ridge, slope and valley, and seedling densities varied across topography and time, partly explained by species composition and recruitment periodicity. These results strongly indicate that traditional sylvicultural treatments of fixed selective cuttings of 40–60 years at set diameter limits (p. 366) will inadequately support regeneration because of variation in advance regeneration and species composition across the landscape.

Forest variation with nutrients. Ordination can establish spatial patterns and correlations between floristic and habitat patterns, and it can suggest explanatory hypotheses, but it doesn't reveal causes. Direct experimentation is needed. A 52 ha ForestGEO plot at Lambir National Park in Sarawak has the geological and soil diversity to make analysis possible. The plot includes, along its 1.04 km length, a ridge with the sandstone strata overlaying early Miocene shale. The shale is exposed on the ridge at its lower southern end (fig. 9.4). The two dominant *Dryobalanops* there are correlated with differences in both topography and soil. Ordination of 200 subplots produced two axes of floristic variation diagonal to the two primary axes of the ordination itself. To the left, plots containing species of clay loam soils clustered on upper slopes and ridges were aligned along a gradient of increasing elevation from slopes to ridges. Canopy gaps along ridges are small, predominantly due to individual standing emergent tree mortality (p. 130). Valley gaps, in contrast, are larger, more variable in area, and more sun-exposed, mostly caused by single or group wind-throws or occasional slips, as at Sinharaja. Plots scattered on yellow humult sandy soils sorted along a nutrient gradient. Plots with sandy humult soils were at one end of the combined gradients, and udult clay loam soils and species at the other. Remarkably, only 13% of the 764 more common species (out of 1,192) used in the Lambir ordinations proved to be generalists whose distributions were *not* associated with the soil categories. Such soil, topographic, and nutrient associations have been observed at other sites in Borneo.

Nutrients versus water stress. It is difficult from site comparisons alone to conclude whether nutrient or soil-water economy is the principal mediator of species distributions, although, in Brunei, dipterocarp species' distributions have varied with soil water and nutrients independently. Bunyavejchewin and colleagues, with a small-scale soils survey and detailed nutrient analysis along a granite catena in seasonal evergreen dipterocarp forest in Thailand were able to discriminate between influences of nutrients and topography on floristic variation.[6] Frequency of water stress and the topography across which it varies there, by influencing canopy height and structure, influences subcanopy light (which is also influenced by lateral exposure

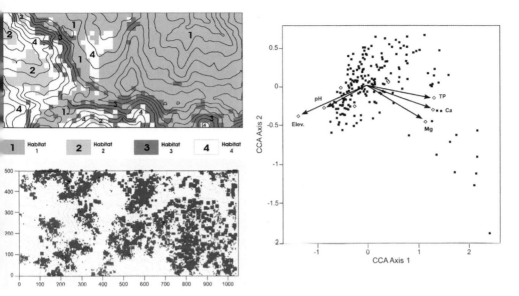

Fig. 9.4 Soil nutrients and species distributions at the Lambir ForestGEO plot. *Top left,* main habitat types: 1, humult sandy loam; 2, humult loam; 3, udult fine loam; 4, udult clay loam (S. R.; see chap. 9, n. 9). *Bottom left,* distributions of *Dryobalanops lanceolata* (blue) and *D. aromatica* (red) in the plot (H. L.; see H. S. Lee et al., *The 52-hectare Forest Research Plot at Lambir Hills, Sarawak, Malaysia: Tree Distribution Maps, Diameter tables and Species Documentation* [Kuching: Sarawak Forest Department, Arnold Arboretum of Harvard University & The Smithsonian Tropical Research Institute, 2002]). *Right,* canonical correspondence analysis (CCA) of 200 0.256 ha plots within the Lambir ForestGEO plot, based on 776 species with at least 50 individuals, trends in physical habitat variables superimposed, including soil nutrients (TP—total phosphorus, Ca, Mg), elevation, and soil Ph (S. D.; see Stuart Davies, S. I. Tan, J. V. La Frankie, and M. D. Potts, "Soil related floristic variation in a hyperdiverse dipterocarp forest," in *Pollination Ecology and the Rainforest,* ed. D. J. Roubik, S. Sakai, and A. A. Hamid [New York: Springer, 2005], 22–334).

along ridges), thereby influencing species composition, as at Sinharaja, where soil nutrient variation is minimal, and at Lambir on udult clay loams.

Canopy height is controlled by soil-water stress in windless climates (p. 116), and low-stature, even-canopied forests (forests with the structure of *kerangas,* p. 140) can occur on a variety of shallow soils. Nevertheless, the *floristic composition* of such short-statured MDFs on shallow, yellow-red soils correlates more with their soil nutrient status than with their structure. Where substrates and overlying soils change abruptly from yellow (with available phosphorus) to white mineral soils (bereft of nutrients), the floristic transition from *kerangas* to MDF is immediate, independent of canopy stature.

Floristic compositions of swamp and upland forest also vary. Periodically flooded alluvium at the edge of the Borneo swamps is generally rich in clay, and soils are only oxygenated near the surface. The tree flora differs little from that on freely draining upland clays except that it has a higher proportion of early successional species. This is commensurate with the forest's shallow rooting habit and consequently greater dynamism through windthrow (p. 140). Beyond the reach of floodwater, shallow peat forms with a permanently high-water table, and the species change completely.

Exceptions to the topography-mediated soil-water and nutrient correlations occur when ridges and slopes overlying different soils are compared. Sometimes strata with different mineralogy and soils that differ in clay/sand ratios may be interbedded along the same ridge.

At very low levels, in *kerangas* and peat swamps, nutrients minimally influence floristic variation, which is instead mediated by soil-water economy. As they increase in total availability, nutrients become correlated with increasing spatial diversity (that is, ß-diversity, p. 13) in tree species composition. Above a certain nutrient threshold, which may vary among regions, the correlation is once again lost. Nutrients no longer constrain competitive interactions between species above a threshold, while soil-water deficits may continue to do so. Above this threshold, which in Borneo appears to be at about 250 ppm total phosphorus and 800 ppm magnesium, floristic composition of a forest varies less, and mainly in relation to topography. Below it, composition varies more, and mainly in relation to soil nutrients.

The reduced correlation between species composition and soil nutrients on higher-nutrient and udult soils in part may explain why researchers in Peninsular Malaysia took so long to recognize that MDF does vary floristically, and that its individual species do have restricted edaphic ranges. Currently, conclusions can only be inferred there from local, plot-based research at one site, Pasoh Forest Reserve (West Malaysia). That forest lies on low rounded hills amid broad flat valleys with alluvium sediments at two levels: (1) an upper level, well-drained, rarely flooded, and usually sandy but with yellow soils, deposited before the later Holocene epoch and drier period; and (2) the current alluvium, generally flooded at least once a year, bearing clay and shallow peat in the lowest areas. In the ForestGEO 50 ha plot at Pasoh, the flora vary between hills and alluvium, and between the old alluvium, the young alluvium along streams, and the peat. There are sharp *overall* floristic changes at hill bases, where surface disturbance by rooting pigs and former elephant wallows is concentrated (fig. 9.5). In the ForestGEO

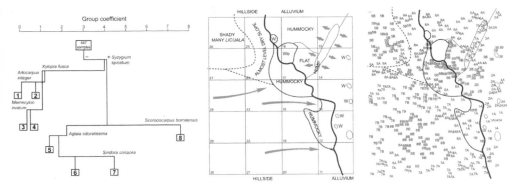

Fig. 9.5 Classification diagram of clusters of individual trees (447 over 9.7 DBH and their 35 nearest neighbors, in a 100 × 80 m plot at the Pasoh Forest Reserve, Peninsular Malaysia (P. A.; see Peter Ashton, "Mixed dipterocarp forest and its variation with habitat in the Malayan lowlands: A re-evaluation at Pasoh," *Malaysian Forester* 39 [1976]: 56–72). *Left*, eight species clusters, with prominent species listed; *center*, habitats mapped on the plot with locations of wallows of pigs (w) and pig or elephant (we), arrows indicate slopes; *right*, map of the plot with species clusters Numbers refer to the clusters in the classification diagram; the letters *A* and *B*, to the final segregates in each.

50 ha plot at Pasoh, 129 of the 617 species within the plot are confined to the alluvium. However, all but five of these occur at densities below 1/ ha.

Differences in nutrient response among species must be tested by experimental observations of performance. Attempts to relate performance of individual species to concentrations of exchangeable soil nutrients have inevitably been focused on seedlings in pot experiments. Seedlings of four red merantis—*Shorea leprosula* (sect. *Mutica*), *S. fallax* (as *S. oleosa*), *S. johorensis* (both sect. *Brachyptera*), and *Dryobalanops lanceolata* were raised in udult clay soils with light level at maximum photosynthetic capacity, at Danum (Sabah). All species responded to added nitrogen with increased growth and photosynthetic capacity, and this increase was greater than the variance in capacity between them. Only *D. lanceolata* responded under low nitrogen. Seedlings of *S. johorensis* increased most in light-saturated rates of photosynthesis and chlorophyll content in the enhanced light levels of logged forest. Thus, *S. johorensis* has the competitive advantage on higher-nutrient soils, and *D. lanceolata* has the advantage on lower-nutrient soils. Nitrogen availability may affect seedling competition more than inherited differences in light response. *Shorea johorensis* and *S. fallax* are most abundant in regional forests on udult clay soils of moderate fertility derived from shale, but *S. fallax* is a dark red meranti of relatively high specific gravity, while *S. johorensis* has low specific gravity. This implies different growth rates and different light responses in nature, at least beyond the seedling stage. *S. leprosula* has wide, edaphic distribution, occurring on relatively low udult sandy clay and leached illitic clay soils. It is the most widespread of the four species, being ubiquitous in the zonal MDF throughout Sundaland. *S. johorensis* is almost equally widespread but is local and confined to higher-nutrient soils. All four species occur on the highest-nutrient, basic volcanic soils, but in low densities. The exception is *Dryobalanops lanceolata,* which may dominate on those soils, with growth in height and diameter comparable to the fastest-growing canopy pioneers. Clearly, factors other than nutrients are involved in mediating competitive advantage, even in relation to such larger-scale soil diversity. Among these factors, the extraordinary plasticity of *Dryobalanops lanceolata* juveniles in relation to light may be crucial. The possibility, too, of changes in rank order of response to light during ontogeny cannot be ruled out. This can only be tested by long-term observation, as in the ForestGEO plots.

Three widespread floristic associations. We can recognize three of these associations in Sunda MDF (see fig. 9.1). (1) The first is an association formerly continuous over vast areas of the lowlands, rich in species but with low endemicity, on udult sandy clay and clay loams of intermediate fertility on a variety of substrates. This *Shorea parvifolia* association approximates to the red meranti-keruing forest type of Wyatt-Smith (p. 139). (2) The second is a variably rich association in which medium-to-heavy hardwood dipterocarps and the Riau Pocket flora (p. 312) are best represented and show high local to regional endemism. This is found on freely draining, leached, siliceous, variably humult sandy clay loams over sandstone or siliceous igneous and volcanic rocks on drier sites. Within Borneo, this is the *Anisoptera grossivenia* association; in Peninsular Malaysia, where it is represented by the *hill* and *coastal hill* dipterocarp forests of Symington, and in the Riau archipelago, it might appropriately be named after the sister species, *A. curtisii.* (3) The third association is dominated by light hardwood dipterocarps, relatively poor in species but rich in local to regional endemics, especially in Borneo. It is found on friable udult clay loam soils over argyllic and basic volcanic substrates. Within Borneo this is the *Dryobalanops lanceolata* association, but MDFs on base-rich soils in the south of Sumatra, and down its west coast, with *Shorea retinodes* and *S. javanica* locally dominant in the south, is analogous. Dipterocarp species endemism varies among Sunda land masses: 10% for Sumatra, 20% for Peninsular Malaysia, and 59% for Borneo.

Other MDF floristic associations. Analogous floristic patterns occur in Sri Lankan *Dun MDF* where, on the sandy humult soils that extend down slopes in Kanneliya forest, species confined to ridges in Sinharaja, described above, such as *S. worthingtonii* and *Axinandra zeylanica,* descend to the valleys. Seventy percent of the tree flora is endemic (p. 171).

In the Philippines, *lauan MDF,* on soils almost universally influenced by base-rich lava and volcanic ash, is analogous to Borneo's *D. lanceolata* association, but is distinguished by a gradient in rainfall seasonality to 5 dry months associated with declining species diversity, and another gradient in cyclone frequency, which increases from south to north and is also associated with declining species diversity. The tree flora is less rich than on Sundaland, with 50 dipterocarp species, but with high endemism, 46% among dipterocarps.

East of Wallace's Line (p. 289), little is known about patterns of floristic variation within the major lowland forest formations. Species richness declines, except in New Guinea, with richness particularly reduced in tree families with poor seed dispersal: 31 species of Dipterocarpaceae, all but two of which are endemic, and 17 Fagaceae, all but 3 endemic.

Other forests of everwet climates. A floristically distinct *kerangas* occupies the same upland habitat as MDF, but where soil-nutrient levels are inadequate to support MDF (p. 140; fig. 2.7). The oligotrophic *kerangas* ecosystem may also be limited by water stress. *Kerangas* is most widespread on the sandstone ridges and plateaus of inland Borneo, on raised beaches up the east coast of Peninsular Malaysia and Southwest Cambodia, and possibly along the foot of the central range of Irian Jaya.

The change in flora between MDF and *kerangas* is striking, enhanced by the usual change in forest structure. The *kerangas* tree flora is poorer than that of MDF. The same families are included, but some (notably Dipterocarpaceae, Meliaceae, and Sapindaceae) are represented by few species, while others (notably Myrtaceae, Clusiaceae, and Rubiaceae) are relatively species-rich. Conifers (*Agathis, Dacrydium, Podocarpus*) and other Australian elements—including *Gymnostoma nobile* (Casuarinaceae), several Myrtaceae, and *Styphelia* (Epacridaceae)—are well represented.

Several forest types in the Asian tropics are defined by periodic inundation, independent of the regime of steady precipitation (fig. 2.7). *Riparian forests* establish on the margins of fast-moving streams and rivers and are influenced by brief flooding from heavy rainfall. Many species, both early successional and mature phase, are specialized for survival in these demanding environments. Shrubs and juveniles of trees in the flora below the flood line have evolved traits, such as pliable twigs, long narrow leaves, and precociously germinating seeds, that are typical of these "rheophytes" throughout the tropics.

Floodplain forests. These hardly survive in the seasonal tropics, having long ago been converted to rice cultivation. In Sundaland, particularly northwest Borneo, aseasonal flooding prevents irrigation, while excessive rain leads to leaching and peat accumulation. In such areas, floodplain forests still occupy narrow belts between river banks and the vast peat swamps, becoming wider only where the rivers issue from the hills and flood most extensively. The great rivers to the east and south of Borneo, and elsewhere around the land masses of Malesia, flood seasonally as well as intermittently, and may have broader floodplains continuously enriched by fine sediments. Shallow rooting on these floodplains is associated with high frequencies of tip-ups and an irregular canopy with many large gaps. Tall pioneers and large strangling figs are locally abundant, interspersed with scattered individuals of late successional species of upland clay loams. In more seasonal regions of India, the elegant, layered crowns of *Terminalia arjuna* grace such sites.

Peats only form in the everwet Asian tropics, reaching their continental limits in southernmost peninsular Thailand. Peat swamp forests cover ~225,000 km^2 overall, down the coast of East Sumatra, on the southern, southeastern and northeastern coasts of Borneo, and on more limited areas of both coasts of Peninsular Malaysia. They are also found on Sulawesi, Ceram, and on the southern coast of Papua.

Most Far Eastern peat swamps were formed over sediments along the coasts of epicontinental seas, following slight declines in sea levels (p. 000). Peats accumulated beginning ~9 Ka and resumed ~6.5 Ka, as sea levels declined. In Central Borneo, peat swamps, or *kerapa,* have also formed on gentle uplands above maximum Pleistocene sea levels. These are more ancient, here and in Java going back to the beginning of the everwet climate there, at the start of the Holocene epoch, 11 Ka (p. 64).

Peat establishes domes, and these spread into massive bog systems up to 60 km in diameter, with peat to 20m in depth, and the mineral sediment base is often below the current sea level. The tops of the domes have the most variable water table; here, *kerangas* elements are most evident (fig. 2.7). Fewer trees grow in these forests compared to the MDF: some 250 species in northern Borneo, of which one-fifth are confined to peat swamps and one-third are endemic to that region. These late Holocene communities may be the least diverse in Sundaland.

Another forest type, occurring at the mouths of rivers and nearby coasts and influenced by the inflow of tidal salt water, is *mangrove*. As much marine as terrestrial, we exclude its discussion in this book, and refer you to the excellent book by Barry Tomlinson.[7] *Littoral forests* establish along beaches, less directly influenced by salt. In coastal regions subject to cyclonic storms, *typhoon forests* with canopies shaped by winds occur, as in the Philippines and Taiwan (fig. 2.7).

Floristic composition and soil in the seasonal tropics. The edaphic influence of soils on tree floras in seasonal evergreen and semi-evergreen forest echoes that in MDF and still mainly derives from casual field observation. Differentiation between forests on sandy and clay soils is clear, even though raw humus and peat fail to accumulate in climates with more than two dry months.

In general, reduced species diversity and specially adapted tree species correlate with increasing dry season length and the associated increased fire frequency and intensity. The distribution of forest types and species is often the result of fire history—and human disturbance—creating mosaics of forest types (and species) within small areas (p. 149).

The core seasonal evergreen dipterocarp forest of Indo-Burma, on clay loam soils, has emergent dipterocarps, including *Parashorea stellata, Dipterocarpus hasseltii*, and *Hopea helferi*, whereas more sandy soils, especially on coastal hills, bear dominant stands of *Dipterocarpus costatus*, which otherwise grows on high ridges, and *Shorea hypochra*. Based on floristics, two forest types, *northern* and *southern* have been identified. In Xishuangbanna, such remnant forests feature magnificent stands of *Parashorea chinensis* (fig. 2.3).

Semi-evergreen forest overlying sandy soils, as at Huai Kha Khaeng, is dominated by *Dipterocarpus alatus*, whereas the predominant clay loams of Indo-Burmese semi-evergreen forest support stands of *Dipterocarpus turbinatus* (fig. 2.4), *Anisoptera costata*, also deciduous *Xylia xylocarpa*, and teak, *Tectona grandis*.

Semi-evergreen forests also occur in south Asia, particularly in the coastal lowlands of western Peninsular India. In them, tree species diversity declines from south to north. Toward the north, there is an association of *Machilus-Holigarna-Diospyros*. North of Goa, it is replaced, as deciduousness increases, with an association of *Tectona-Lagerstromia-Terminalia*.

Northern semi-evergreen forests. A distinct semi-evergreen forest ranges from southern Yunnan to Central Nepal, easily recognized by the prevalence of *Terminalia myriocarpa* with its brilliant red inflorescences and copper-hued senescing leaves (fig. 2.3). That community forms in drier conditions, but in the wettest localities has replaced the dipterocarp-dominated seasonal evergreen forest following shifting agriculture. There it is associated with late successional tropical evergreen lowland species, including *Pometia pinnata* and *Duabanga grandiflora*.

Deciduous forests. The floristic transition from evergreen to deciduous forests (by way of semi-evergreen forests) shifts family dominance from Dipterocarpaceae to Lythraceae, Combretaceae, Verbenaceae, and Fabaceae. Tall deciduous forests retain evergreen trees and shrubs in the understory. Changes to short forests include the replacement of the conspicuous (though scattered) pale-barked canopy tree of tall deciduous forest, *Anogeissus latifolia*, by the often-dominant *A. pendula*; of *Lagerstroemia lanceolata* by *L. parviflora*; and of *Dalbergia latifolia* by *D. lanceolaria*.

Throughout Indo-Burma, freely draining soils along ridges or sandy soils support a suite of five more or less tardily deciduous, light-demanding dipterocarp species, which are fire resistant and require bare soil or a ground fire prior to establishment. Occurring separately or in mixture, each occupies a distinct edaphic range (fig. 9.6). Uniquely, this deciduous dipterocarp forest transcends climate-induced drought, occurring from the climatically dry, short deciduous region of the Irrawaddy valley to the edaphically dry karst of Langkawi Island, which has less than 2 dry months. Their associated flora changes with adjacent forest formations, and from partially evergreen to entirely deciduous (p. 149); only if fire is excluded will this forest return to an adjacent formation on similar soil (p. 149; fig. 8.4).

Teak and sal. The well-documented Indian deciduous forests are clearly divided into those dominated by sal, *Shorea robusta*, on sandy soils, and teak, *Tectona grandis*, on clays derived from metamorphic rocks and basalt (fig. 2.1). The two species rarely co-occur, contrasting with most other species. Both are light-demanders, and they are the two most economically important tree species in the Indian subcontinent.

Teak (*Tectonia grandis*, Verbenaceae) is a deciduous timber tree known for its medium density, durable, and attractive timber (fig. 2.5). It extends westward into Gujarat, with 8 dry months and 850 mm annual rainfall, and eastward into Indo-China in climates with up to 1,800 mm annual rainfall and 5–6 dry months (fig. 4.1). Teak fruit is a 15 mm diameter globe with a single dormant seed. It germinates when the pericarp rots off or is abraded, and fire increases the germination rate. Teak forests are associated with tall bamboos. Teak, and the mixed forests with which it is associated, follows

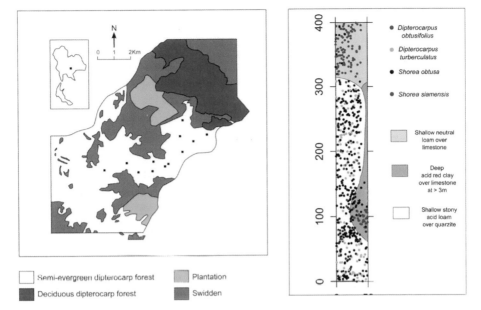

Fig. 9.6 Floristic variation in deciduous dipterocarp forests in Thailand (I. B., after S. Bunyavejchewin). *Left,* A patch of forest embedded within a landscape on sandy soils, Sakhaerat, Korat Plateau. *Right,* distribution of dipterocarp individuals in a forest at Mae Ping National Park, scale in meters, with substrates indicated (S. Bunyavejchewin, "Ecological studies of tropical semi-evergreen rainforest at Sakhaerat, Nakhorn Ratchasima, northeast Thailand. I. Vegetation patterns").

the basaltic trap clay and calcium-rich sedimentary rocks, on red-brown fertile clay loams (p. 80).

Sal (*Shorea robusta*, Dipterocarpaceae) is a tardily deciduous timber tree, known not only for its dense, durable construction timber, but also for its aromatic resin and seed oil. It grows to 35 m throughout the north of India (fig. 2.5). Sal forests are a floristically impoverished deciduous dipterocarp forest. They establish on sandy, freely draining, low-nutrient soils on siliceous granite, metamorphic schists, and sandstones, particularly in the north and east (p. 87). The distributions of sal and teak overlap in northeastern Andhra Pradesh (fig. 2.1). Nowadays, wherever teak or sal occur in nature, each dominates overwhelmingly. This is a consequence of silvicultural manipulation, even though both species perform better and are less attacked by pests when mixed with other species. Both occur over a wide range of rainfall seasonality regimes. Sal fruit are winged, the pericarp is lignified and resinous, the single seed lacks dormancy, and it is killed by fire, as are the young seedlings. Establishment is impeded by litter, especially when waterlogged; ground fire early in the monsoon and prior to seed fall promotes establishment. Sal forests are often found with short bamboo species, grasses, and cycads.

With sufficient moisture, tall forests establish within the distribution of sal and teak: *tall deciduous sal forest* and *tall deciduous teak forest*. Further east, into Burma and Thailand, sal forest is replaced by more species-rich deciduous dipterocarp forest, and a different teak-bearing forest establishes: *the Indo-Burmese tall deciduous teak forest*. It is richer in species within *Lagerstroemia, Dalbergia,* and *Terminalia* than its Indian equivalent, and includes endemic sister species of *Xylia* and *Pterocarpus*. In southern India, but beyond the distributions of teak and sal, *Indian tall deciduous forests* occur, well stocked with *Lagerstroemia lanceolata*. Throughout Indo-Burma, there is a similar forest, with *L. calyculata* or *L. tomentosa*. Scattered individuals of sal and teak in semi-evergreen forests betray former establishment in a firebreak, subsequently reinvaded by evergreen canopy species in the absence of fire.

Short (dry) deciduous forests occur in dry seasons of 6–9 months, are strongly influenced by crown fires, and produce canopies less than 20 m high. Where teak is distributed, the forests are *short deciduous teak forest,* and for sal, *short deciduous sal forest* (fig. 2.5). In regions south of the sal and teak distributions in India, there are areas of *southern short deciduous forest,* with *Cochlospermum religiosum, Pterocarparpus santalinus, Acacia amara,* and *A. planifrons*.

Javanese tall deciduous teak forests may be the result of the introduction of teak by forest farmers well over a millennium ago. They are normally highly degraded from overgrazing, although small trees may establish when protected by spiny shrubs.

In the diversely forested landscapes of seasonal tropical Asia, human history, soil, and topography influence species diversity and distributions. In Thailand's Khorat Plateau, deciduous dipterocarp forest dominated by *Dipterocarpus obtusifolius* can be embedded within a landscape of semi-evergreen dipterocarp forest on sandy soils, dominated by *Shorea henryana* and *Hopea ferrea* (fig. 9.6). Historically this region was semi-evergreen dipterocarp forest.

Thorn woodland is most widespread in South Asia, occurring in climates with 8–11 dry months (fig. 2.6). Spiny trees include *Acacia, Zyziphus,* and cactus-like *Euphorbia*. These are the *acacia thorn woodlands*. In similar dry climates in Burma, *Burmese deciduous thorn woodlands* are present, with *Acacia catechu, A. leucophloea, Dalbergia lanceolaria,* and *Bauhinia vahlii.*

The stability of floristic associations. The continuous leaching of clay and nutrients in stable soils within hot, everwet climates removes nutrients over the span of centuries (p. 87). Low-nutrient soils, protected from physical erosion by their root-matted surface raw humus, will be particularly prone, and may only be rejuvenated over prolonged periods by the periodic collapse and retreat of scarps and the banks of streams issuing from the hills. Leaching may often be actively promoted by the tree flora itself; phenolic compounds in solution can slow litter decomposition. Species with leaves rich in phenols are more abundant in low-nutrient, humult soils—including coastal dipterocarp forest soils—where leaf nutrients, notably nitrogen, are low. Thanks to landslides and surface deposition, the mosaic of soils where clay prevails is locally dynamic and may change during the life of a climax tree. In contrast, change on freely draining sandy soils takes place over millennia, and at larger spatial scales correlated with the substrate. On such soils, rejuvenation may be mediated as much by orogeny and by climate change as by erosion.[8]

Light, competition, and survival. Competition influences species composition within habitats, through differential survival. In addition to physical growth and survival in natural forest, floristic composition within specific habitats is clearly differentiated by competition. At Gunung Palung National Park (West Kalimantan), in 17 of 49 species, adult trees were associated more closely with habitats than were seedlings (5 of 22), implying that selective mortality was operating prior to achieving reproductive maturity. The five species of Sterculiaceae co-occurring in the ForestGEO plot at Lambir (Sarawak) have different soil preferences and heights at maturity, and soil conditions become more restrictive as the trees increase in size.

The diversity of understory light regimes across topography correlated with floristic differences among congeneric canopy species at Sinharaja, seen in the light-responses of their seedlings (p. 179). The same influence may also operate within the dynamic mosaic of a single forest community, permitting the coexistence of several species with similar nutrient responses and explaining the frequently wide overlaps in their ecological distributions. But competitive success is determined as much by survivorship as by performance. Either may be determined by other factors besides nutrients, especially light, soils, and herbivory.

Trade-offs between rapid growth and higher survivorship may permit the coexistence of related species within MDF communities. Differences in competitiveness can be mediated by differences in the light and soil-water regimes of adjacent habitats. The *Shorea* species at Sinharaja forest live on soils with moderate nutrient levels and with little change in topography, other than soil depth. Here, species performance was influenced by variation in the understory light regime. Ridges and spurs had much lateral light exposure and a diffuse canopy, compared with the contrasting deep shade and bright light of large gaps in the lower slopes and valleys resulting from poor drainage and windthrows or landslides (p. 171).

In the ForestGEO plot at Lambir, Sabrina Russo studied growth and mortality among trees, comparing two censuses five years apart in a search for correlations of growth and mortality with soil-nutrient levels (fig. 9.7). The soils (fig. 9.4) varied in nutrients (3x for P, 5x for Mg, and 2.5x for Ca) and in moisture-storage capacity. Growth rates of pioneers were lowest on the least fertile and most drought-prone soils, but growth rates of climax species did not differ significantly. Mortality rates correlated with fertility among both pioneers and climax species, grouped according to soil preferences. The species restricted to the least fertile soils grew less and died less frequently—that is, among the specialists to a soil type. Mortality differed as tree diameter increased. This led to more frequent loss of generalists than specialists on all soil types, and consequently produced the floristic patterns observed in the landscape. Mortality was the main mediator of eventual species composition, whereas on more fertile soils, differential growth rates were also important. This trade-off was steepest toward the lower end of the nutrient gradient. Coexistence and differentiation in competitiveness along the habitat gradient must be governed by multiple inherited adaptations. Russo and collaborators concluded that adaptation for fast growth may entail a greater mortality risk among inherently fast-growing species when soil resources are scarce. Slow-growing species benefit from lower mortality when soil resources are limited but are outcompeted when they are less so.[9] However, such trade-offs may not occur in more disturbed forests. Plant life-history strategies can influence species distributions *across* resource gradients. The diversity of life histories can also therefore potentially allow species to coexist in floristic associations within which several resource gradients may be varying independently. Performance of closely related species may not differ sufficiently across any edaphic or topographic gradient to support competition gradients among them *within* a uniform geological substrate or topography, although Ash-

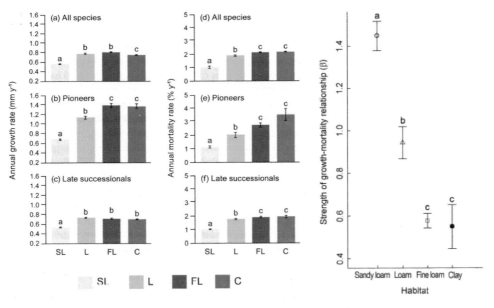

Fig. 9.7 Comparison of growth and mortality of tree species, grouped by their soil preferences, across four soil types within the ForestGEO plot at Lambir, Sarawak (S. R.; see chap. 9, n. 9). Soil types (in order by increase in nutrients and moisture capacity): SL = humult sandy loam, L = thinly humult loam, FL = fine udult loam, and C = udult clay loam. *Left,* mean annual growth for all species, pioneers, and late successional plants (shared letters indicate no significant difference); *right,* a statistical analysis of relationships between growth and mortality on all soil types.

ton's monitoring of seedling performance, and especially mortality, of four climax *Shorea* species at Sinharaja is consistent with Russo's Lambir results (p. 171).

Forests on unusual geological substrates. In everwet forests on soils dramatically affected by peat formation or flooding, or by geological substrates, distinct floras have evolved, with different regional patterns in diversity, often on local scales (fig. 9.1). Such diversity is also seen in more seasonal climates.

Forests on limestone. Limestone in the wet tropics of Asia is generally ~35 Ma or older (p. 29). The rock is hard, creating the dramatic karst topography seen in classical Chinese paintings. Limestone karst, that is freestanding rock pillars, is absent from South Asia, but occurs as archipelagos in several regions of Indo-Burma and Malesia to New Guinea. In everwet northern Borneo, there is no mineral soil overlying the karst. In the seasonal dry tropics and at the tropical margins of northeastern Indo-Burma, udult red loam soils may form from limestone impurities, and the topography beneath can become more rounded.

Karst limestone hills include four distinct topographic elements, each with a flora linked to communities in other habitats. Around the base, a calcium-rich alluvium supports tall forest typical of MDF and floodplains. The karst towers are surrounded by prominent talus boulder-slopes, often steeper than 30°. These generate little surface mineral soil, but the tree flora is nevertheless very similar to that on the friable udult red-brown loams of basalt hills, with few dipterocarps. The rocks and humus-filled fissures support a rich, often endemic herbaceous flora.

Cliffs rise vertically above the talus, sometimes exceeding 500 m. Largely bare, their fissures carry a rich, herbaceous, full-sun-adapted flora, including many local endemics, with a few trees and shrubs of the summit flora. On summit plateaus, the hard limestone is dissolved by rainwater into a honeycomb of sharp-edged, vertical surfaces, sometimes as angular pinnacles up to 100 m in height, as on Gunung Api, Mulu (Sarawak, fig. 2.8). The dry limestone summit plateaus allow, against all odds, the establishment of a few tree saplings. In regions of high humidity, especially on higher mountains, acid raw humus accumulates between and even over these rocks to depths of up to 50 cm. This supports a *kerangas* flora (p. 176). Dipterocarps are represented as *kerangas* elements on peat-bearing summits, including *Cotylelobium lanceolatum* and *Shorea multiflora*. These drought-prone karst tops are especially prone to lightning-induced fires (Api means 'fire'), which destroy the raw humus, leaving bare rock and even more drought-prone conditions. Human-induced fire has resulted in prevalence of such conditions elsewhere in Sundaland.

Forests on ophiolite. Although the influence of such high cation concentrations on tropical, "ultrabasic" soils has been little investigated, species adapted to ultramafic soils synthesize chelating compounds, which inactivate unusual concentrations of heavy metal ions and accumulate them in leaf tissue.

Some ultramafic soils are reddish clay loams, bearing tall MDF floristically like those on basalt-derived soils (p. 85). Such soils and forests are at the base of Kinabalu (fig. 9.1). More often, the soils are coarsely sandy and freely draining; blackwater streams, rich in humic acids, emanate from them. Under these conditions, the vegetation resembles *kerangas* in structure and physiognomy, while the ultramafic forest is often even shorter than most *kerangas,* the short trunks branching low (fig. 2.8). In Borneo, it includes many *kerangas* species and many endemic species with the leaf and branch physiognomy of *kerangas.*

A Naturalist's Eye

The exploring naturalist discovers not only that forest structure constantly varies with habitat within a landscape, but that this variation is repeated in similar habitats, on similar landforms, and in similar climates regionally. This predictability of structure is explained by the constraints that habitats variously impose—through soil-nutrient limitations and proneness to water deficits—on plant function, on species' life cycles, and on the dynamics of the forest. The species vary with those habitats.

Notes

1. Beccari, O., *Wanderings in the Great Forests of Borneo* (London: Constable, 1904).

2. Richards, Paul W., *The Tropical Rainforest: An Ecological Study,* 2nd ed. (Cambridge: Cambridge University Press, 1996).

3. Wyatt-Smith, J., *Ecological Studies on Malayan Forests* (Kuala Lumpur: Malayan Forest Department Research Pamphlet 52, 1966).

4. Symington, C. F. (revised by P. S. Ashton & S. Appanah), *Forester's Manual of Dipterocarps,* Malayan Forest Records 16 (Kepong, Malaysia: Forest Research Institute of Malaysia & Malayan Nature Society, 1943 & 2004).

5. Ashton, Mark S., Elain R. Hooper, Balangoda Singhakumara, and Sisira Ediriweera, "Regeneration recruitment and survival in an Asian tropical rain forest: Implications for sustainable management," *Ecosphere* 9 (2018): e02098.

6. Bunyavejchewin, S., "Ecological studies of tropical semi-evergreen rainforest at Sakhaerat, Nakhorn Ratchasima, northeast Thailand. I. Vegetation patterns," *Natural Historical Bulletin, Siam Society* 34 (1986): 35–57; Bunyavejchewin, S., A. Sinbumroong, B. L. Turner, and S. J. Davies, "Natural disturbance and soils drive diversity and dynamics of seasonal dipterocarp forest in Southern Thailand," *Journal of Tropical Ecology* 35 (2019): 95–107; Sukri, R. S., R. A. Wahab, K. A. Salim, and David Burslem, "Habitat association and community structure of dipterocarps in response to environment and soil conditions in Brunei Darassalam, Northwest Borneo," *Biotropica* 44 (2012): 595–605.

7. Tomlinson, P. Barry, *The Botany of Mangroves,* 2nd ed. (Cambridge: Cambridge University Press, 2016).

8. Jucker, Tommaso, et al., "Topography shapes the structure, composition and function of tropical forest landscapes," *Ecology Letters* 21 (2018): 989–1000; Vincent, John B., Benjamin L. Turner, Clant Alok, Vojtech Novotny, George D. Weiblen, and Timothy J. S.

Whitfeld, "Tropical forest dynamics in unstable terrain: A case study from New Guinea," *Journal of Tropical Ecology* 34 (2018): 157–75.

9. Russo, Sabrina E., P. Brown, S. Tan, and S. J. Davies, "Interspecific demographic trade-offs and soil-related habitat associations of tree species along resource gradients," *Journal of Ecology* 96 (2008): 192–203; Russo, Sabrina E., et al., "The interspecific growth–mortality trade-off is not a general framework for tropical forest community structure," *Nature Ecology and Evolution* (November 2020): DOI: 10.1038/s41559-020 -01340-9.

On my first Brunei forest sojourn, I explored Gunung Pagon Periok: Saucepan Mountain. Though it was but fifty miles from the capital as the crow flies, we spent two weeks reaching it, and were away for two months. We approached from the south, up the Limbang River. We first scaled a steep spur onto a long undulating ridge. Gradually ascending, we reached Ubin Hill, a knoll at 750 m where the forest became shorter, its canopy more even, and unfamiliar dipterocarps indicated we had reached upper dipterocarp forest. We had escaped the hot muggy lowlands for a paradise of cool mountain breezes, the sound of distant cataracts far below, and the occasional hysterical cackle of a helmeted hornbill. The ground then rose sharply, up 500 m through a shorter forest on peaty soil, mixed with oaks and chestnuts, Syzygia, and the first mountain conifers, all new to me: a lower montane kerangas. *Scaling a sharp crest, we entered a thicket so dense and embedded in mossiness that we could hardly tell whether we were near the ground or in the canopy. And the fog: dense, so quiet that sound was muffled but for the constant drip off everything. We had reached the "cloud," or upper montane forest. As evening approached, the fog lifted, revealing before us the base of Pagon Periok's precipitous sandstone slab. For nearly a month, we explored the slopes and the shrubby, knife-edge defile of the summit. Our long descent was perilous, following the source of the Temburong River, and another story (p. 70).* — P. A.

10

ABODE OF THE CLOUDS

We have long been fascinated by mountain forests in the tropics, a reassuring mix of the temperate climes of most scientists, and the otherworldly vegetation of a mossy forest. Despite long fascination, we have much to learn about them. Are these changes with altitude in forest structure and composition gradual or continuous? What physical characteristics influence these changes? What are the roles of competition, physiological constraints, or catastrophe in limiting the altitudinal distributions of tree species? Since the mean annual temperature declines with altitude at a similar rate throughout the wet equatorial tropics, at a rate of about 0.6° C per 100 m (p. 60), do forest ecotones always occur at the same altitudes?

Our approach to answering these questions is primarily biogeographical and comparative.[1] Others have named and described forest types along altitudinal gradients in Asia and elsewhere in the tropics, but these categories have not been generalizable to all the tropics. The physical and structural characteristics of these forests should correspond reasonably well to floristic categories.

Four forest zones. Relying mainly on field observations and the altitudinal ranges of observed species, four forest zones have been recognized on the higher tropical mountains up to the tree line: *lowland, lower montane, upper montane* and *subalpine* (table 10.1).[2] This zonation scheme has been recognized globally in equatorial and oceanic climates, though such forests differ in species composition. During the ascent of Kinabalu (fig. 10.1), the most studied mountain in tropical Asia, all forest zones are encountered,

Table 10.1 Characteristics that define the widely accepted zonation of tropical montane forests found in Asia

Formation	Lowland forest	Lower montane forest	Upper montane forest	Subalpine thicket
Maximum canopy height	25–70 m	15–35 m	1.5–20 m	1.5–5 m
Emergent trees	Present to 70 m, or absent	Mostly absent, <50 m	Rare, <20 m	Rare, <5 m
Pinnately compound leaves	Frequent	Less frequent, a few species common	Uncommon	Absent
Dominant woody plant leaf size	Mesophyll or notophyll	Notophyll (microphyll)	Microphyll	Microphyll and smaller, cupped, shiny
Buttresses	Usually frequent, large	Uncommon, small	Absent	Absent
Cauliflory	Frequent	Uncommon	Absent	Absent
Big, woody vines	Uncommon to abundant	Rare to common	Absent	Vines absent
Bole climbers	Locally common	Locally common	Rare	Absent
Vascular epiphytes	Usually uncommon	Abundant, species-rich	Abundant, less species-rich	Few
Nonvascular epiphytes	Local, species-poor	Frequent, increasing with altitude, species-rich	Abundant in equatorial regions and coastal, less so in seasonal forests, less species-rich, especially where seasonal	Sparse

Source: Tim C. Whitmore, *Tropical Rainforests of the Far East*, 2nd ed. (Oxford: Clarendon Press, 1984).

ending in a glacially polished summit plateau. In Bhutan, there are only two tropical zones, lowland and lower montane forest (fig. 10.2).

Floristic peaks and ecotones. Within Kinabalu National Park, the number of species initially declines slightly from the lowland estimate, then increases slightly at 1,000–1,399 m. In Bhutan, species number increases to a maximum at 1,000–1,299 m, where the greatest number of high and low range limits—or the highest rate of species turnover—is found (fig. 10.3). This pattern reflects the relatively impoverished flora in Bhutan's lowland deciduous forest. On Kinabalu, the *upper* limits of most species ranges cluster around 1,300–1,600 m, while fewer species' *lower* range limits are encountered in ascending. From 1,600–2,500 m, species present and new "arrivals" and "departures" drop greatly. In Bhutan, as elsewhere in the Eastern

Fig. 10.1 Views of Mount Kinabalu. *Top left,* mountain view in 1974 (D. L.); *top center,* lower montane forest at 1,500 m dominated by oaks and laurels, with crown of *Agathis borneensis* on right, park headquarters (D. L.); *top right,* upper montane mossy forest, at 2,100 m (D. L.); *bottom left,* looking down from 3,700 m in subalpine thicket (D. L); *bottom center,* at 3,800 m, looking up past subalpine woody vegetation toward summit (P. A.); *bottom right,* temperate sedges and grasses, plus shrub (*Rhododendron ericoides*), and glacial polish of quartzite intrusion, both on summit plateau at 4,000 m (D. L.).

Himalayas, there is a further peak in "arrivals" and "departures" (though not in total numbers present) at 1,600–2,500 m. There, this elevational band marks the transition from mostly frost-free to entirely frost-prone sites, from a tropical to a rich, warm temperate Himalayan flora. There are no temperate tree species on Kinabalu, though there are some frost-hardy herbs of widespread temperate genera, with occasional frost above ~3,100 m.

The Lowland to Montane Forest Transition

Symington was the first to distinguish altitudinal forest zones in Sundaland,[3] based primarily on dipterocarp distributions in Peninsular Malaysia, with lowland dipterocarp forest terminating at ~300 m. Above, he described a separate "hill dipterocarp forest" (p. 171). Many of the lowland dipterocarps are represented in the hill forests, although they become scarce toward the upper limits, but many species appear in the hill forests that only exceptionally occur in lowland forests, and then near the coast. This small

Fig. 10.2 View of montane forest in Bhutan, from Dotula Pass, at 2,800 m, inner valleys and frontier peaks; cool temperate forest is on ridges, warm temperate broadleaf forest in foreground, and tropical lower montane forest beneath that. Distant clouds above 3,000 m peaks and ridges (H. H.).

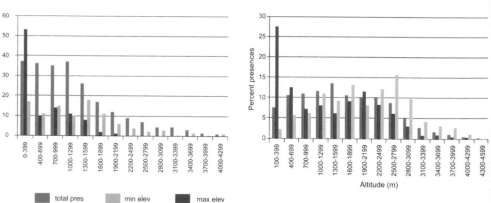

Fig. 10.3 Species zonation (percentages) on two tropical Asian mountains; *left* is Mount Kinabalu, and *right* is the Bhutan and Sikkim Himalaya (I. B., after P. A.; see chap. 10, n. 9).

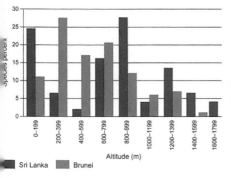

Fig. 10.4 Dipterocarps at higher elevation. *Left*, altitudinal range maxima of dipterocarps in Brunei and Sri Lanka (I. B., from P. A.); *right*, *Shorea gardneri* in lower montane forest, Peak Sanctuary, 1,600 m, Sri Lanka (H. H.).

elevation increase would not affect temperature, so other factors must account for such altitudinal zonation. In Borneo, this Peninsular Malaysian hill flora is most widespread in the lowlands, in MDF on humult yellow sandy soils.

The stepped elevational ranges of Brunei's dipterocarps (fig. 10.4) occur throughout everwet Sundaland. The altitudinal ranges of dipterocarps and other trees in everwet Sri Lanka are similarly influenced by their edaphic ranges, although dipterocarps are better represented there at altitude for historical reasons (chapter 13). Thus, the floristic differences between "lowland" and "coastal hill" and "hill" dipterocarp forests are mediated by soil, not by temperature, and the ecotone at 200–300 m is attributable to a widespread change from undulating to steep topography, and sometimes to changes in geology.

Lower Montane Forests

Lower montane forests were, until recently, the least studied of tropical forests, and there is still much that we do not understand about them (fig. 10.5).

Structure. Like lowland forests, lower montane forests vary in structure with altitude, topography, and soils. The mature phase has two strata, a canopy lacking emergents and a subcanopy, although scattered emergent species do occur at its lower altitudes, especially in sheltered valleys. The main canopy is dense; the predominantly notophyll leaves are densely clustered. The canopy varies in height from 20-40 m among different habitats, but

Fig. 10.5 Examples of lower montane forest. *Top left*, forest profile, Kinabalu (I. B., after Abu Salim); *top right*, upper limits of lower montane oak-laurel forest, Zhemgang, Bhutan, 1,900 m (H. H.); *bottom left*, *Castanopsis* crowns on ridge, lower boundary in oak-laurel forest, ~800 m, Mengla, Xishuangbanna, Yunnan (Z. H.); *bottom center*, oak-laurel forest, Pinosok Plateau, 1,500 m, Kinabalu (D. L.); *bottom right*, looking up into microphyll-notophyll canopy of oak-laurel forest, Doi Inthanon, 1,700 m (H. H.).

it is relatively even in height within any one habitat, increasingly so with altitude. Large-leaved pioneers are rare except on the steep earthquake- and landslide-prone New Guinea mountains.

Lacking buttresses, lower montane oak-laurel forest trees develop deep, coarse roots, with a diffuse horizon of fine roots within 15 cm of the soil surface. Fine roots are generally sparse beneath the litter, often from a dense clay surface horizon eroded and trampled by boar and deer.

Vines decrease in size and abundance with altitude, peaking in richness at the lowland-lower montane forest ecotone in seasonal tropical Asia. A few species extend to the upper limits of forest. Vines may be abundant in disturbed or secondary forests at lower altitudes.

Tree ferns (*Cyathea* and other, less species-rich genera) are rare in the Asian lowland tropics but common in lower montane forest (fig. 10.6). They even reach the upper montane secondary grassland in New Guinea

(3,800 m). Ground herbs and epiphytes, including ferns, urticaceous herbs, and orchids (fig. 10.6) increase in abundance and diversity with altitude.

Flora. The floristic transition between lowland and lower montane forests is substantial. On both Kinabalu and in the eastern Himalayas, we find many species (and a few genera) whose lower limit is 1,200 m. A few start at 700 m (fig. 10.3). On Kinabalu, 499 tree species are present at 1,000–1,900 m. These include 35 of the 61 Fagaceae and 56 of the 106 Lauraceae known from the massif, whose abundance in both canopy and subcanopy at these altitudes makes them *lower montane oak-laurel forest.* Fagaceae, absent in peninsular South Asia, are replaced in everwet SW Sri Lanka by Dipterocarpaceae (p. 172).[4]

Of the more than 100 lowland dipterocarps known from Kinabalu and adjacent hills, only 19 reach 1,300 m, and 1 reaches 1,700 m. Of the 1,442 tree species in all families known on Kinabalu, 979 are confined below

Fig. 10.6 Plants characteristic of lower montane forest. *Left*, tree fern, *Cyathea obscura*, Genting Ridge, West Malaysia, 1,600 m (D. L); *upper center*, *Strobilanthes capitata*, Gelephu, Bhutan, 865 m (H. H.), *upper right*, *Elatostema macintyrei*, Doi Inthanon, 1,800 m (H. H.); *lower right*, cemara (*Casuarina junghuhniana*) near Gunung Batur, Bali ca. 1,400 m (D. L.).

1,000 m. The decline of the MDF flora accelerates beyond 700 m, and characteristic lower montane forest hardly appears below 750 m.

At the northern margin of Asia's tropics, in the Bhutan and Sikkim Himalayas, of the 503 known tree *species,* only 110 are confined to the lowlands below 1,300 m, reflecting both the species-poor deciduous forest in the lowlands and the rich, tropical lower montane and warm temperate, mostly evergreen, tree floras above. Lowland species drop out with altitude. There is a zone of floristic impoverishment in the everwet tropics at ~400–850 m (fig. 10.3). In both cases, lowland mixed forest species fail to extend beyond the lower montane forest; some *kerangas* species (p. 176) are exceptions.

Lower montane forests are the centers of diversity of epiphytes throughout the humid tropics. In tropical Asia, vascular cryptogams and orchids are dominant; Melastomaceae and Ericaceae are less abundant, although characteristic. Among other plants important and distinctive in these forests, the spectacular shrub *Strobilanthes* (Acanthaceae) is found particularly in the seasonal tropics (fig. 10.6), especially at upper elevations. With more than 200 species, most are monocarpic (like bamboos), coming into gregarious flower at intervals of 5–15 years, then dying. Pollinated by honeybees and bumblebees, their copious production of small dry seeds attracts birds and small mammals. Seedlings of shade-tolerant trees survive beneath *Strobilanthes,* to emerge after their next flowering.

Physical correlates. Remarkably, the transition from lowland to lower montane forest, both structurally and floristically, generally occurs over a gradual ecotone at 750–1,300 m whether on the equator or at the margin of the tropics, whether on Mount Kinabalu, beyond the geographical tropics in Bhutan at 27° N, or on the slopes of the vast massifs of New Guinea.

The lowland/lower montane forest ecotone is correlated with the elevation of the diurnal cloud base, including seasonal tropical Asia during the wet summer monsoon. This ecotone is influenced by the climatic convergence zone (p. 56), whose seasonal oscillation is responsible for most of the rainfall throughout the region. Much speculation has arisen about possible physical correlates of these forest boundaries, arguments bedeviled by poor data, particularly for climate. There is no long-term monitoring of climate at any stations along an altitudinal transect in Asia. Forest zonation in tropical Asia could correlate with Kira's index of warmth,[5] which is the sum of mean monthly temperatures above 5° C. This index declines more rapidly at higher than at lower latitudes in humid tropical climates. Within the tropics, summer temperatures do not decline with latitude, but winter night temperatures do. Nevertheless, the ecotone from lowland to lower montane forest remains remarkably constant, and temperature cannot be its major determinant.

Fog does not regularly penetrate the canopy of lower montane forest during the heat of the day. It is frequent within it at higher altitudes, and correlates there with increasing mossiness on trunks and branches. Cloud as fog is generally excluded from the forest interior by the diurnal upwelling of hot humid air following the canopy surface until rain commences. Massing around higher slopes and peaks, clouds shade forests beneath, lowering daily temperatures. The ecotone from lowland into lower montane forest, being universally at the same elevation, is thus caused by an environmental change mediated by the one climatic factor common to all latitudes, and from coast inland: equatorial mean weather conditions and those of the wet summer monsoon.

Soils. Soils provide the most direct correlation with change in forest species at the lowland/lower montane forest ecotone. In the understory, temperatures decline continuously with higher altitude, as do the rates of chemical processes, including litter decomposition. However, litterfall does not change. At the ecotone, rates of litterfall come into balance with rates of litter decomposition, and udult soils then come to resemble their temperate counterparts, with humus discoloring the mineral soil deep within the profile. There, earthworms replace termites. Unlike termites, earthworms actively draw decomposing litter down into the soil (p. 77). Along road cuts, fine tree roots follow the humus down through the mineral profile in the mull soils of lower montane oak-laurel forest, just as they do in temperate forests. This pattern contrasts with the concentration of organic matter and finer roots near the surface in lowland tropical yellow/red soils, and in humult leached ultisols and podzols at all altitudes. The change in forest root systems with altitude is poorly understood.

The soils of lower montane forests have been favored by swiddening communities (p. 350), and lately for growing commercial tea, whose traditional cultivation coincides with the distribution of the *warm temperate* oak-laurel forests to the north of the lower montane forests, extending to northeast Honshu (fig. 10.7).There are two ecotypes of domestic tea (*Camellia sinensis*): (1) the Chinese var. *sinenesis,* a short canopy tree (to 6 m) of warm temperate forests, with microphyll leaves, upward-tilted and shiny; and (2) the Assam-Upper Burma form, var. *assamica,* which is a slender subcanopy tree (to ~15 m) of tropical lower montane forest, in which

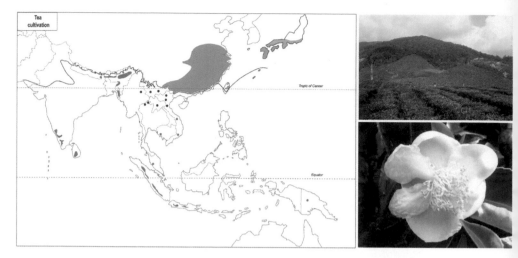

Fig. 10.7 Tea cultivation. *Left*, the distribution of indigenous and cultivated tea in Asia: black dots indicate the natural distribution of the Assam variety, in tropical lower montane forest; the yellow areas indicate the Chinese variety, both cultivated and wild, within the natural range of warm temperate forest (I. B.); *top right*, Boh Tea Estate, Cameron Highlands, West Malaysia (D. L.); *bottom right* hybrid tea flower, Munnar, India (L. E.).

the notophyll leaves are larger, hardly tilted, spreading, and often dull in appearance.

Kerangas. Ascending Pagon Periok (p. 192), one would note that the lower montane forest was very different from these descriptions, as would also be observed in the ascent of Gunung Mulu (2,376 m), a sandstone mountain in northeast Sarawak. At 1,200 m on the first true ridge, the forest changes to resemble upper montane forest in its short stature and smaller leaf sizes. However, this forest lacks the characteristic species of upper montane forest, with 15–20 m, dense, mostly straight stems; it resembles *kerangas: lower montane kerangas.* The mineral soil is shallow, a leached, greyish sandy loam, and bears a horizon of raw humus, dense with fine roots. Mosses and liverworts cover the trunks patchily, but there are no tussocks until the high ridges are reached (above 1,900 m). Similar forests occur elsewhere, as on Kinabalu and in the mountains of southern Cambodia and southeastern Thailand.

Still other forests. Symington regarded upper dipterocarp forest in Peninsular Malaysia as a category of lower montane forest, forming a distinct zone within its lower ecotone below the oak-laurel forest throughout everwet Sundaland. It has the structure and dipterocarp dominance of short-stature

MDF, in which emergent dipterocarp species hardly exceed the main canopy, but it includes some species common in lower montane forest, such as *Schima wallichii*. Some ridges at ~1,000 m in Indo-Burma bear local stands of massive, gregarious *Dipterocarpus costatus*, a species occurring in upper dipterocarp forest in Peninsular Malaysia. Similar are *Hopea forbesii* in New Guinea and *Shorea gardneri* in Sri Lanka (fig. 10.4).

On the everwet outer slopes of the New Guinea mountains, lower montane forest is initially dominated by *Lithocarpus* spp. and *Castanopsis acuminatissima* (Fagaceae), which also occur on Kinabalu. *Araucaria cunninghamii* (var. *papuana*) forms gregarious, emergent stands on ridges and rocky defiles, while *A. hunsteinii* may dominate on former landslides down into the lowlands. The Australasian *Nothofagus* may dominate on limestone. Rainfall-seasonal inner valleys and basins support mixed lower montane forests that lack dominant species; they are rich in Lauraceae, Cunoniaceae, Elaeocarpaceae, and conifers, but include few Fagaceae. In these lower montane forests, there is a high proportion of species with bird-dispersed seeds.

On the wet west-facing slopes of the Western Ghats, with a 3–4-month dry season, *mixed montane sholas* establish; these lack emergents and Fagaceae, and are rich in Lauraceae and *Magnolia nilagirica*. The lower ecotone is dominated by *Palaquium ellipticum* and *Cullenia exarillata* or *Mesua ferrea*. On the drier eastern slopes, leeward of the southwest monsoon, lowland forests are deciduous below ~1,200 m.

Upper Montane Forests

With a narrow lower ecotone, the transition from lower montane to upper montane forest is generally distinct on the ground, particularly in rugged topography along ridge crests. It also varies greatly in altitude, both on different mountains and on different sites on the same mountain (fig. 10.8).

On Kinabalu, upper montane forest, whether structurally, physiognomically, or floristically defined, is ubiquitous by 2,200 m (fig. 10.1). However, the interdigitation of lower and upper montane forests, in valleys and on ridges respectively, blurs species ranges in identifying ecotones on Kinabalu and other mountains.

Structure and physiognomy. Upper montane forests are often popularly called "elfin woodland" (fig. 10.9). They are single-canopied and low in stature, due to the decline and eventual disappearance of the woody subcanopy of trees or subordinate shrubs. Leaves are microphyll and smaller sizes

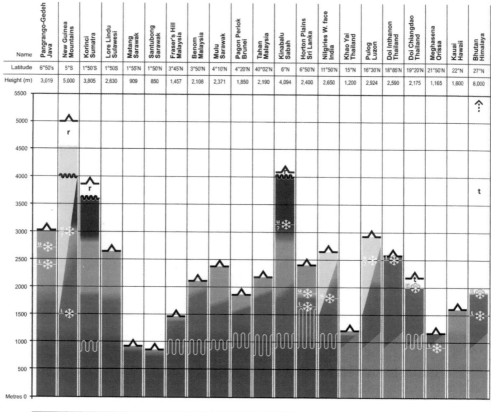

Name	Pangrango-Gedeh Java	New Guinea Mountains	Korinci Sumatra	Lore Lindu Sulawesi	Matang Sarawak	Santubong Sarawak	Fraser's Hill Malaysia	Benom Malaysia	Mulu Sarawak	Pagon Perlok Brunei	Tahan Malaysia	Kinabalu Sabah	Horton Plains Sri Lanka	Nilgiries W. face India	Khao Yai Thailand	Pulog Luzon	Doi Inthanon Thailand	Doi Chiangdao Thailand	Meghasena Orissa	Kauai Hawaii	Bhutan Himalaya
Latitude	6°50's	5°S	1°50'S	1°50S	1°55'N	1°50'N	3°45'N	3°50'N	4°10'N	4°20'N	40°02'N	6°N	6°50'N	11°50'N	15°N	16°30'N	18°85'N	19°20'N	21°50'N	22°N	27°N
Height (m)	3,019	5,000	3,805	2,630	909	850	1,457	2,108	2,371	1,850	2,190	4,094	2,400	2,650	1,200	2,924	2,590	2,175	1,165	1,600	8,000

Fig. 10.8 Forest zonation on the major mountains of the wet Asian tropics. Note: oblique ecotones indicate coexistence of two forest zones, only indicated on mountains observed by P. A. Horton Plains includes, for lower altitudes, the surviving sequence on the adjacent Eastern Peak Sanctuary (I. B., from P. A.).

(table 6.1), often fibrous, thick, and without drip tips—perhaps adaptations to peaty soils. Brilliant red, young leaves among canopy species may protect against high morning solar irradiance. Boles are generally unbuttressed and contorted from the base, and twigs with short internodes and dense, twiggy branching crowns bear a compact layer of leaves.

Upper montane forests vary in stature as much as other tropical forests.

Fig. 10.9 Upper montane forest. *Top left*, canopy diagram from Kinabalu (I. B., after Abu Salim); *top center*, Horton Plains, Sri Lanka, 2,100 m, *Eleocarpus glandulifer* with flat crowns (H. H.); *top right*, tall forest, Mount Albert Edward, New Guinea, 2,500 m (P. A.); *bottom left*, Gunung Mulu, Sarawak, ~2,200 m (H. H.); *bottom center*, emergent *Calophyllum walkeri*, Horton Plains, 2,100 m (H. H.), *bottom right*, two living fossil forns, *Dipteris conjugata* (wide lobes) and *Matonia pectinata* (narrow lobes) characteristic of openings in forest, 2,100 m, Ulu Kali, West Malaysia (D. L.).

At lower altitudes on Kinabalu and other large massifs (and occasionally to 3,000 m), forests up to 18 m tall and without dense epiphytic cryptogams—but with more crooked boles and a mixture of lower and upper montane genera and species—occur on shallow acid organic soils over bedrock. With their frequently tall stature there, upper montane forests are often regarded as starting at higher altitudes in New Guinea than elsewhere (see cloud truncation, p. 200). Southern beech (*Nothofagus*) species are dominant, particularly above 2,600 and as high as 2,800–3,000 m with canopies to 30 m. At the higher altitudes of upper montane forest on Kinabalu and other Borneo mountains, forest in moist swales reaches 15 m in height, whereas on nearby ridges it hardly exceeds 5 m and may often be reduced to a 2 m, dense scrub. The canopy of upper montane forest is diffuse, bearing mostly microphyll and even smaller leaves (table 6.1), inclined upward or sometimes hanging. Catastrophic canopy openings, such as those caused by landslides, are infrequent, mitigated by densely intertwined roots, and single-tree mortality prevails. In both South and East Asia, the generally smooth upper montane and subalpine forest canopy is broken by one or more taller species with mature crowns. These include *Agathis*, Podocarpaceae (*Dacrydium pectinatum*), *Leptospermum recurvum*, and *L. javanicum* on Kinabalu, and *Calophyllum walkeri* in Sri Lanka.

Upper montane forests lack woody climbers but carry an abundance of vascular epiphytes of few species. Epiphytes are more abundant in upper than in lower montane forests, but species diversity of both epiphytic orchids and pteridophytes reach their peak at ~1,600 m in the lower to middle zones of lower montane forest on Kinabalu.

Mossiness. These forests also support a diverse cryptogamic flora, particularly mosses, filmy ferns, and liverworts, which swathe the trunks and major branches as sleeves. Upper montane forest is often named *mossy* forest (fig. 10.1), or "cloud forest," especially in the Neotropics. Pervasive mossiness is associated with the upper montane tree flora of the equatorial tropics, along with abundant vascular epiphytes. On the mountains of Borneo, extreme mossiness is found on sheltered slopes and lesser ridges, while the short scrub at the summits is less mossy. In such continuously foggy sites, mosses form characteristic mounds around the bases of trees and extend along the main branches as sleeves resembling pipe insulation. In extreme cases, these associations (rich in liverworts and filmy ferns) may extend into the branches so densely that the visitor cannot easily determine whether he/she is standing on the ground or in the canopy (p. 193)!

Dense tussock moss occurs at 2,200–2,700 m on exposed ridges of the high equatorial mountains of New Guinea and Kinabalu (fig. 10.1), and typically at 1,400–2,300 m on the lower summits in Borneo and Peninsular Malaysia. Thick blankets and tussocks of moss can also be found in narrow misty lowland valleys emerging from higher mountains, as in New Guinea and Borneo.

The flora. The vascular flora of upper montane forest is relatively impoverished, even on the great mountains. On Kinabalu, there are only 232 woody dicotyledonous species above the limit of lower montane forest. This number declines with increasing altitude and forest stature, as does the number of new species (fig. 10.3). At 3,100 m, there are only 73 tree species, including seven Fagaceae and nine Lauraceae. The same families prevail in New Guinean upper montane forests, but no Fagaceae and poor representation of Melastomaceae.

Physical correlates. Climate. The upper montane forest climate is one of dense fog during the heat of the day, when transpiration and therefore photosynthesis are curtailed, alternating with spells of direct sunlight and high surface temperatures. Diurnal fog is not merely associated with upper montane forest, but it is part of the mechanism that distributes moisture to the soil surface and thus contributes to the conditions that produce the highly leached humult podzols that, in part, determine the structure and species compositions of these forests. Extensive moss and cryptogam cover coincides with consistent penetration of fog within the canopy during the heat of the day, increasing canopy area relative to its volume and thereby increasing capacity to trap fog vapor as drip. Fog penetration of the forest canopy coincides with many upper montane forest tree species and their associated canopy structure and leaf physiognomy, but not consistently with marked reduction of stature.

Fogginess. Fog (and possibly wind) likely influence both leaf size and soil in upper montane forests, but not necessarily floristics *per se.* The presence of much of the upper montane forest tree flora, including tree stature, appears instead to be correlated with accumulation of raw humus. Climate (fogginess) and raw humus accumulation co-vary. But there is no exact correlation between the floristic and the structural-physiognomic transitions from lower to upper montane flora.

Soils. The upper montane woody taxa correlate with the first appearance of raw humus on the soil surface on Mount Kinabalu and high Bornean sandstone ridges. Soil surface horizons of mossy upper montane forests in everwet climates are generally acid, with deep raw humus, though also high total nitrogen and extractable phosphorus. Upper montane forest soils are frequently waterlogged, but short stature upper montane ridge forests suffer occasional drought as well. As in *kerangas*—but in contrast with other lower montane forests—rooting in upper montane elfin woodland is mostly shallow and confined to organic horizons. The presence of upper montane woody species within the upper reaches of lower montane forest on Kinabalu is partially explained by soil. The mixing of lower and upper montane forest floras between ~1,400–2,200 m within individual stands is particularly evident on the ultramafic substrate on Kinabalu, but it is also widespread on siliceous substrates in the mountains of Borneo. On Gunung Jerai (1,200 m), in coastal Kedah, there is a similar paucity of lower montane forest flora, but an abundance of heath and upper montane forest families on the upper slopes.

Upper montane forest and kerangas. Lowland *kerangas* (p. 140) shares several attributes with upper montane forest, including smooth canopy, sclerophylly, small leaf size, and lack of drip tips. However, trees in both low-land and lower montane *kerangas* are straight-boled, internodes are longer, and mosses, though locally abundant in "premontane" habitats, do not form tall hummocks or sleeves along branches. Lower montane *kerangas* merges with upper montane forest in a gradual ecotone. As in *kerangas,* the leaves of upper montane tree species are rich in tannins that inhibit bacterial decomposition. These depress rates of litter humification and increase the accumulation of raw humus.

Tropical Subalpine Thickets

Subalpine thickets are confined to those equatorial mountains high enough to support them (fig. 10.8). In New Guinea, subalpine thickets appear at 3,650 m (fig. 10.10); they are above 2,900 m on Kinabalu in Sabah (fig. 10.1), and they are on the tallest volcanoes of West Java and Sumatra at 2,200–2,700 m. Leaf sizes reduce to nanophyll (table 6.1). Many species have cupped leaves with the upper surface convex and shiny, and very short internodes. On exposed narrow ridges below those altitudes, upper montane forest is short, its canopy exceptionally dense and manifestly wind-pruned; but floras remain upper montane. Cryptogamic epiphytes, particu-

Fig. 10.10 Tropical subalpine thicket. *Top left*, canopy diagram from Mount Kinabalu (I. B., after Abu Salim); *right*, Mount Wilhelm (Enduwa Kombuglu), Papua New Guinea, lower lake and grassland below tree line and peak (4,509 m; A. S.). *Bottom*, two species endemic to Kinabalu: *left and center left*, *Leptospermum recurvum*, tomentose and recurved leaves, and flower detail (D. L.); *center right*, *Rhododendron ericoides* (D. L.).

larly mosses and liverworts, are sparser at higher altitudes. Mossiness is less dense on both ground and branches, although scattered clumps mixed with lichens continue to the highest altitudes.

Flora. Subalpine thickets are distinguished on Kinabalu and Kerinci (Sumatra, 3,805 m), by a distinct shrub flora, including specialist rhododendrons and other Ericaceae, as well as on New Guinea, where southern elements are also present: individuals of Australasian taxa become frequent in upper montane forest, but diversity is greatest in subalpine thickets from New Guinea to Kinabalu.[6] Between 2,700–3,200 m, new tree taxa, including sister species or subspecies of taxa at lower altitudes, appear. Such pairs are found within large genera, including *Syzygium* (Myrtaceae), *Memecylon* (Melastomaceae), *Rhododendron* and *Vaccinium*, *Elaeocarpus* (Elaeocarpaceae), and *Ilex* (Aquifoliaceae)—also *Leptospermum javanicum* with its subalpine and ultramafic sister, *L. recurvum*, both abundant on Kinabalu (fig. 10.10).

On Kinabalu, mossiness declines gradually above ~2,900 m (fig. 10.1). Above 3,000 m, trailing strands of the drought-tolerant lichen *Usnea* are common. It is also abundant on the drier Javanese summits. It festoons the upper montane canopy on the Horton Plains (Sri Lanka), the summits of the Western Ghats, and lower montane forest at the summit of Doi Inthanon, Thailand.

Temperate herbaceous genera appear, including violets, buttercups, cinquefoils, primulas, and grasses. These often first appear in open sites in the upper reaches of upper montane forest, but especially in valleys receiving cold nocturnal downdrafts, as in New Guinea and the grassy caldera bottoms of Javanese volcanoes.

Physical environment. The transition from upper montane to subalpine thicket is gradual, correlated with the cloud cover. Ultraviolet radiation is intense during the frequent sunny intervals at highest elevations. The small leaves, highly inclined against incident radiation and with shiny upper surfaces, reduce penetration of ultraviolet light. Many upper montane and subalpine species bear vivid anthocyanin-crimson photo-protective flushes.

Upper elevations are prone both to squalls and high winds, occasionally associated with extreme droughts in ENSO years. Soils on Kinabalu are increasingly thin above ~2,700 m, though deep humic muck soils occur in swales up to 3,100 m. Earthworms there are abundant almost to the summit.

The altitudinal limits of forest in equatorial Asia. The transition to subalpine thicket appears to correlate with the onset of occasional winter frost (although we lack good meteorological data). Frost may occur when skies remain clear for prolonged periods in ENSO years. Valley frost descends in valley grasslands as low as 1,500 m, even in New Guinea. However, closed forests extend on slopes well above the lower limits of occasional killing frost in valley flats, and they may exclude frost from within their canopy. Buds in the canopy remain dormant during the cold, dry season in mildly seasonal climates, and species often flush and flower after the frosts as the wet monsoon approaches. Trees may thus escape nocturnal freezing and delay diurnal warming, reducing destructive impacts of drought and frost.

The tree line appears uncorrelated, then, with frost incidence—which must be widespread over the canopy at these high altitudes—and frost does not therefore represent a potential altitudinal limit of forest on equatorial mountains. Only on the rocky, high mountains of New Guinea, and probably on Kinabalu, does forest reach its altitudinal limit for equatorial mountains. The physiological and environmental determinants of the altitudinal limits of subalpine trees on tropical mountains need study.

The tree line on tropical mountains is often sharp but varies with elevation and topography. Kinabalu's summit (4,094 m) is exceptional in its lack of soil except in seepages and fissures, and in the absence of human impact. The tree limit there is unclear; a few shrubby trees extend almost to the summit along soil-filled fissures. On summits with soil cover, forest is replaced by alpine dwarf shrublands or, more frequently, by grasslands. Grasslands attract ungulates and their predators, and people have brought cattle for centuries and have lowered the altitude of the tree line by fire. In continental Asia, this ecotone is fringed by frost-hardy dwarf bamboos and woody pioneer shrubs and treelets.

Canopy death near the equatorial tree line. Trees currently suffer mortality near the upper limits of forest in many parts of the tropics, probably due to frost and/or drought. On Kinabalu, larger individuals of subalpine *Leptospermum* died after an intense drought in 1995–1996, and others earlier, but pubescence under partial genetic control decreased damage (fig. 10.10).[7] This mortality could be caused by the drought, which promotes freezing that accompanies a periodic decline in atmospheric humidity (which also might be caused by lowland deforestation). We need long-term observation of winter temperatures, leaf flush phenology, and the causes of bud mortality.

Alpine grasslands. Mountain grasslands, dominated by narrow-leaved, temperate grasses and herbs of temperate genera and families, are widespread on tropical Asian mountains. The lowest natural alpine grassland

communities are confined to valley bottoms; their margins are sharply defined, and the adjacent woody pioneer vegetation is periodically scorched
by frost. Flats with temperate, narrow-leaved grassland genera occur as low
as 900 m in Javanese calderas at 7° S latitude, at 2,000 m in valley bottoms
on Sri Lanka's Horton Plains (8° N), and lower in the Palni and Nilgiri
mountains of the Western Ghats (to 12° N).[8]

Physical determinants of the tree line. The natural tree line marks the altitude at which the production gains from photosynthesis are matched by
the maintenance costs of respiration. The growing-season temperature at
which this occurs has been estimated, worldwide, at 5° C.[9] As root/shoot
ratios of trees increase with altitude, respiration increases. Persistent diurnal
fog may reduce rates of photosynthesis. However, fog is intermittent near
the tree line, and diurnal surface temperatures are frequently high. Although
gnarled, mossy, upper montane forest is associated with continuous diurnal
fog, which may affect growth, the decline in stature and increase in gnarled
structure with altitude may partly be due to the frosting of buds, but it will
also be partly due to increase in respiratory costs relative to photosynthetic
gains. Diurnal temperatures decrease with elevation, thereby explaining the
upper montane structure in the absence of dense mossiness, which implies
infrequent fogginess.

Climatic Factors That Narrow Zonation

Frost and drought on equatorial mountains. The natural distribution of montane grasslands correlates with frost-ponding in climates with a marked dry
season, but it has been greatly extended by human-induced fire for pasture,
as on eastern slopes in the Western Ghats. This extension continues down
to the upper limits of lowland deciduous forest on the dry Peninsular Indian
mountains.

Frost occurs in the subalpine zone of New Guinea above 3,600 m, and is
an occasional hazard to agriculture in valleys there down to 1,500 m. Early
morning hoarfrosts occur on the grasslands of the East Javan Diëng Plateau
(~2,000 m), Java, and the once-forested grasslands on South Asian mountains; yet the frost-sensitive trees in adjacent forests appear unaffected.[10] In
South Asia, tree rhododendrons *R. nilagiricum* and *R. zeylanicum* resist both
fire and frost, and extend into the grassland as scattered individuals.

Toward the latitudinal tropical margin, winter frost increasingly reduces
the upper elevations of inland tropical forest zonation, initially by eliminat-

ing the subalpine and tropical upper montane forest habitats. Seasonal variation in diurnal temperature and fog reduces mossiness. Nevertheless, elements of the equatorial upper montane tree *generic* flora remain, even above the tropical limits on the eastern Himalayas.

Transition to temperate evergreen montane forest at higher latitudes. Annual ground frost and occasional air frost within the forest canopy mark the climatic margin of the tropics. This margin forms an ecotone between largely evergreen, notophyll, tropical lower montane forest and warm temperate (subtropical), moist evergreen forest, and transition from tropical to temperate canopy Fagaceae, also predominantly notophyll and microphyll, but frost-hardy and partly winter-deciduous. This forest is not upper montane, because it gives way at higher altitude to temperate conifer, then cool temperate deciduous forest at the tree line.

Transition of tropical to temperate evergreen forests in East Asia. Tropical forests extend further north than the geographic margin of the tropics (23.26° N) in the eastern Himalayas and northern Burma (p. 60), but their limits coincide with the Tropic of Cancer where they descend to the lowlands of South China. Some tropical tree species grow on continental islands south of Japan to 30° N. The southerly ocean current warms the coast as far north as Sendai (northeast Honshu), where northern mountains shield from the winter northeasterlies and mark the limit of temperate evergreen forests. The forests of the northern tropical frontier remain mostly intact in the Bhutan Himalayas at 27° N (fig. 10.2). In these mountains, winter hoarfrost descends in valley stubble down to ~1,500 m, the same altitude as in equatorial New Guinea.

The altitude of greatest tree species richness in Bhutan (1,900–2,200 m) coincides with the upper limits of frost-free habitat, and in South China coincides with the northern frost limit down to the lowlands. The transition (ecotone) to warm temperate evergreen forest is marked by a changeover in canopy species, notably *Castanopsis* (Fagaceae), but the tropical woody subcanopy flora is only gradually replaced by warm temperate as frost incidence and intensity increase. This is associated in East Asia with the penetration of the summer monsoon north to the central Chinese mountains and the Japanese alps, and is unique to Asia. The ecotone between 1,800–2,200 m in Bhutan is also the altitude where diurnal fog penetrates during the wet summer monsoon, and thin but conspicuous mossiness of limbs prevails. The ecotone witnesses the upper limit of tree families and genera confined to tropical climates, a major turnover of *species*, especially in those major canopy genera well-represented across the ecotone, and is accompanied by an increase of deciduous species belonging to temperate genera, with toothed leaf margins. Unlike the classic, tropical upper montane forests, these forests are not significantly shorter than the tropical lower montane forest below 1,800 m. Both floristically and structurally, this is an ecotone to warm temperate forest, still predominantly evergreen.[11]

In the lowlands, the ecotone changes floristically from northern seasonal evergreen dipterocarp forest to tropical lower montane forest. It is distinct in tree genera and families, lacks the emergent canopy stratum and prominent bole buttresses, has a canopy dominated by notophylls instead of mesophylls, and grows on temperate soils. Many subcanopy tropical elements persist, gradually replaced by warm temperate species as the winter climate intensifies with altitude or latitude. This transition occurs over ~500 m elevation in the Himalayas, and gradually between ~10° to ~23°–30° N, for more than 1,000 km longitude in lowland China. There, the Qinling Shan mountains halt the East Asian tropical summer monsoon northward, while reducing the impact of the cold, winter northeast monsoon from the north. In China, tropical elements mostly halt at the Qinling; a few species gradually disappear north of these mountains and toward the drier west. They continue northward on the coastal Ryukyu Archipelago to halt at the Tokara Gap, a deep oceanic trough that marks a biogeographic barrier between those low islands and the temperate continental uplands of Yakushima Island and Kyushu. The lowland evergreen forests on the Ryukus have a partially tropical understory and are designated "subtropical" by Chinese and Japanese ecologists, and "warm-temperate" elsewhere.[12]

Some warm temperate evergreen species at all altitudes, growing beneath a deciduous canopy and often related to lower montane congeners, produce leaves with shiny upper surfaces and cupped blades, resembling species of equatorial subalpine thickets (p. 208). Some subtropical forest taxa are related to those in Far Eastern tropical lower montane forests, even in the lowlands. Some are frost-tolerant (to −5° to 10° C). Their shiny leaf surfaces reflect early morning winter sun, perhaps slowing warming and reducing embolism in the frozen xylem in leaves. This may promote species persistence in killing frosts at higher altitudes and longitudes—a possibility worthy of physiological ecological research.

Fog and the mountain mass effect. Condensation—cloud formation—occurs as temperature declines with increasing altitude. In the everwet equatorial convergence zone, diurnal humidity saturation is reached at al-

Fig. 10.11 The mountain mass effect. Pulau Tioman is a small, high island off the east coast of West Malaysia. *Left*, view of mountains from the south end; pale crowns are *Shorea curtisii*; *right*, the summit of Gunung Kajang, 1,038 m, upper montane forest with *Leptospermum flavescens* about 3 m high, E. Soepadmo providing scale, May 1974 (D. L).

titudes as low as ~800 m from humid air over the sea and along coasts. It rises to ~1,000–1,200 m on low inland mountains, but higher on longer slopes where the heated morning canopy warms a greater mass of humid air. Mossy upper montane forest appears where the canopy cools and diurnal fog penetrates. If it were due to temperature differences alone, this variation in ecotone elevation would require a doubling of the lapse rate in some cases, which is highly improbable. Thus, variation must be explained by other factors.

The ecotones between lower and upper montane forest are higher on larger and more continental mountains (fig. 10.8). Moss and upper montane forest cover the summit of Gunung Kajang, at 1,000 m (fig. 10.11). On the other hand, on the large equatorial massifs of Kinabalu and New Guinea, mossy upper montane forest starts at 1,700 m on steep spurs and ridges, becoming prevalent by 2,200 m, again coinciding with fog penetration of the canopy. Dense tussock moss is restricted to mountains that are bathed in diurnal fog in all seasons; it is thus confined in the tropics to oceanic climates, everwet at least in the upper montane zone.

Upper montane forests disappear on inland, tropical Asian mountains at higher latitudes where diurnal penetrating fog is interrupted by the dry season. There, lower montane forest continues to altitudes prone to frost and also within rain-shadow valleys and basins of the great East-West equatorial massif of Papua New Guinea.[13] On the Western Ghats, upper montane forests clothe western slopes and summit ridges exposed to the winds of the southwest monsoon, receiving orographic clouds during the dry season. The eastern slopes, with an intermittently foggy dry season, support a short but straight-boled woodland of lower montane physiognomy and flora, though impoverished in epiphytes, including cryptogams.

Tropical Gymnosperms

Gymnosperms, including the cone-bearing conifers, are an important component of tropical montane forests, particularly in Asia (fig. 10.12). Geographically, conifers are viewed from their origins, either from the south of Pangaea (p. 34), around 250 Ma, which later split as Gondwana, or from the north, later becoming Laurasia. The northern families are the Pinaceae, with pines and firs, and some members of the Cupressaceae (*Thuja, Juniperus*) and Taxaceae (*Taxus* and *Torreya*). The latter two families also include southern genera. The southern families are the Podocarpaceae (originating ~200 Ma), with *Podocarpus* (originating ~63 Ma) and many other genera,[14] and the ancient Araucariaceae, with *Araucaria* and *Agathis*.

Among the northern conifers, besides the pines, only the drupe-bearing *Taxus wallichiana* occurs in lower montane forest of the Far Eastern tropics.

North: Pines. Dominating much of the savanna woodlands in which they occur, pines play a special role in the drier forests of seasonal tropical East Asia. All species require fire for their establishment (fig. 10.12).

Pinus merkusii occurs widely in lowland Indo-Burma, also along the western slopes of Sumatra's Barisan Range and the western lowlands of Mindoro (Philippines). It reaches 1,200 m in the hills of northern Indo-Burma.

Pinus kesiya savanna occurs on sunny south-facing and rain shadow slopes throughout Indo-Burma eastwards to the central massif of Luzon, within the lower montane forest zone. Frequent fire helps establish *P. kesiya* savannas. It descends to 1,000 m in northern Indo-Burma, joining lowland deciduous forests and often mixing with *P. merkusii* in short deciduous dipterocarp forest at 850–1,200 m. It occurs at 1,400–2,200 m in the central mountains of Luzon. In China, *P. kesiya* is confined to the extreme southwest, in the tropical lower montane evergreen forest zone. The related *P. yunnanensis* is a warm temperate species that replaces *P. kesiya* to its northeast.

Pinus roxburghii (*chir*) occupies the habitat of *P. kesiya* from northeast

Fig. 10.12 Conifers. *Top*, North: pines; *left*, distribution map in tropical Asia (blue = *P. roxburghii*, green = *P. kesiya*, purple = *P. yunnanensis*, red = *P. merkusii*); *center, P. kesiya* at 1,200 m, Mountain Province, Luzon, Philippines (P. A.); *right, P. roxburghii* at 600 m along ridge, Sankhosh Valley, Bhutan (H. H.). *Bottom*, South: podocarps and more, all on Kinabalu; *left, Agathis lenticula* (A. F.); *middle left, Phyllocladus hypophyllus*, 3,350 m (D. L.); *middle right, Dacrycarpus imbricatus*,1,500 m (D. L.); and *right, Dacrydium beccarii*, 3,350 m (D. L.).

India to the western end of the Himalayas in the tropical lower montane forest zone. Being mildly frost-hardy, it extends into temperate winter-dry forests west to Afghanistan. Toward the front ranges of the eastern Himalaya, *chir* savanna may be sandwiched between lowland semi-evergreen forest and lower montane oak-laurel forests. *Chir* is replaced there by *P. wallichiana* at warm temperate altitudes.

South: Podocarps and more. The southern conifers prevail in lower montane forest (fig. 10.12), where *Agathis* and *Araucaria* are most important. Both have dry light seeds They mostly occur on acid humult soils towards the upper elevations, also east of Wallace's Line, contrasting with the fleshy drupes of the podocarps and *Taxus*.

On Kinabalu, there are three *Agathis* species, up to 2,800 m. The podocarps are well-represented on Kinabalu, often dominant at 2,800–3,400 m, and on other East Asian mountains. Of the seventeen species, *Dacrycarpus imbricatus, D. gracile* and *Dacrydium elatum* are quite common in Sunda lower montane forest, while *D. elatum* dominates lower montane *kerangas* of Khao Yai in southern Thailand, and the coastal mountains of Cambodia. *Phyllocladus hypophyllus* has the greatest vertical range on Mount Kinabalu (1,200–4,000 m)—large trees lower down and dwarf shrubs near the summit plateau. In Indo-Burma west to Nepal, there is a single, wide-ranging southern species, *Podocarpus neriifolius*, with some *Syzygium* among the very few Australasian trees that have readapted to the light frosts of warm, temperate South China.

South of the equator, in East Java and the Lesser Sunda Islands, where the dry season exceeds four months, pines are replaced by the southern angiosperm cemara (*Casuarina junghuhniana*; fig. 10.6). Cemara occurs above 1,300 m and can be found as high as the subalpine zone at 3,100 m. It also requires full sun and open ground for germination and establishment. Leaf scales on slender green shoots function like needles, and rough, flaky, persistent bark attracts epiphytes in sheltered habitats. Fire, originally often naturally volcanic in origin, is a regeneration requirement.

Epilogue Forest structure, physiognomy, and flora on tropical mountains is distinctly zoned, but also changes continuously with elevation within each zone (figure 10.13). Kira[15] proposed that zonation correlates with an index of warmth, but the realization that the lowland-lower montane forest ecotone remains at a constant altitude irrespective of latitude or seasonality, whereas that of lower and upper montane forest varies even locally with the size of a mountain and its isolation, defies that explanation. Zonal ecotones coincide with sudden changes in the physical environment, which provide factors that limit which species survive above them. The mediator of the lowland-lower montane ecotone is a change from tropical to temperate soils; the lower to upper montane ecotone, the incidence of fog; the upper to subalpine thicket ecotone, the incidence of occasional frost or, on large mountain massifs especially, the impact of increasing respiratory cost on that of photosynthetic gain.

However, the continuous altitudinal change within each zone is attributable to the influence of the gradual decline in temperature on the ratio between these physiological costs and gains on interspecific competition, and floristic impoverishment as this ratio imposes increasingly harsher conditions for survival.

Alt. (m)	Climate				Soil		Forest Formation	
	Summer monthly T°C	Everwet/summer rain	Winter monthly T°C	>2 dry months	Everwet/summer rain	>2 dry months	Everwet/summer rain	>2 dry months
4,000	0	Annual frost	0	Winter cloud base	Skeletal soils, mires		Tree line	Tree line
		Occasional frost					Gradient to subalpine thicket	
3,000	8	Intermittent clouds		Summer air frost / Winter annual frost	Ferric podzols	Increasingly acidic humic mulls		Warm temperate forest (>20° N). Gradual transition to upper montane species/structure (New Guinea rain shadow)
		Daily diurnal canopy fog						
2,000	14	Isolation and altitude bring increasing daily canopy fog. / Fog free / Daily shadow		Increasing seasonal daily canopy fog / Fog free / Daily shadow	Humic podzols (siliceous substrates) / Humic mulls	Humic mulls	Upper montane forest	Lower montane forest
1,500			Lowest valley frost			Humic ultisols/latosols	Lower montane forest	Deciduous forest/pine savanna
1,200								
1,000	21	Daily inland cloud base		Summer cloud base	Ecotone	Ecotone	Ecotone	Ecotone
750					Tropical lowland udult/humult ultisols/podzols	Tropical lowland udult/humult ultisols/latosols	Lowland forest	Lowland forest
0	27		15					

Figure 10.13 Summary of elevational zonal change in climate, soils, and forests in the humid tropics. Habitat changes with altitude, and those changes are driven by temperature decline (~0.65°/100 m of ascent on tropical mountains) and its interaction with atmospheric humidity to produce cloud (fog) and consequent precipitation. These factors of climate, along with substrate, determine soil formation in interaction with forest cover and litter production. Absence or presence of cloud/fog results in different soil and forest zonation. *Source:* Peter Ashton, "Patterns of variation among forests of tropical Asian mountains, with some explanatory hypotheses," *Plant Ecology & Diversity* 10 (2017): 361–77.

Notes

1. *OTFTA,* condensed from chapter 4 (241–307); Ashton, Peter S., "Patterns of variation among forests of tropical Asian mountains, with some explanatory hypotheses," *Plant Ecology & Diversity* 10 (2017): 361–77.

2. Whitmore, Tim C., *Tropical Rainforests of the Far East,* 2nd ed. (Oxford: Clarendon Press, 1984); Muellner-Riehl, Alexandra N., et al., "Origins of global mountain plant biodiversity: Testing the 'mountain-geobiodiversity hypothesis,'" *Journal of Biogeography* 46 (2019): 2826–38.

3. Symington, C. F. (revised by P. S. Ashton and S. Appanah), *Forester's Manual of Dipterocarps,* Malayan Forest Records 16 (Kepong, Malaysia: Forest Research Institute of Malaysia & Malayan Nature Society, 1943 and 2004).

4. Ediriweera, Sisira, et al., "Changes in tree structure, composition, and diversity of a mixed-dipterocarp rainforest over a 40-year period," *Forest Ecology and Management* 458 (2020): 117764.

5. Kira, T., "A climatological interpretation of Japanese vegetation zones," in *Vegetation Science and Environmental Protection,* ed. A. Mayawaki and R, Tüxen (Tokyo: Maruzen, 1978), 21–30.

6. Brambach, Fabian, Christoph Leuschne, Aiyen Tjoa, and Heike Culmsee, "Predominant colonization of Malesian mountains by Australian tree lineages," *Journal of Biogeography* 47 (2020): 355–70.

7. Rehm, Evan M., and Kenneth J. Feeley, "Freezing temperatures as a limit to forest recruitment above tropical Andean tree lines," *Ecology* 96 (2015): 1856–65; Lee, David W., and J. Brian Lowry, "Plant speciation on tropical mountains: *Leptospermum* (Myrtaceae) on Mount Kinabalu, Borneo," *Botanical Journal of the Linnean Society* 80 (1980): 223–42; Ando, Soichi, Yuji Isagi, and Kanehiro Kitayama, "Genecology and ecophysiology of the maintenance of foliar phenotypic polymorphisms of *Leptospermum recurvum* (Myrtaceae) under oscillating atmospheric desiccation in the tropical-subalpine zone of Mount Kinabalu, Borneo," *Ecological Research* 35 (2020): 792–806.

8. Kluge, Jurgen, et al., "Elevational seed plants richness patterns in Bhutan, Eastern Himalaya," *Journal of Biogeography* 44 (2017): 1711–22.

9. Körner, C., "A reassessment of high elevation tree line positions and their explanation," *Oecologia* 115 (1998): 445–59; Ashton, Peter S., "Floristic zonation on tropical mountains revisited," *Perspectives in Plant Ecology, Evolution & Systematics* 6 (2003): 87–104.

10. Joshi, Atul A., Jayashree Ratnam, and Mahesh Sankaran, "Frost maintains forests and grasslands as alternate states in a montane tropical forest–grassland mosaic, but alien tree invasion and warming can disrupt this balance," *Journal of Ecology* 108 (2020): 122–32.

11. Kluge, et al., "Elevational seed plants richness patterns; Ashton, P. S., and H. Zhou, H., "The tropical-subtropical evergreen forest transition in East Asia: An exploration," *Plant Diversity* 42 (2020): 255–80.

12. Aiba, S.-I., et al., "Latitudinal and altitudinal variations across temperate to

subtropical forests from southern Kyushu to the northern Ryukyu Archipelago, Japan," *Journal of Forest Research* 26 (2021): doi.org/10.1080/13416979.2020.1854065.

13. Ashton, "Patterns of variation among forests of tropical Asian mountains."

14. Kitayama, Kanehiro, Shin-ichiro Aiba, Masayuki Ushio, Tatsuyuki Seino, and Yasuto Fujiki, "The ecology of podocarps in tropical montane forests of Borneo: Distribution, population dynamics, and soil nutrient acquisition," *Smithsonian Contributions to Botany* 95 (2011): 101–17; Quiroga, Maria Paula, Paula Mathiasen, Ari Iglesias, Robert R. Mill, and Andrea C. Premoli, "Molecular and fossil evidence disentangle the biogeographical history of *Podocarpus*, a key genus in plant geography," *Journal of Biogeography* 43 (2016): 372–83; Cernusak, Lucas A., et al., "Podocarpaceae in tropical forests: A synthesis," *Smithsonian Contributions to Botany* 95 (2011): 189–95.

15. Kira, "Climatological interpretation of Japanese vegetation zones."

*Simmathiri Appanah and Chan Hung Tuck revealed
the marvelous story of pollination among* Shorea *(sect.
Mutica), thanks to their careful observations at Pasoh
Forest, Malaysia. Awaiting mass flowering to complete
their dissertations, in February 1976 they were alerted
by the expanded racemes and elongating buds of a
S. hemsleyana tree in the Kepong Arboretum. Taking
some buds back to their laboratory at the university,
they were surprised the following morning to find the
still-closed petals punctured by minute, neat round
holes, and tiny insects trapped within the bag: thrips!
Could these prove to be the long-sought pollinators?*

When the populations of the six Shorea *species in
the Pasoh MDF successively came into flower, no other
floral visitor was found. The flowers opened during the
evening and dropped the following morning, carpeting
the forest floor and permeating the shade with their
heady scent. On board were thrips in all developmental
stages. A haze of mature individuals rose erratically
into the canopy each morning—the pollination mystery solved.* — P. A.

11

TREES AND THEIR MOBILE LINKS: POLLINATION

In tropical forests, the tree species may drastically exceed the maximum number of neighbors allowed to any given individual by space and geometry. Consequently, opportunities for consistent, direct physical contact and competition between species are reduced. Interspecific competition for space must generally be relaxed, with different species functioning with ecological equivalence. Competition must be mediated by third parties, or *mobile* links—mobile organisms that themselves exploit a species, thereby positively or negatively influencing their competitiveness. In the case of plants, some mobile links *increase* fitness, especially pollen dispersers.[1] Continuously warm and humid climates reduce the physical challenges, such as drought and temperature extreme, minimizing the influence of the physical environment on plant fitness and competitiveness. These mobile links can exert strong—though indirect—influences on plant diversity.

The study of those mobile links that disperse by air is demanding. Much of the action takes place in the forest canopy, where our own, non-brachiating species is awkward and insecure (fig. 11.1). By the time hints of flowering have been detected, few days remain before close observation must begin. It may be necessary to access the outermost twigs of canopy species for flower observations, especially for cross-pollination experiments. Consequently, the number of detailed studies is still limited to a few species, and it is too early to draw general conclusions regarding many issues. The species richness of pollinator groups is less in Asian than in Neotropical rainforest communities, including MDF, as a comparison between two major research

Fig. 11.1 Studying pollinators in tropical rainforests. *Top left*, frequency of pollinators encountered at Lambir (Sarawak) and La Selva (Costa Rica); wind was studied at La Selva (I. B.; see Kuniyasu Momose and A. Hamid Karin, "The plant-pollinator community in a lowland dipterocarp forest," in *Pollination Ecology and the Rainforest*, ed. D. J. Roubik, S. Sakai, and A. Hamid [New York: Springer, 2003], 65–72.); *top right*, pollinating a tree in Sri Lanka (N. G.); *bottom left*, using a tree boom at Pasoh, West Malaysia (© Mark Moffat); and *bottom right*, Le Radeau des Cîmes (Canopy Raft), Gabon (D. L.).

sites revealed (fig. 11.1). Within individual forest formations, the number of species of birds, bats, bees, and butterflies is substantially fewer in the Asian tropics (p. 110). On the other hand, certain other plant-dependent groups are particularly species-rich in tropical Asia, including the scale insects and fruit flies.

Breeding Systems in Hyperdiverse Forests

Charles Darwin recognized that heritable diversity was required among members of species for evolution to occur through natural selection. For genetic diversity to be sustained at high levels in populations, genes must be continuously exchanged among genetically different individuals. Less exchange among different individuals means less genetic diversity in these local populations. The exchange takes place by the meeting and fusion of pollen grains (mobile and male) and the egg (sessile and female), thereby making a new whole, a zygote. The ovule remains within the parent flower, housing the embryo as it develops inside the seed.

Among most north temperate canopy trees, pollen is carried by wind, only randomly reaching fertile stigmas. Since only a tiny proportion ever reaches its destination, this is an enormously wasteful process. Flowers are among the most nutrient-rich plant organs, and their production is energetically costly. Pollen is rich in amino acids, therefore expensive in the nitrogen-starved environments of high temperatures and rainfall-leached soils. Many trees hardly grow during flowering and fruiting years. In these species-rich but nutrient-poor rainforests, how is pollen transferred from one distant individual to another, and how is nutrient loss minimized?

Pollen waste is inevitably great with cross-pollination between individuals separated by the dense foliage of a host of other species. In this setting, attracting a pollinator requires sugar-rich nectar, or even the sacrifice of some portion of the pollen itself as food to lure the disperser. It may also entail the synthesis of pollinator-specific aromatic compounds, attracting moths, bats, and beetles.

Importance of outbreeding. Rainforest trees were once considered self-compatible and overwhelmingly self-pollinated. Were this so, local populations would become genetically more uniform and less variable because this would reduce change in gene frequencies—and evolutionary change. If the rainforest environment itself changed little, natural selection would be effectively neutral. This situation would favor self-pollination, minimizing waste of pollen and ovules without trading off adaptability. While pollen dispersed to the nearest conspecific individuals may result in increased gene fixation, only that dispersed in quantity to greater distances will maintain genetic variability at the highest levels. Breeding systems are the primary determinants of genetic variability. The great genetic diversity in such an environment—where each seed successfully produced is costly in both energy requirements and nutrients—would be evidence for selection in response to competition between organisms.

In evergreen forests in Central America, Kamal Bawa and colleagues discovered a pattern of dominant outbreeding among canopy species.[2] Obligate outcrossing among bisexual tree species appears rare, and partial autogamy and a variety of monoecious systems are widespread. However, outbreeding dominates even where self-compatibility in stigma or style is prevalent. This is also true in tropical Asia.

The evidence of molecular data shows that overall levels of genetic diversity among dipterocarp populations are comparable to those of Neotropical and temperate zone tree species, indicating high levels of outbreeding. Phenological and pollination studies in Asia and the Neotropics indicate processes of surprising precision, implying intense selection for cross-pollination and maintenance of genetic variability.

Wind pollination. Wind pollination is rare in the Asian tropics, including seasonal regions with prevailing winds. The most archaic surviving species in the Asian rainforest jackfruit genus *Artocarpus,* in Borneo, are said to be partially wind-pollinated.[3] In everwet Malesia, wind pollination is confined to coasts with diffuse-canopy forests; *Casuarina equisetifolia* and *Podocarpus polystachyus* are common. Both produce copious male reproductive cones (cone-like floral inflorescences in *Casuarina*) and female structures on the same plant. Near-coastal *kerangas* and peat swamps affected by daily on-shore breezes feature podocarps with male cones, *Quercus kerangasensis* with male catkins, and *Gymnostoma nobile* (p. 287). Exposed riverbanks feature *Octomeles sumatrana* with dense flower clusters, male and female on separate plants. High ridges in lower montane forest also feature plants with adaptations for wind pollination, including *Quercus* and *Engelhardia* (Juglandaceae).

Bisexual (perfect) flowers and their pollinators. Vertically stratified forest species guilds vary by where they display flowers and fruits (p. 121). In the open landscape and full sun of the canopy, visual cues are important. Brightly colored flowers and fruits in Asian forests are found on canopy, gap, or riverside tree species, or on exposed vine and epiphyte species. Flower colors, shapes, and odors attract different pollinators, and the arrangements of stamens and stigmas promote efficient pollen transfer and outcrossing (fig. 11.2). Species with such flowers have a variety of mechanisms to promote outcrossing: in *Polyalthia,* for example, anther development may precede that of pistils (protandry); specialized oil secretions attract different beetles; and flower timing prevents crossing between species.

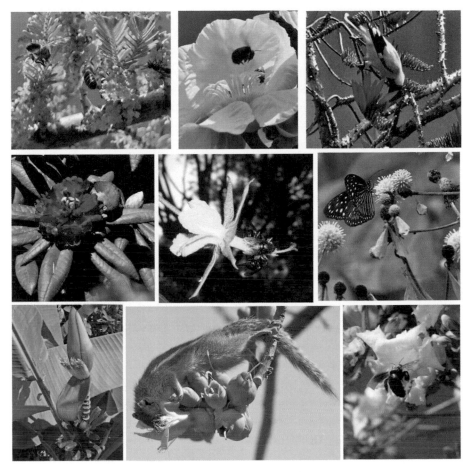

Fig. 11.2 Flower characters and pollinators. *Top left,* oriental white eye on *Canthium dioicum,* Mudumalai (H. H); *top center, Xylocopa* bumblebee on *Dillenia obovata* (H. H.); *top right,* Indian black-headed warbler on *Erythrina stricta,* Northwest Thailand (H. H.). *Middle left,* Ceylon white-eye on *Rhododendron zeylanicum,* Horton Plains, Sri Lanka (H. H.). *middle center,* flies on *Eriolaena candollei* (H. H.); *middle right, Euploea* butterfly on *Mitragyna rotundifolia,* Mae Ping National Park (H. H.). *Bottom left,* bat-pollinated *Musa violescens,* Peninsular Malaysia (D. L.); *bottom center,* squirrel on *Careya arborea* (Lecythidaceae), Mudumalai, India; *bottom right, Xylocopa* on *Stereospermum chelonoides* (Bignoniaceae), Mudumalai. (H. H.)

Monoecy and dioecy. The emergent canopy of Asian MDF is comprised almost entirely of species with bisexual flowers. In the main canopy and beneath it—as throughout Central America, where a distinct and continuous emergent stratum is unusual—the production of separate male and female (or unisexual) flowers on individuals is prevalent (*monoecy*). *Dioecy* is the separation of male and female sexual functions in flowers of different individuals. For several reasons, dioecy may be an inefficient means of reproduction in sedentary organisms. The sex of individuals remains constant, so the number of seed-bearing individuals is effectively halved within the mature population, and the next generation depends entirely on the success of the pollination process. Reduced fertility and population bottlenecks from constraints of cross-pollination often reduce reproduction in dioecious species.

It is therefore surprising that dioecy is so pervasive in evergreen rainforest. In Asia, dioecy is concentrated overwhelmingly beneath the canopy and in many of the largest genera with co-occurring species (fig. 11.3). As many as a third of subcanopy species in MDF are dioecious. The height of the MDF canopy has permitted the evolution and coexistence of congeneric species-series that reach reproductive maturity at different heights within the canopy. Many of these species share pollinators. The prevalence of dioecy among most or all species in a genus, and even whole families, implies that the condition is generally ancient and stable (assuming it was derived from the bisexual state in the remote past). What then, are the strong selective advantages of dioecy?

Concentrating the colorful, juicy fruit, dispersed by primates and birds, into a few individuals, may produce a tempting show in the sea of green. Dioecy may also be advantageous where pollinators are unspecialized and have restricted foraging ranges. It is striking, though, that these species are obligate outcrossers in the understory—precisely the habitat where visual attraction of pollinators is most difficult. Reducing the energetic costs of reproduction may also be advantageous. The cost, in energy and nutrients, of post-anthesis ovule or seed abortion can become vastly higher than that of producing sufficient pollen to ensure sufficient seed production to sustain the population. Dioecy also permits flowering of lower-cost male plants in smaller individuals and earlier in the season.[4]

Androecy, male and bisexual flowers on different individuals, is rare. *Xerospermum intermedium* (and most members of its tree family, Sapindaceae) is androdioecious. It is more abundant than any other tree species at Pasoh. As with monoecy, it may permit more production of male gametes, where successful outcrossing is difficult and extra pollen therefore at a premium, in plants whose flowers have an inflexible pollen/ovule ratio. Androdioecy may be advantageous in species with low selfing rates (since male reproductive allocation thereby increases relative to the overall resource devoted to flowering), or where the relative cost of ovules per flower is high and the chances of setting fruit from a pistillate flower are thereby increased.

Fig. 11.3 Monoecy and dioecy. *Left,* breeding systems by tree size at Bukit Raya MDF, Sarawak (I. B., from P. A.; see Peter Ashton, "Speciation among tropical forest trees: Some deductions in light of recent evidence," in *Speciation in Tropical Environments*, ed. R. H. Lowe-McConnell [London: Academic Press, 1969], 155–96); *right, Bischofia javanica,* a dioecious tree common in mixed evergreen forests throughout tropical Asia; *top center,* female flowers, ~2 mm across; *bottom center,* male flowers, ~2 mm across (all photos D. L.).

Hybridization. Equatorial rainforest tree species are remarkably constant in those characters—many of them seemingly trivial—by which related species can be distinguished. Mature individuals rarely blend the physical characters of two co-occurring species, implying that hybridization is rare or that hybrid individuals are unfit to survive. Morphological uniformity is particularly noticeable within populations, even over vast distances. However, three of the four species of *Shorea* section *Mutica* co-occurring in the Bukit Timah forest fragment in Singapore yield hybrid seedlings. *S. curtisii,* a more shade-tolerant ridgetop humult soil specialist and the commonest species, readily hybridizes there with *S. leprosula,* a light-demanding, faster-growing, hillside udult soil specialist. They share completely overlapping flowering times. Their potted hybrid seedlings, including back-crosses with flowering F1 individuals, had higher growth rates and survivorship, and larger leaves under high light, than did parental seedlings, demonstrating introgression. But hybrid wildlings only performed at levels intermediate between the parent seedlings and shared similar survivorship. Hybrid survival to reproductive maturity is rare in this mature stand, implying that selective mortality during growth favors habitat specialization.[5] Introgressive hybridization among co- occurring species series under reduced competition has grave implications for the future of species diversity in forests opened and degraded by logging.

Morphologically intermediate individuals are markedly more common in the seasonal tropics—for example, among *Dipterocarpus* species in the seasonal evergreen and semi-evergreen forests of Burma and Thailand. These are most frequent between semi-evergreen forest species (especially *D. turbinatus*) and deciduous dipterocarp forest species. Lower hybrid fitness, if finally shown to be due to competitive mortality, will provide an ultimate test of niche specificity.

Apomixis. When breeding tests reveal self-compatibility, embryos may form asexually from maternal tissue of an unfertilized ovule or surrounding tissues. This adventive embryony is a form of asexual reproduction, or *apomixis.* Surprisingly, it is common among rainforest tree species, which often produce multiple embryos. It occurs in Asian fruit tree genera, such as *Mangifera, Garcinia, Lansium,* and *Citrus*—and in at least a few dipterocarps.

It is hard to explain how populations could remain both apomictic and competitive over evolutionary time, since persistence of large numbers of identical genotypes would expose them to infection. Apomixis may promote "firework displays," the genetic diversification of populations into niches that are temporary in evolutionary time, while the core, outbreeding metapopulation—often a widespread sister species—persists. Whether those congeneric species series so characteristic of forests within the everwet tropics of Asia are thus evolved is an intriguing research subject.

Pollinators and Reproductive Ecology

Given the importance of tree reproductive biology for understanding the maintenance of species diversity, most research into the reproductive biology of tropical trees in Asia has focused on the species-diverse MDF, where many canopy species, and most dipterocarps, flower supra-annually in ENSO drought years. Seasonal forest trees differ notably in annual flowering, although the reproduction may vary between years. Flowering in seasonal tropical Asia occurs during the dry season—especially in deciduous forests. Studies in the Neotropics have been more extensive than those in Asia, with similar results but little research on apomixis and none on the influence of ENSO on reproductive biology.

Cross-pollination despite uniform flower morphology. The diversity of seed plants is partly explained by the co-evolution of their elaborate and versatile flowers with a diverse suite of pollinators.[6] Curiously, the basic ground plans of flower structure are generally consistent within rainforest tree taxa; species within speciose genera and even some speciose families share details of flower and fruit morphology, and share the same families of pollen and seed dispersers. These are ancient characters. Insects and flowering plants diversified rapidly during the Cretaceous period, and insects evolved in intimate communion with the proliferation of chemical attractants—and defenses—among flowering plants. Co-evolutionary relationships are often conservative. Many basal flowering plant families remain associated with primitive pollinators, such as Annonaceae and Myristicaceae with beetles, and Dilleniaceae with flies. Among congeneric series of tropical trees, most share the same pollinator groups, and frequently species. In larger genera with single, dominant pollinators, including *Shorea* and *Macaranga,* subgenera attract different dominant pollinators. Figs are a major exception. How then did these species-series, which contribute so greatly to overall tree species diversity, arise, and how are they sustained? How do these series of often ecologically sympatric tree species avoid pollen wastage and pollution of their generally tiny stigmatic surfaces with foreign pollen?

Everwet forests. Species within large genera of more advanced families, as in *Dipterocarpus, Shorea* (sect. *Mutica*), and *Syzygium,* share pollinators. Sequential yet overlapping flowering times within such co-occurring species suggest that competition for pollinators could, through differentiation of flowering times, sustain species diversity through the mutual advantage in maintaining pollinator numbers. These co-occurring tree species could either be ecologically complementary, or they could differ in other ways, such as light response and height at reproductive maturity. Sequential flowering has been observed in several dipterocarp genera and *Parkia,* but there is little evidence so far of overlapping sequential flowering among the many other series of co-occurring congeneric species sharing floral colors and morphologies.

Flowering phenology. Although differences in the time of fruit fall may be caused by heritable differences in the rate of fruit development, and can be modified in individual seasons by weather, the triggers to reproduction overall must precede flowering. Once begun, the events cannot be stopped except from catastrophic abortion by weather or predation.

Flower triggers. The climatic triggers to flowering remain obscure. For dipterocarps, it is a trigger for cell division to begin forming an inflorescence primordium.[7] In other families, such as in some Meliaceae and Sapindaceae, the trigger leads to expansion of an already formed, dormant inflorescence for which there may or may not have been a prior climatic trigger. Water stress is correlated with initiation of reproduction in the everwet tropics. However, it is unlikely to generate the precision for precise intraspecific synchrony and consistent sequential interspecific anthesis, irrespective of topography and geology (and therefore soil-moisture variation). Mass flowerings have been observed in peat swamp forests with permanently saturated soils in the same years as upland forest mass flowerings. A likely principal trigger to canopy flowering in the seasonal tropics, as well as to mass flowering in the everwet Sundalands, is a depression of night temperature to below 20° C for several days (despite some lack of consistency in meteorological records), but this cannot be the sole trigger.[8] Mass flowerings occur, perhaps invariably, in years of ENSO drought, when there are dry periods in January, the middle of the northeast, and July, the middle of the southwest monsoons. North of the equator, as in North Sumatra, Peninsular Malaysia, and northern Borneo, the trigger is usually set in January; south of the equator, as in Palembang, South Sumatra, and even at Purukjau, 0.36° S latitude in the South Borneo foothills, it is set in July, where it correlates with a drop in minimum night temperature to 18° C for several nights.

Mass flowering. The everwet regions of the Sunda Shelf are celebrated for periodic mass flowering of the dipterocarps (p. 103). Malaysia (particularly at Pasoh forest, Negeri Sembilan, and Sepilok Forest, Sabah) and Gunung Palung National Park (West Kalimantan) have been sites of the most research. Mass flowering periodicity is correlated with ENSO drought years and therefore has a regular periodicity of about five years. Some species also flower in the year preceding or following a mass flowering. Minor mass flowerings may occur that are uncorrelated with ENSO events and may vary in their timing even within small regions: relief for those starving vertebrates foraging over long distances.

Community mass flowering in the Sundalands is concentrated in the forest canopy, and to a lesser extent in other habitats exposed to direct sunlight. It is not a highly standardized event. Most canopy species flower with exceptional intensity during these events, but some dipterocarp species rarely flower *exclusively* in those years—and even those species do not come into flower during *every* mass flowering year. Subcanopy and understory shrub and tree populations flower annually but may flower somewhat less during mass flowerings. The geographical extent of mass flowering in the Far East remains unknown, as we lack records from New Guinea, and those from the Philippines are inadequate.

The carbon-cost of mass flowering—estimated at 3% of net primary production—is greater than that of wind pollination, but there is negligeable reproductive cost in intervening years. Mass reproduction events reduce later tree growth. Onset of reproductive maturity in tropical trees may permanently reduce leaf size, implying a permanent change in carbon allocation from growth to reproduction.

Semelparous gregarious (once in a lifetime) reproduction is widespread in the Asian tropics, prevalent in the omnipresent bamboos and *Strobilanthes* (p. 199). Like the dipterocarps, these genera produce abundant, undefended seeds which, by their occasional heavy production, saturate the appetite of the seed predators in their habitat. Different species of *Strobilanthes* have life cycles of 4–14 years; however, like some dipterocarps and bamboos, a few may flower in the year prior to mass flowering. In this case, the sparse seed is eliminated by predators.

The extreme dominance of the dipterocarps in the MDF emergent canopy, whose pollination and seed dispersal do not depend on arboreal vertebrates, likely partially explains the comparative poverty there of both vertebrate species diversity and population sizes. However, the ENSO-induced cycle of multiyear famine between single years of excess must also play a major part.

Foraging patterns. The foraging behavior of pollinators influences the distance over which pollen is dispersed and the effectiveness of cross-pollination, thereby influencing the genetic structure of plant populations. In closed-canopy rainforests, tree pollination is accomplished almost entirely by animals—mostly insects (fig. 11.2). In the everwet tropics, relatively few tree species are pollinated by birds or bats, but there are exclusively bird-pollinated canopy species in semi-evergreen and deciduous forests where flowering occurs when trees are leafless. Other trees are principally pollinated by insects but are also visited by birds. Most tropical trees are visited by a diversity of potential pollinators. Bat flowers may be visited by birds, others by various butterflies. Most flowers with spreading petals are visited by a variety of bees. Promiscuous pollination is particularly common among species with excess pollen as a lure, an expensive strategy most often found on the fertile soils of the seasonal tropics and on river banks. Most—perhaps all—tree species have a "lead pollinator" species distinguished by its faithfulness during foraging excursions. Some trees attract a single, dominant visitor almost exclusively, as in the archaic Myristicaceae and Annonaceae.

It is difficult to determine which floral visitors are the most effective pollinators. They must carry pollen between trees and between flowers so that it reaches the stigma. They must demonstrate high consistency to a single host species during each flight. Among pollinators that respond to visual cues, bees are exceptional, although they are also pollen robbers. Other such species that may be effective pollinators include bats and Danaid butterflies, which also "trapline." This foraging pattern increases the likelihood of visiting subsequent trees of the same species as the animal retraces the same or a similar circuit over the course of a day. Many beetles are pollen robbers whose clumsy activities within the flower and random foraging patterns may reduce cross-pollination. Beetles are often attracted by odors of a single species.

The seasonal tropics. Flowering throughout tropical Asia north of the equator usually begins as the wet southwest monsoon ends, whether it was rainier, as in Sunda, or less so, as in SW Sri Lanka. It reaches a peak toward the end of the cool dry season, March–April. Canopy flowering generally peaks after leaf fall, with fruit fall toward the second half of the southwest monsoon. Flowering times differ somewhat among species visited by different pollinators. Synchronicity is loose in many deciduous species, and trees may flower and fruit simultaneously during the dry season. Seeds break dormancy and germinate following the onset of the rains. By that time, nutrients will have been reabsorbed from the senescing leaves of many deciduous species (p. 152).

In the open landscape and full sun of the canopy, plants attract with visual cues. Most bright-colored flowers and many such fruits in Asian forests are found on canopy, gap, or riverside trees and vines. Deciduous trees often produce large flowers, visually attractive during leaflessness (fig. 11.4). Several, such as the *mahwa* (*Madhuca longifolia*: Sapotaceae), present flowers rich in nectar, with fleshy corollas ripening at the same time as the stamens. They attract various vertebrates and honeybees, which may act mainly as pollen robbers (and then there are humans who make country liquor). Some genera with species whose mature crowns differ widely in height, such as *Syzygium* and *Canarium* and other Burseraceae, have flowers that attract a wide range of visitors, mostly insects, or are pollinated by generalist pollinators, such as flies and beetles.

In seasonal regions, several tree species, including *Butea, Erythrina,* and

Fig. 11.4 Flowers of seasonal forests. *Left, Careya arborea* (Lecythidaceae) Maharashtra, India; *left center, Helicteres isora* (Malvaceae), Maharashtra, India; *right center, Bombax ceiba* (Malvaceae), India; *right, Madhuca longifolia* (Sapotaceae), Maharashtra, India (all D. L.).

Pterocymbium—all deciduous and with brilliant coral red or orange (on *Firmiana*), flowers—attract a wide diversity of feeding, generalist passerine birds, including mynahs, orioles, and bulbuls, as well as sunbirds and spiderhunters. Such trees are rare in everwet climates, occurring only occasionally on nutrient-rich soils, where these birds visit epiphyte and mistletoe flowers.

Pollinators in Asia and the Neotropics. Much has been made of the low diversity of pollinators in the Asian tropics, especially compared with their number in the Neotropics, where there is greater diversity among vertebrates and the promiscuously foraging euglossine bees. About 2,000 bird species are said to regularly visit flowers in the Neotropics, of which two-thirds are flower specialists. These are dominated by the hummingbirds (Trochilidae), 97 species of which occur in Venezuela alone. Besides sunbirds, spider-hunters and flowerpeckers, the main groups likely to pollinate in the Asian tropics are bulbuls and minivets, though parakeets and others visit flowers. The total number in tropical Asia is in the hundreds. Perhaps the far fewer epiphytes (p. 109) and the few flowers between mass flowering years in everwet regions reduce food supplies for birds.

Tropical Asia is particularly rich in pollinators from archaic insect taxa, notably beetles but also flies (fig. 11.2). This may reflect the relatively recent origin of extensive, wet equatorial lowlands in Asia, and constraints imposed by ENSO at the western end of the widest ocean (chapters 4, 13). In the seasonal tropics of Asia, wide-ranging pollinators, including honeybees, xylocopid and *Amegilla* bees, and a diversity of generalized pollinating birds are more abundant.

Compared with the Neotropical hummingbirds, the nectarivorous Asian sunbirds and flowerpeckers are poorly diversified and less host-specific, and they visit few tree species. The Australasian lorikeets are represented in

Sunda MDF by only one species, the blue-crowned hanging parrot (*Loriculus galgulus*), although more species occur east of Wallace's Line.

Pollinators that forage over long distances or migrate may sustain populations in everwet regions by visiting flowers in mass flowering years. Pollinators of more restricted range must either have alternate food sources, as have leaf beetles (which visit flowers), or have high fecundity and short life cycles, as do thrips.

Momose and colleagues observed the pollinators of all plants in the mixed dipterocarp forest at Lambir National Park (northeast Sarawak) over 53 months, including a mass flowering, in the most comprehensive study of its kind.[9] Compared with a Central American site, Lambir had fewer species of far-ranging solitary bees (*Amegilla* substituting for euglossines), birds and bats, and Lepidoptera; but it had more frequent visitation by other bees, beetles, and by a diversity of other insect visitors. All but one of the 19 species of *Hopea* and *Shorea* were visited by leaf beetles.

Vertical stratification of mature forests depends in part on where trees present their flowers and fruits (p. 121). A major ecological difference between the Asian tropics and the Neotropics is that vertebrate pollinators in Asia, including the macroglossine bats, and honeybees, visually respond at distance, and thus forage almost entirely in the canopy, in large gaps, and along forest edges and riverbanks. Only the Meliponid sweat bees forage in the shady subcanopy, *Apis* and *Xylocopa* being primarily canopy, gap, or forest-edge feeders. Compared with Neotropical forests, with their *Passiflora* and the bright-flowered Acanthaceae and Rubiaceae, the Asian MDF subcanopy is a uniform green. There are exceptions: swallowtail butterflies pollinate the gleaming, brick-red flowers of *Ixora* (Rubiaceae) and gingers of the genus *Globba*. Danaid butterflies are important flower visitors and presumed pollinators in Asian forests, though they do not compare in diversity with the Neotropical Ithomiidae and Heliconiidae. The subcanopy of Asian rainforest especially contains pollinators that respond to olfactory chemical cues. In contrast, Asian pioneer vegetation and semi-evergreen and tall deciduous forest understories are often rich in Zingiberaceae and other tall herbs.

Though hawk (sphinx) moths are not important tree pollinators, moth-pollinated trees are well-represented in certain genera, especially in the Sapotaceae and *Diospyros* (Ebenaceae). Parakeets and flying foxes (*Pteropus* bats), most common in the seasonal tropics, feed on the flowers themselves as well as their pollen and nectar, which they inadvertently cross-pollinate.

Case Studies of Tree Pollination in the Asian Tropics

Although there is some work of high quality, there is much less detailed research in pollination ecology in tropical Asia, compared with that done in the Neotropics, particularly Central America.

The dipterocarps. Given their ecological importance in forests in the Asian tropics (p. 104), it is more than useful to summarize the research on this important family. Dipterocarp flowers are bisexual, with a basic floral morphology for the family. However, flowers are produced in many sizes (fig. 11.5). Species in seasonal forests and more exposed in the canopy or as emergents tend to produce flowers that are larger (measured by calyx diameter) than the very small flowers in MDF understory.

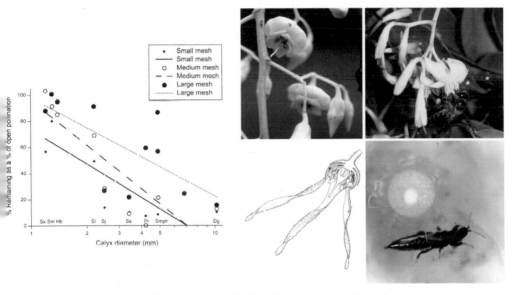

Fig. 11.5 Dipterocarp flower size and pollination. *Left,* plot of percentage of open pollination versus flower size (calyx diameter) in inflorescences covered by small, medium, and large mesh bags. Smaller flowers are more effectively pollinated by small insects passing through fine mesh bags (I. B., from C. J. Kettle, "Flower size and differential pollination success in tropical trees," ATBC annual meeting, Denpassar, Bali, 2012 [unpublished]). Hb = *Hopea beccariana,* Sj = *Shorea johorensis,* Ss = *Shorea superba,* Pt = *Parashorea tomentella,* Smph = *Shorea agamii,* Dg = *Dipterocarpus grandifloras,* sx = *S. xanthophylla;* m = *S. macroptera;* l = *S. leprosula. Top center,* large flower of *Shorea siamensis* (H. H.); *top right,* large flower of *Dryobalanops aromatica* with *Apis dorsata* (T. I.); *bottom center, Shorea* sect. *Mutica* flower, showing positions of thrip egg deposition (L. S.; see E. Hagerup, and O. Hagerup, "Thrips pollination in *Erica tetralix,*" New Phytologist 52 [1953]: 1–5); *bottom right,* thrips on *Hopea nutans* flower (© Mark Moffat).

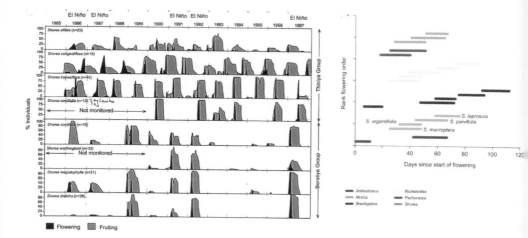

Fig. 11.6 Flowering patterns in dipterocarps. *Left*, reproductive phenology of eight co-occurring *Shorea* sect. *Doona* species over 12 years at Sinharaja, Sri Lanka; *right*, flowering phenology of co-occurring dipterocarps at Danum and Sepilok, Sabah; colors indicate species within sections of *Shorea*. (both I. B., from S. Dayanandan; see S. Dayanandan, D. N. C. Attygalla, A. W. W. L. Abeygunasekara, I. A. U. N. Gunatilleke, and C. V. S. Gunatilleke, "Phenological and floral morphology in relation to pollination of some Sri Lankan dipterocarps," in *Reproductive Ecology of Tropical Forest Plants*, ed. K. A. Bawa and M. Hadley [Paris: UNESCO, 1990], 103–33).

In the seasonal tropics, dipterocarp species flower annually with varying intensity, or supra-annually but independently and without familial or generic synchronicity. The best-known dipterocarp, the gregarious sal (*Shorea robusta*) of Indian deciduous forests, flowers annually from February well into April, with extra intensity every few years, and heaviest flowering every decade. In seasonal climates, dipterocarps may flower for one month or more within the cool dry season.

In everwet climates, most canopy dipterocarps flower over a season en masse as a family. Trees flower at about every five years, broadly coinciding with the drought years occasioned by the ENSO (p. 9; fig. 11.6). Flower timing varies regionally in relation to the season in which drought occurs. At least since systematic observations began at Pasoh in 1958, heavy flowering years have tended to alternate with weaker years in the five-year cycle. Other canopy families similarly flower heavily in these years, but also more frequently in the years intervening. The climatic trigger, which leads to development of reproductive primordia in the bud, is becoming better understood. It is broadly correlated with periods of drought (p. 137), but not directly with soil-water deficit, as flowering does not vary with local topography. Flowering may be promoted by several days of night temperatures below 20° C, but only if trees have enough sugar reserves. It is preceded by increased activity of certain genes. Canopy dipterocarps only flower once their crowns are exposed.

Where several sister species co-occur, such as *Shorea* within one section, they flower in overlapping sequence, each flowering alone for a period (fig. 11.6). Appanah and Chan (p. 222) found at Pasoh that populations within sect. *Mutica* flower with high synchronicity, the first for about two weeks, then successive populations over an increasing period to the sixth and last, which flowered for 3.5 weeks. The differentials in flowering times imply that each species has a different rate of floral development, following simultaneous triggering of a flowering gene. The dates provided the means, by extrapolating backward, to identify the time at which all these developmental trajectories originated—and therefore to search for any climatic event correlated with it. The flowering sequence has subsequently been confirmed in other seasons and sites, where species co-occurring in different regions within Sundaland flower in sequence. Similar results were seen in eight wild, honeybee-pollinated *Shorea* (sect. *Doona*) in MDF at Sinharaja (southwest Sri Lanka). Four species mass-flower at intervals of about five years, and four flower more often (fig. 11.6). Both East and South Asian supra-annually flowering *Shorea* species series flower in overlapping sequence, but with much greater overlap in sect. *Doona* than in sect. *Mutica*. Individual *Doona* species populations never flowered alone. Clearly, different climate triggers must be involved in the two *Doona* groups. Individuals of some taxa in the Far East, notably bee-pollinated *Dryobalanops* species and *Dipterocarpus crinitus*, may flower more frequently than others in the intervening years between mass flowerings. Other species series within small and large-flowered dipterocarp genera flower in overlapping sequences.

Since dipterocarps are predominantly outbreeders, the integrity of species in such extreme diversity can only be sustained where there is little or no introgressive hybridization.

Among the sequentially flowering species and genera of dipterocarps in the Far Eastern everwet tropics, mass flowering is followed by mast fruiting over a period of no more than four weeks; species series that had flowered in sequence fruit synchronously. Those species which develop from expanding bud to anthesis in the shortest time take the longest time to ripen their fruit, and vice versa.

Species within each dipterocarp genus (and within each section of the large genus *Shorea*) share the same principal pollinator. The flowers of most dipterocarps open early in the morning, and corollas start to fall in large numbers around noon. The isolated and basal genus *Dipterocarpus* is pollinated by Lepidoptera. Pollination-exclusion experiments confirmed that, though all dipterocarp taxa are visited by diverse insects, flower size and morphology (especially of the androecium) select different predominantly effective pollinators for each genus and infrageneric species series (fig. 11.5).

The flowers of Asian dipterocarps are found in two forms: large, and with larger, narrowly oblong and yellow anthers, containing many pollen grains and bearing short, stout, hairless appendages; or small, with small ellipsoid and cream-colored anthers, with fewer grains, bearing slender and often bristly appendages.

The flowers of all dipterocarps attract Chrysomelid beetles. *Dipterocarpus* flowers are visited by butterflies, and the nocturnally flowering species *D. tempehes* is visited by moths. Others that have been observed (*Vateria, Dryobalanops, Neobalanocarpus,* and *Shorea* sections *Doona* and *Pentacme,* all of which are canopy trees) are pollinated by honeybees, although many other insects visit in small numbers, including Scarabid beetles (possibly Chrysomelids too), and *Trigona* bees in other taxa when honeybees are absent.

The Asian dipterocarps with smaller white anthers include most emergent and all subcanopy dipterocarp taxa in MDF, in many genera and sections. Thrips and Chrysomelid beetles are pervasive in dipterocarp flowers, and thrips are the likely pollinators in the small-flowered dipterocarps throughout the region (p. 222). At the mercy of convectional breezes that can carry them beyond individual crowns, these minute insects bear oar-like wings that allow them to land directionally. The small-flowered trees produce enormous quantities of flowers.

Fruit set is substantially increased when trees are adjacent to others of the same species. Reproductive success as a function of flower density is generally greater in larger flowered species whose pollinators disperse pollen at greater distances than the thrips pollinating the small-flowered species.

The dipterocarp genera and *Shorea* sections with large, usually yellow anthers and short, hairless appendages are phylogenetically primitive within their higher taxa.[10] They are restricted to the less diverse forests of Sri Lanka, peninsular India and the Philippines, low diversity habitats (*kerangas,* peat swamps, on base-rich soils), or are taxa that flower more frequently after mast years, or have resinous pericarps and lower seed predation, and occur in gregarious "drifts." Therefore, fecund insects with short life cycles pollinate more effectively in highly diverse forests than larger insects with life cycles longer than a mass flowering, and these insects may have been more general in Asian rainforests when the everwet Sunda lowlands expanded 20 Ma (p. 45).

The many dipterocarp species in hyperdiverse forests, such as Lambir in Sarawak, appear to be sustained by (1) soil specificity; (2), by growth responses to light combined with survivorship in shade; and (3) by their place at reproductive maturity in the vertical structure of the forest. In some cases, though—including several *Shorea* (sect. *Mutica*) species at Pasoh and elsewhere in Peninsular Malaysian MDF—species appear to be at least partly ecologically complementary in these respects. Such populations of species in co-occurring, congeneric series may then be differentiated in niche occupancy by a fourth factor: sustained competition for a shared pollinator. These species, at least, could be physiologically complementary, but their numbers are checked by competition for a shared pollinator. In this respect the spatial predictions of neutral theory (p. 241) may prevail among their populations. Yet they are interdependent in that they also rely on each other to build pollinator numbers.

Fruit bats and their flowers. All bat flowers, Neo- and Paleotropical, have several common characteristics. Many contain sulphur in their nectar, imparting the characteristic "sour milk" odor. Some have mushroom-like smells from fatty acid derivatives. Flowers are large, pale colored, open in the evening, produce copious nectar, present a silhouette away from the crown, and lose the corolla the following morning. Bat flowers (fig. 11.7) are of two basic forms in Asia: "shaving brush-like" and "campanulate." Shaving-brush flowers present abundant pollen on stamens with long spreading filaments, which are offered as part of the lure. The campanulate flowers, typical of Bignoniaceae, only offer nectar, and place the few stamens to avoid self-pollination. Usually, a few flowers open each night on any tree. There are many exceptions. Bat-pollinated wild banana flowers remain open during the day, attracting sunbirds. Durian flowers combine many stamens with a campanulate corolla shape. Also, bats sometimes visit flowers lacking any of these characteristics: mango (*Mangifera*), wild jack (*Artocarpus*), *Rhizophora* mangrove, coconut, and occasionally *Arenga* palms (fig. 11.8).

Bat-pollinated plants either flower continuously or seasonally, once a year or more often. These include the mangrove *Sonneratia*; and the wild

Fig. 11.7 Bats and flowers. *Top left*, the bat *Eonycteris spelaea* feeds on *Durio zibethinus* (Malvaceae) flowers (© Merlin Tuttle); *top center*, *Macroglossus minimus* feeds from a banana flower in Sarawak (© Chien Lee); *top right*, *Ptenochinus jagori* consumes a fig, Luzon (© Tim Laman); *bottom left*, protruding brush flowers of *Duabanga grandiflora* (Lythraceae) (H. H.); *bottom center*, *Parkia javanica* (Leguminosae) inflorescence, *Eonycteris spelaea* feeds on the nectar from male flowers and pollinates the bisexual flowers covering the bulb (H. H.); *bottom right*, a fulvous fruit bat feeds on durian buds (© Tim Laman).

bananas (*Musa*) in small groves, forest gaps, and secondary vegetation. *S. alba* and *S. ovata* flower in intermittent, overlapping flushes throughout most of the year. The estuarine *S. caseolaris* populations present some flowers all year. Banana shoots develop, flower, and die continuously within a grove. Bat-pollinated bananas bear hanging inflorescences, and the bats land on the inflorescence bud after a bract curls back in late afternoon, exposing the large oblong anthers at its base and allowing the copious nectar to flow down the bud. Megachiropteran bats are confined to the Old World

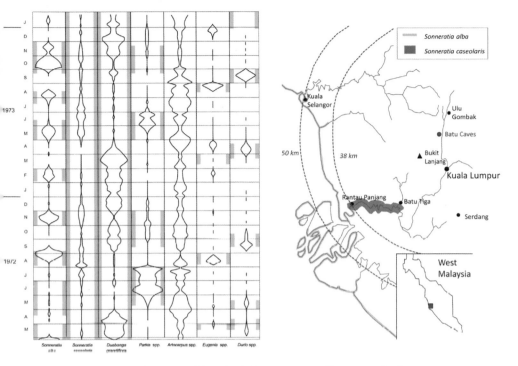

Fig. 11.8 Bat pollination in West Malaysia. Left, kite diagram of frequency of pollen from tree species visited by *Eonycteris spelaea*, from guano collected weekly from Batu Caves, Selangor; *right*, map of the area of Selangor, West Malaysia, showing the range of *Eonecteris spelaea* flying from Batu Caves, and the distribution of two mangrove species visited by the bats (both I. B., from A. Start; see chap. 11, n. 11).

(p. 242). They are best known for the large fruit bats (*Pteropus*), which forage and roost in flocks like parakeets. They act as pollen robbers, often damaging flowers by eating their soft parts and accidentally pollinating the flowers. The Megachiroptera also includes several genera of small bats, including all the nectar- and pollen-feeding bats of the Old World. There are three exclusively nectarivorous genera of small Megachiropteran bats: *Eonycteris, Macroglossus,* and *Syconycteris*. Like flower-visiting Asian birds, but unlike Neotropical Microchiropteran nectarivorous bats and hummingbirds, they do not hover—instead, they land on flowers to feed. They feed exclusively on nectar and pollen, which they ingest by rapidly flicking a long, narrow, hairy tongue in and out.

Unlike Microchiroptera, Megachiroptera have good eyesight; they navigate to food and roosts by eye. They therefore feed in the evening by moonlight, mostly foraging in the forest canopy, edge, riverbanks, and large gaps. None regularly enters beneath the canopy.

Eonycteris occurs from South Asia to Malesia west of Wallace's Line, with one species in Peninsular Malaysia, *E. spelaea*. It roosts by the thousands in the roofs of caves. Tony Start collected fresh guano from individuals and from roosting groups in Batu Caves, near Kuala Lumpur, ascertaining that different bat groups fed on the same tree species nightly (fig. 11.8).[11] Some interchange among groups may serve to communicate the availability of newly flowering trees by the pollen on their coats. *Eonycteris spelaea* is medium sized, flies fast, but can only land on a flower by braking and leave it by somersaulting. It visits all flowers foraged by Peninsular Malaysian bats, and it feeds primarily on trees that flower seasonally, preferring certain flowers (fig. 11.8). Bats only visited the continuously flowering *Sonneratia* species in the coastal mangroves more than 38 km from Batu Caves, when closer seasonal forest sources were unavailable. Bats in another cave, 60 km inland, never visited this major, continuous food source, indicating that they can survive without it and revealing the limits to their foraging distance. Trips to the mangroves take at least an hour, by which time night has fallen; the bats must then rely on moonlight to locate flowers visually. Bats repeat these long-distance journeys, and they may therefore be trapliners (p. 233). In contrast to primates and bees, which are heavily dependent on masting years, *Eonycteris spelaea* breeds continually. Successively flowering trees sustain relatively stable bat populations.

Two species of *Macroglossus* live in Peninsular Malaysia (fig 11.7). These tiny bats are very different from *Eonycteris*. They are solitary, roosting alone near their food sources, which flower continuously, although they occasionally visit nearby seasonally flowering species. They occupy overlapping territories of no more than 1 km in diameter. Their flight is slow and fluttering but agile, on relatively broad wings. On the peninsula, *Macroglossus minimus* feeds almost exclusively on the flowers of the three *Sonneratia* species of the coastal mangrove, but it occasionally visits cultivated banana, coconut, and other village bat flowers; the inland *M. sobrinus* concentrates on wild bananas.

Pteropus, the true fruit bats, are pollinators of the brush-flowers visited by *Eonycteris*, including durians. *Pteropus* feed on a wider range of flowers than *E. spelaea*, eating the stamens, fleshy bracts, and perianths, and often damaging the pistil—but pollinating flowers.

These natural histories of pollination, elegantly derived from such sim-

Fig. 11.9 *Dendrocalamus pendulus* in Hulu Gombak, West Malaysia, 2019. *Left*, clump in senescence after flowering; *right*, dense flower clusters (W. K).

ple techniques as collecting guano, have helped give these mammals a better reputation and have helped in their conservation—for example, by preserving their roosting places in the case of *Eonycteris*.

Bamboos. Bamboos are perennial woody grasses, extremely important in tropical Asia (p. 106). Many bamboos flower annually, but some gregarious taxa flower communally over whole landscapes at intervals of many years (up to 55 years in *Bambusa bambos*) or patchily in clumps (fig. 11.9). Like masting trees such as oaks and dipterocarps, bamboos yield vast quantities of nutritious seed. This feast attracts small vertebrates, especially rodents and birds. Despite this predation, vast numbers of seedlings, whose foliage is often poisonous, manage to establish. Famines among human populations once followed the flowering of bamboo over large areas when rats, having multiplied in the year of fruiting and abundance, moved into villages and grain stores. Although the selective advantage of mass flowering in bamboos is probably the satiation of seed predators that allows more individuals to survive over the long run, even 40–50 years, it may also promote reproductive success in this wind-pollinated grass. The real puzzle, with not a viable hypothesis proposed, is what triggers such flowering. Another example of such flowering is the genus *Strobilanthes* (p. 199).

Figs. Ficus, with ~800 species, is among the largest genera of woody plants. The majority occur in tropical Asia, and their range extends to Australia and the Pacific islands. *Ficus* is perhaps the most diverse in form among the woody plant genera, with free-standing pioneer trees, successional and climax trees, stranglers (banyans), vines (many with aerial roots), understory shrubby species (including some that bear figs on specialized rhizomes as much as 10 cm below the soil surface), small climbers, and even six rheophytes, including the Bornean *F. macrostyla*, which presents its figs below normal water level among rocks in white-water rivers.

All figs produce an inverted inflorescence resembling a fruit, hollow inside where the many flowers are attached to the inner face: the syconium (fig. 11.10). The pollinators, specialized fig wasps of the family Agaonidae, enter through a terminal pore, the osteole. Figs are unique in their pollination biology and remarkable among woody plants for the near-absolute specificity of their pollinators. Four subgenera have been recognized on morphological evidence, three of which are monoecious, bearing male and female flowers within each fig. The female flowers vary in the length of their styles. The female fig wasps have specially adapted needle-like ovipositors with which they penetrate the style. They prefer flowers with shorter styles, which thereafter become sterile gall figs, the ovule nevertheless swelling and providing food for the wasp larvae, which bite their way out of the ovary at maturity. When a male wasp emerges within a syconium, it loses its rudimentary wings, mates with one or more females, and dies. The female wasps escape through the scale-lined osteole, and their bodies pick up pollen from the male flowers that mature just as the females are preparing to leave. Figs develop synchronously on each tree, populations reproduce continuously, and different individuals simultaneously produce gall-figs as well as seed-figs (fig. 11.10).

Figs originated about 110 Ma. In a molecular phylogeny, George Weiblen found substantial congruence between phylogenies of figs and their wasps, a stringent demonstration of co-evolution (fig. 11.11).[12] Wasps of the large monoecious strangling figs fly at full canopy height in low densities. Caught in prevailing breezes they can be carried many tens of kilometers. Some monoecious figs are among the most wide-ranging woody plants in tropical Asia; few are local endemics. Monoecious figs were among the early immigrants and reproducing colonizers of Krakatau, which is more than 50 km from both Java and Sumatra, following its devastating eruption in 1884. Dioecious figs (male and female syconia on separate plants) have evolved more than once from monoecious ancestors. The wasps of dioecious figs are found mainly low in the canopy, where their hosts' crowns occur. Their host fig species tend to cluster in local populations, and many are local endemics.

Honeybee pollination. As with birds, the Old World lacks the euglossine

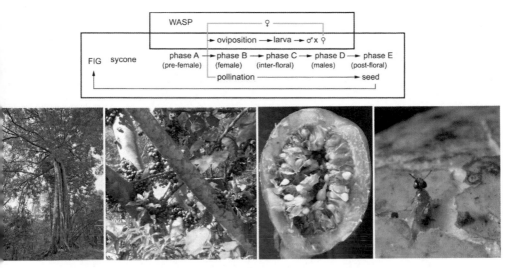

Fig. 11.10 Fig life cycle. *Top,* life cycle diagram of a monoecious fig, showing the life history stages of the pollinating wasp and floral development within the host syconium (I. B., from E. Herre, K. C. Jandér, and A. Machado, "Evolutionary ecology of figs and their associates: Recent progress and outstanding puzzles," *Annual Review of Ecology and Systematics* 39 [2008]: 439–58). *Bottom,* illustrations of life history stages: *left,* adult strangling fig at Hua Kha Khaeng, Thailand (H. H.); *left center,* F. *variegata,* Sabah (P. A.); *center right,* syconium interior, F. *retusa,* Florida (D. L.); and *right,* female agaonid wasp emerging from syconium of F. *crassiramea,* W. Kalimantan (© Tim Laman).

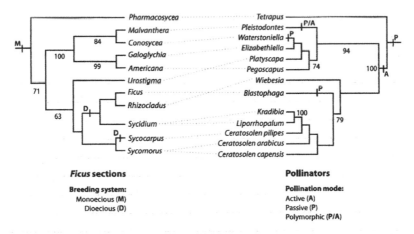

Fig. 11.11 Co-phylogenetic history of main sections of *Ficus* and genera and species of their wasp pollinators. Numbers by branches are bootstrap values, indicating the strengths of relationships (E. Herre; see the article cited in the caption for fig. 11.10).

Fig. 11.12 Bees and pollination. *Top, Apis dorsata* (the giant honeybee): *left,* nests on main branches of *Koompassia excelsa* crown, a common sight in Malaysia and Borneo (Wikimedia: T. R. Shankar Raman); *center,* nest close-up, ~1 m long (Wikimedia: Sean Hoyland); *right,* pollinating *Phyllanthus emblica,* Mudumalai (H. H.). *Bottom, Trigona* bee pollination; *left,* bee approaching *Shorea siamensis* flowers (H. H.); *bottom center,* nest (~1 cm dia. entrance), Pasoh, W. Malaysia (S. L.); and *right,* pollination of *Nephelium* sp. (S. L.)

and other bees whose diversity co-evolved with important plant groups in the Neotropics. The closest equivalent in Asia is the small, solitary bee genus *Amegilla,* whose species forage long distances but are not exclusive pollinators of any tree. Asian pollinating bees are overwhelmingly in three categories: Xylocopidae (pantropical bumblebees), Apidae (the honeybees, which are indigenous to the Old World), and Meliponidae (the pantropical sweat bees). None of these are truly species-rich in Asia.

Bees, like bats, feed on nectar but also rob pollen to feed their larvae. Bee flowers offer excess yellow pollen as an attractant (fig. 11.12). The honeybees (*Apis*) and the big xylocopid bees forage in full sunlight, although some xylocopids may be nocturnal. They have eyesight of appropriate spectral sensitivity.

Xylocopa visit flowers along riversides and roadsides, in secondary forest and large canopy gaps, and among the crowns of canopy species. They are solitary foragers and are said to range several kilometers on a flight. They visit a narrower range of flowers than honeybees, at intervals, and they probably trapline. They are fond of large, mauve, white, and yellow-orange flowers that liberally yield nectar. During mass flowering events, they desert the roadsides for the canopy.

Honeybees visit a wide range of canopy flowers of varying size, many of which offer less nectar than xylocopid flowers. They occasionally visit bat- and bird-pollinated flowers, but they are unlikely to pollinate these effectively. They are ever-present above and around the forest, but the giant honeybee, *Apis dorsata,* becomes particularly prevalent during mass flowering years and during the flowering season in the seasonal humid tropics, after which its nests are replete with pollen and honey. Honeybees are opportunistic pollen and nectar foragers. It takes time for a hive to build up workers in response to increased floral resources—much longer than it takes a population of thrips to expand. Honeybees, like wild boar (p. 252), appear in unusual numbers when food becomes abundant—in this case, during mass flowerings. Hives quickly move to locales with many trees in flower, but it is unlikely that their numbers can be sufficiently increased by reproduction alone within the duration of these events to qualify them as lead pollinators.

The sweat bees, Meliponidae, are less diverse in Asian than in Neotropical forests, being represented by few genera, of which *Trigona* is the most important (29 species in Peninsular Malaysia and Sarawak). They are even more opportunistic feeders than honeybees. Although they are most numerous below the forest canopy, they also visit honeybee canopy flowers in smaller numbers, including those of emergent dipterocarps. Individual species appear to have some floral preferences. Their nests, entered by a resin vestibule, characteristic in form to each species, are built up in layers, each layer studded with the pollen grains currently being collected (fig. 11.12). Careful examination of this pollen might reveal the seasonal floral preferences of each bee species. Sweat bees are generally weak flyers, and workers of each colony probably stay within a restricted area.

Rainforest tree species maintain high genetic diversity within continuous habitats in their populations, even when restricted to ecological ranges of extraordinary narrowness and precision Natural selection has, through pollination, played a dominant role in the dynamics and structure of the forest and in the evolution of its species. There is much to learn, but opportunities to rigorously investigate these interrelationships are fast declining as lowland forests are fragmented and destroyed.

Notes

1. *OTFTA*, chap. 5, 309–60.

2. Bawa, Kamal S., "Breeding systems of tree species of a lowland tropical community," *Evolution* 28 (1974): 85–92; Bawa, Kamal S., D. R. Perry, and J. H. Beach, "Reproductive biology of tropical rainforest trees. 1. Sexual systems and incompatibility mechanisms," *American Journal of Botany* 72 (1982): 331–45.

3. Williams, Evelyn W., Elliot M. Gardner, Robert Harris III, Arunrat Chaveerach, Joan T. Pereira, and Nyree J. C. Zerega, "Out of Borneo: Biogeography, phylogeny and divergence date estimates of *Artocarpus* (Moraceae)," *Annals of Botany* 119 (2017): 611–27.

4. Bruijning, Marjolein, et al., "Surviving in a cosexual world: A cost-benefit analysis in tropical trees," *American Naturalist* 189 (2017): 297–314.

5. Kenzo, Tanaka, et al., "Overlapping flowering periods among *Shorea* species and high growth performance of hybrid seedlings promote hybridization and introgression in a tropical rainforest of Singapore," *Forest Ecology and Management* 435 (2019): 38–44; Kamiya, K, Y. Y. Gan, K. Y. Lim, M. S. Khoo, S. C. Chuah, and N. H. Faizu, "Morphological and molecular evidence of natural hybridization in *Shorea* (Dipterocarpaceae)," *Tree Genetics & Genomes* 7 (2011): 297–306; Lum, Shawn, and Ngo Kang Min, "Lessons in ecology and conservation from a tropical forest fragment in Singapore," *Biological Conservation* 254 (2021): 108847.

6. Fleming, Ted H., and W. John Kress, *The Ornaments of Life: Coevolution and Conservation in the Tropics* (Chicago: University of Chicago Press, 2013).

7. Kobayashi, Masaki L., Yayoi Takeuchi, Tanaka Kenta, Tominori Kume, Bibian Diway, and Kentaro K. Shimizu, "Mass flowering of the tropical tree *Shorea beccariana* was preceded by expression changes in flowering and drought-responsive genes," *Molecular Ecology* 22 (2013): 4767–82; Weston, David J., Stan D. Wullschleger, and Gerald A. Tuskan, "Extending the *Arabidopsis* flowering paradigm to a mass flowering phenomenon in the tropics," *Molecular Ecology* 22 (2013): 4603–05.

8. Kurten, Eric L., Sarayudh Bunyavejchewin, and Stuart J. Davies, "Phenology of a dipterocarp forest with seasonal drought: Insights into the origin of general flowering," *Journal of Ecology* 106 (2017): 126–36; Satake, Akiko, Yu-Yun Chen, Christine Fletcher, and Yoshiko Kosugi, "Drought and cool temperature cue general flowering synergistically in the aseasonal tropical forests of Southeast Asia," *Ecological Research* 34 (2019): DOI: 10.1111/1440–1703.1012; Chen, Yu-Yun, et al., "Species-specific flowering cues among general flowering *Shorea* species at the Pasoh Research Forest, Malaysia," *Journal of Ecology* 106 (2017): 586–98; Ashton, Peter S., Jacqueline Heckenhauer, and V. Prasad, "The magnificent Dipterocarps: Précis for an epitaph?" *Kew Bulletin* 76 (2021): 87–125.

9. Momose, Kuniyasu, et al., "Pollination biology in a lowland dipterocarp forest in

Sarawak, Malaysia: Characteristics of the plant-pollinator community in a lowland dipterocarp forest," *American Journal of Botany* 85 (1998): 1477–1501.

10. Heckenhauer, Jacqueline, et al., "Phylogenetic analyses of plastid DNA suggest a different interpretation of morphological evolution than those used as the basis for previous classifications of Dipterocarpaceae (Malvales)," *Botanical Journal of the Linnean Society* 185 (2017): 1–26; Heckenhauer, Jacqueline, R. Samuel, P. S. Ashton, K. Abu Salim, and O. Paun, "Phylogenomics resolves evolutionary relationships and provides first insights into floral evolution in the tribe Shoreeae (Dipterocarpaceae)," *Molecular Phylogenetics and Evolution* 127 (2018): 1–13.

11. Start, Anthony, "The feeding biology in relation to food source in nectivorous bats (Chiroptera: Macroglossinae) in Malaysia," PhD thesis, University of Aberdeen, 1974.

12. Weiblen, George. D., "Correlated evolution in fig pollination," *Systematic Biology* 50 (2004): 128–39.

The bearded pig of Borneo (Sus barbatus) was cele-brated for its migrations in search of seed crops, which it consumed and dispersed. To observe hundreds of these stocky creatures, the boars with fearsome incisors, hurtling down a steep slope and across the rocky bed of a narrow inland torrent—like a crowd trying to catch the last bus—was one of the most thrilling and signal sights of inland Borneo. Now, however, they are in pathetic decline, as the forests have been converted or trashed by loggers and overhunted. Whereas the lowland forests experience mast fruiting years, the lower montane forests are less affected. From March-May the acorns fall abundantly in the hills. Hordes of pigs migrate tens or hundreds of kilometers up from the lowlands, fording the cataracts in their quest for a feast. Senkelang told me that this is the time to shoot the sows, who are waxing fat and nubile as their scrawny lovers give them the best Fagaceous treats. But when the sows are pregnant, less lovely and scrawnier, their former admirers lose interest ("just like us"): it is their turn to go partying, but only, it seems, if there is a mast. — P. A.

12

TREES AND THEIR MOBILE LINKS: DISPERSAL AND SURVIVAL

Mobile links that increase plant fitness include organisms that disperse seeds and enhance their survival. Other mobile links, including organisms that feed on seedlings and cause disease, may decrease fitness.[1] In the high diversity of the tropics, the rarity of many species and their separation from congeners (p. 223) affect these mobile links. The equable climate and lack of physical constraints on seedling growth and survival also increase the importance of mobile links. What are they? They are dispersers of seeds, insects that defend them against herbivores, and the spores of symbiotic fungi. Others *reduce* fitness, such as herbivores and predators of seeds or of plant tissues (especially of young plants), and pathogens. These links tend to exert strong—though indirect—influences on plant diversity. As primary producers, plants synthesize a vast array of complex molecules, but they are less diverse (and complex) than the products of fungi and of microbes themselves.[2] These may act as specific attractants to dispersers, or as detractants against herbivores, predators, and pathogens. In this way, animals and microorganisms—dependent on plant biomass for food—may themselves be the major mediators of interspecific interactions among plants in diverse communities.

If, therefore, plant species within diverse communities are niche-specific, the competition between them that defines their niches could largely be mediated by mobile organisms. However, the species richness of many mobile linkage groups is much lower in Asian than Neotropical rainforest communities. This is true even of lowland Asian MDF, where tree species richness is comparable to that of the Neotropics while stature and attendant structural complexity is generally greater.

Certain other plant-dependent groups are particularly species-rich in tropical Asia, including the ectotrophic mycorrhizae. Toward the margins of the tropics, species richness of vertebrate browsers and grazers is greater in Asia than in the Neotropics, and comparable to that in parts of Africa. Dispersing organisms in species-rich tree communities tend to service multiple plant species. Over time, tree species are likely to evolve characteristics, such as phenological differences that attract their undivided attention.

Studies of the dispersal by mobile links are critical, but observations are only feasible where seed numbers are greatest: around the source. The tail of declining abundance from that source is often impossible to document directly. The occasional seed or fungal spore that disperses at an exceptional distance and germinates successfully may redefine the range of a tree species or a gene. These abundances and distances must be inferred from indirect genetic evidence.

Seed Shadows

There is a theme that binds much of this research together: the Janzen-Connell hypothesis. It was first described by J. B. Gillett, and later independently by the entomologist Dan Janzen and marine ecologist Joe Connell. They all hypothesized that seed and seedling predation may be a major cause of tree species diversity in tropical forests.[3] The seed predators and dispersers are specialist natural enemies of the trees and are distance- and density-responsive. They argued that host-specific predators would preferentially consume easily located seeds and seedlings at high density next to the parent tree. Predation would reduce density, opening new vacant spaces for species that they do not attack. The Janzen-Connell hypothesis has also been named the pest pressure and the seed shadow hypothesis. This is one of the most important ideas of the past century in tropical biology and a core concept in this chapter.

Seed Dispersal

The dispersal of a single seed and subsequent successful recruitment of one individual to reproductive maturity is equivalent to the dispersal of millions of pollen grains a similar distance. Seeds of tropical forest trees are dispersed over short distances compared with those of temperate forests. Limited seed dispersal leads to the clumping of juveniles around mother trees. A few of these juveniles survive and become groups of reproductively mature individuals. Among species that have higher fecundity than the average, this can also lead to local aggregations of a species through "reproductive pressure," independent of interspecific competitive advantage in relation to the physical habitat (in *Dryobalanops* for example, see p. 324). To date, few comparative data for seed production among species have been gathered from tropical forests, though some conclusions can be drawn from the relation between the relative abundance of species and their fruiting phenology and effectiveness of seed dispersal.

Tropical versus temperate tree seed dispersal. Hubbell's unified neutral theory of biodiversity and biogeography (p. 308) arose partly in response to the contrasting predominance of light, wind-borne seeds among temperate canopy tree genera, especially conifers, whose seeds are widely dispersed, compared to trees in the tropics.[4]

A major difference between temperate forests and tropical rainforests is the richness of the tropical subcanopy, including understory species. In Asian forests, animal-dispersed fruits and seeds occur predominantly among species of the main canopy and understory. Many subcanopy tree genera bear fleshy fruits of moderate size, and these are dispersed by birds and mammals, but a surprising number are barely fleshy or dry. In the rainforest understory, small fruits with small seeds predominate. The winged fruits of emergent dipterocarps are generally dispersed by gyration, sometimes enhanced by squalls,[5] but often halted in the nearby forest canopy. Some emergent durian species bear fruits beneath their crowns, on the trunks as well as branches. Emergent leguminous trees bear dull-colored pods and seeds; some may be dispersed by ungulates, but *Koompassia* fruits are thinly flanged and gyrate.

The supra-annual mast fruiting of the canopy guilds of MDF are synchronized to the same season in all participating families, unlike the oaks in North American forests. Mast fruiting (p. 103) is weak in the MDF understory, where fruit production, though varying from year to year, does occur annually, and species are less synchronized.

The seed dispersers. The diversity of dispersing animals is, of course, far greater throughout the continental tropics than in temperate climes. In tropical Asia, the predominant frugivores and seed dispersers include elephants, tapirs, ungulates, rhinoceros, bats, monkeys and other primates,

Fig. 12.1 Animal dispersal agents. *Top left*, an orangutan enjoys the fruit of *Polyalthia hypoleuca* (© Tim Laman); *top center*, an Asian koel purloins fruits of *Diospyros montana* (H. H.); *top right*, a young rhinoceros hornbill tosses a fig of *Ficus dubia* into its gullet, Gunung Palung, Kalimantan (© Tim Laman); *bottom left*, a red leaf monkey with a fig of *F. dubia*, Gunung Palung (© Tim Laman); *bottom center*, giant Malabar squirrel on *Grewia tiliifolia*, Mudumalai (H. H.); *bottom right,* a long-tailed macaque feeding on a *Ficus stupenda* crown, Gunung Palung (© Tim Laman).

and civets, and among birds, they include the barbets, broadbills, hornbills, fruit pigeons, corvids, muscicapids, bulbuls, white-eyes, laughing thrushes, and flowerpeckers (fig. 12.1). Less rich in frugivorous birds than the Neotropics,[6] the Asian tropics are better endowed with browsing and seed-dispersing mammals, comparable to Africa. Fish, ants, and even tortoises and monitor lizards have been implicated. Frugivores consume the pulp surrounding seeds but not the generally toxic seeds themselves.

Birds, fish, and primates can sense color, and seeds dispersed by them are often vivid red, blue, or black. The diversity of size and form in tropical forests is extraordinary (fig. 12.2). These colored fruits and seeds stand out against the prevailing green within the canopy, presumably attracting the disperser. Many subcanopy legumes, Euphorbiaceae and others, yield seeds with contrasting black and red coats or attached arils, dangling on long

Fig. 12.2 Examples of fruit displays. *Top left, Myristica fragrans* (Myristicaceae, nutmeg) red aril (mace) on brown seed (Wikimedia); *top center*, iridescence, *Elaeocarpus angustifolius* (Elaeocarpaceae, D. L.); *top right, Harpullia pendula* (Sapindaceae), red fruit and black seed (D. L.); *bottom left, Sterculia ceramica* (Malvaceae) red fruit and black seed (D. L.); *bottom center, Polygala oreotrephes* (Polygalaceae), black seed, orange aril, purple fruit (D. L.); and *bottom right*, blue fruit and red calyx, *Clerodendron minahassae* (Verbenaceae; D. L.).

funicles after the fruit has split open. The brilliant seeds of some *Sterculia* fruit, released from five dehiscing, spreading, satiny vermillion follicles, is named by Iban Dayaks *buah ayan antu sebayan*: "fruit stared at by the spirits of the dead."

Asian rainforests in both everwet and seasonal climates support lower biomasses of vertebrate-dispersed fruits than do forests of the Neotropics. The years between the fruiting that follows mass flowering events are often years of famine for frugivores. One-third of sun bears died during such a famine period in Sabah, while orangutans suffered—and some died. Deposition in dung may improve germination and seedling growth.

Seed dispersal distances. Much more is currently known about dispersal around tropical than temperate trees, particularly because the enormously greater species diversity of lowland tropical rainforest enables easier monitoring of seed rain from isolated trees of individual species.

The average dispersal distance of a species' seed is correlated with the foraging distances of the dispersing agents. The winged fruit of most dipterocarps is not dispersed more than 100 m, a finding supported by genetic analysis. Few will disperse even half that distance in the generally windless climate and dense canopy of MDF. The wingless seed of many dipterocarps, as well as other species that seem to have no consistent disperser, will mostly fall within the area of the maternal crown. The several genera of Euphorbiaceae, mostly of the subcanopy, with hard, shiny seeds bereft of fleshy tissue and dispersed by the explosive opening of the pericarp, are generally dispersed over a few meters, which is reflected in their highly clumped distributions (fig. 12.3). Seeds furnished with elaiosomes, such as those of many *Macaranga*, are dispersed by ants, also limited in distance. Ants disperse seeds of different plant species, according to the size of the seed relative to that of the ant. In contrast, bats and hornbills may easily carry fruit a distance exceeding 1 km to roosting sites, where they eat them and defecate; these germinating seedlings may often be even more densely clumped than if they had fallen directly from the tree crown. Macaques likewise sit in tree crowns with cheek pouches full of fruits, the flesh of which they remove at leisure without leaving the tree. Some seedlings survive to maturity, explaining the frequent clumping of mature trees of the same species in species-rich forests.

David Chivers documented the social and feeding ecology of the siamang gibbon (*Hylobates lar*) at the Krau Game Reserve, Pahang, and one other site in Peninsular Malaysia.[7] Siamang consume fruit, leaves, and occasionally even small animals. Figs comprised 13% of trees visited, but they yielded the dominant food. One gibbon explored ~8 ha of forest each day, with a maximum radius of 700 m. Gibbons and orangutans swallow fruit and travel up to 1 km before defecating; most of those seeds germinate.

Seeds that disperse and germinate over longer distances expand a species range and maintain breeding populations in rich forests and heterogeneous terrain. Elephants, which feed on a variety of fruits, have a vast home range, estimated at 170 km[2] in Peninsular Malaysian MDF. Their food is particularly limiting and widely dispersed, with a ~60 km[2] home range in more productive secondary forest. Rhinoceros ingest the large, dull fruit of *Trewia nudiflora* in tall deciduous forest, potentially dispersing it considerable distances. Occasional successful long-distance dispersal, critical to speciation in continuous landscapes, may be more frequent than supposed. Scatter-hoarding of dipterocarp seed by rodents probably plays a larger part in their dispersal than expected, the seed sometimes being carried tens of

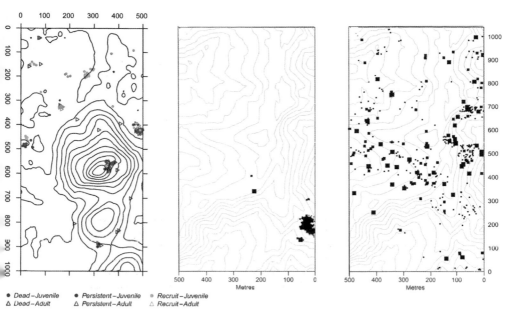

Fig. 12.3 Using plot censuses to document dispersal. *Left*, population distribution of *Koilodepas longifolium* (Euphorbiaceae), with explosively dispersed seeds, ForestGEO plot, Pasoh, shows clumping but some longer range dispersal (P. A ; see N. Manokaran et al., *Stand Tables and Distributions of Species in the 50 ha Research Plot at Pasoh Forest Reserve*, Forest Research Institute Malayan Research Data Series I [Kepong: Forest Research Institute of Malaysia, 1992]); *center*, *Shorea geniculata* at Lambir, Sarawak: an emergent with large, wingless fruit, is clumped (P. A.; see H. S. Lee et al., cited in fig. 9.4 caption); *right*, *S. scrobiculata*, a main canopy species with small, winged fruit, is well dispersed in the plot; dots, trees 1–10 cm; small squares, 10–30 cm; large squares, >30 cm DBH (P. A.; see preceding citation in this caption).

meters from the tree. Many dipterocarp fruits can be surprisingly widely dispersed in squalls by gyration induced by their twisted, wing-like sepal lobes and the large surface of wing relative to seed, but the large majority land near the source (fig. 12.4).

Since chloroplast DNA is solely transmitted by seed, but nuclear markers are transmitted by pollen as well, the relative dispersal distances of seed and pollen can be evaluated. Bee-dispersed pollen increases the distance of fine-scale genetic structure (FSGS) compared to that dispersed by thrips, and species with small, winged seed increase that for female-inherited FSGS relative to species with heavy seed (fig. 12.3).

Certain forms of fruit and seed dispersal correlate with rarity, providing some predictability to the rank order of species abundances in rainforest

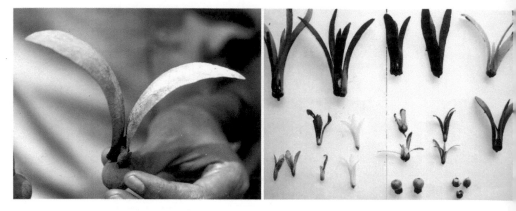

Fig. 12.4 Gyrating dipterocarp fruit. Gyration is dependent upon the total area of wings, in relationship to fruit mass, or wing-loading. Higher values decrease terminal velocities and increase the distance from the parent tree. *Left*, two well-developed wings (~12 cm long) of *Dipterocarpus indicus*, Kerala, India, suggest greater dispersal distance (D. L.); *right*, a sample of dipterocarp fruits from Lambir, in Sarawak; these vary greatly in wing loading, and include some with no wings at all (© Christian Zeigler).

communities, and contrary to some predictions neutral theory (p. 308). Pioneer species with wind-dispersed seed occupy temporary island habitats of low relative area and are therefore consistently in low density in primary forest. Sugary-fruited tree genera of the continuous udult clay and sandy clay soils of the everwet Sundalands are mostly widespread, with low local endemism. Species bearing large fruit with vertebrate-dispersed seeds, including *Mangifera*, many *Durio* species, and the palm *Pholidocarpus*, produce few fruit and seeds, and are consistently at low density in primary MDF (fig. 12.5). Small-fruited species in these same genera, such as *Mangifera parvifolia* and species of *Durio* are equally widespread and locally abundant.

A long-standing debate among biogeographers concerns whether speciation, which is assumed to occur when a barrier divides metapopulations, is most often driven by the occasional dispersal of founder populations to hitherto unoccupied yet suitable habitats, or whether it occurs through vicariance—the division of the habitat itself, as by tectonic drift, sea-level rise, or river capture. In dispersal, we emphasize factors important in determining the first driver, but the second is equally probable when circumstances have favored it (p. 308).

If species within a community are ecologically complementary and enjoy equal reproductive advantage locally, they may spread over time simply by persisting within communities by chance and continually generating new

Fig. 12.5 Large, green, and smelly tropical forest fruit, dispersed by large animals. *Left,* wild durian (*Durio graveolens*, Malvaceae, ~12 cm wide) collected from forest in Selangor, Malaysia; *bottom center*, creamy aril covered seeds exposed, *D. zibethinus*; *top center*, chulta, or elephant apple (*Dillenia Indica*, Dilleniaceae, ~11 cm wide); *right*, jak or jackfruit (*Artocarpus heterophyllus*, Moraceae, a compound fruit ~18 cm wide) represents a genus of 60 tree species widespread in tropical Asian forests (all photos D. L.).

patches of local numerical dominance. Such a pattern of restricted ranges of seed dispersal could facilitate species coexistence, and thus be a leading determinant of floristic structure and species richness. This hypothesis forms the basis of Stephen Hubbell's book *Unified Neutral Theory of Biodiversity and Biogeography* (p. 308). Hubbell has demonstrated that this is both theoretically feasible and consistent with the structure and dynamics of the species-diverse rainforest community samples as large as 50 ha, censused in the ForestGEO plot network (p. 430). According to the theory, seed dispersal limitations should also influence speciation rates and diversification within continuous landscapes. Local fortuitous extinction would exceed competitive extinction in frequency, leading to local populations potentially becoming isolated long enough to become genetically distinct, even to speciate. Such random patterns of species distribution could obscure the coinciding ranges of species limited by common ecological barriers to dispersal. Neutral theory thus predicts that rank order of abundance of species in a community should vary unpredictably. What evidence we have suggests the opposite among some more abundant species, and among some that persist in low population densities.

Dispersal and genetic differentiation. While very small populations suffer inbreeding depression through increasing homozygosity,[8] small populations of freely outcrossing species may maintain high levels of heterozygosity. Isolated but expanding small populations can rapidly diversify, either at random or through selection. Formerly, random effects were predicted to dominate unless selection was stringent, and the likelihood of advantageous genes spreading in natural populations was less than 1%. Selection is now known to be much higher. Trees are characterized by long life spans, high outcrossing rates, and moderately low numbers of highly stable chromosomes (although polyploid series do occur in some rainforest families). Heterozygosity has been found to reach levels at least as high as those in temperate trees.

Other Mobile Links

Organisms influence the establishment of seeds in other ways. Herbivores, large and small, consume seedlings. Other organisms, mainly ants, may defend against the attacks of the herbivores. Spores, passively mobile, may infect seedlings, and fungal spores may establish connections with roots of trees, particularly dipterocarps and legumes, enhancing the survival of seedlings.

Herbivory. Herbivores, from insect larvae to elephants, may influence plant competitiveness by reducing their photosynthetic capacity, but this only affects species competitiveness if the herbivore is relatively specific to it, unless the plant is already rare. Elephants are well known to prefer some browse over others; they can therefore influence the demography of the browsed species (fig. 12.6). Harvesting of tree juveniles by pigs for nest building is a major cause of mortality in some forests but is generally nonspecific. Most specific herbivores are insects. The diversification of the major insect groups from the Cretaceous period onward is broadly correlated with the diversification of flowering plants and of their chemical defenses and attractants.

Insect predation of seeds and seedlings. An overall pattern of species' juvenile mortality positively correlated with their spatial density has been demonstrated in both Neotropical and Asian forests (p. 264). Few vertebrate seed predators or seedling browsers are host-specific. Janzen was studying Bruchid beetles, which are host-specific predators of leguminous trees. So far though, few other host-specific insect seed or seedling predators have been examined. Although they may be expected to exist for tree taxa with chemically defended seeds, their influence on tree population fecundity and

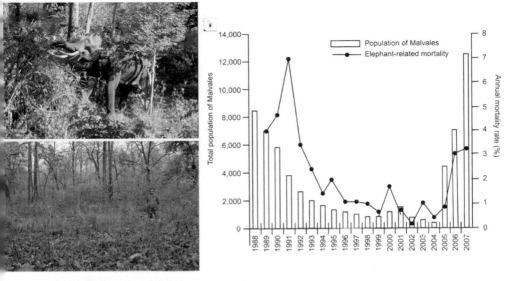

Fig. 12.6 Elephants at Mudumalai, India. *Top left*, browsing in forest understory (P. A.); *bottom left*, *Helicteres isora* regeneration after browsing, (C. Lo.); *right*, annual mortality rates and total reduction in Malvales (*Kydia calycina*, *Helicteres isora*, *Grewia tiliifolia*, and *Eriolaena quinquelocularis*) in the ForestGEO plot (I. B., after H. Suresh; see H. S. Suresh, H. S. Dattaraja, N. Mondal, and R. Sukumar, "Seasonally dry tropical forests in Southern India: An analysis of floristic composition, structure and dynamics," in *The Ecology and Conservation of Seasonally Dry Forest in Asia*, ed. W. J. McShea, S. J. Davies, and N. Bhumpakphan [Washington, DC: Smithsonian Institution Scholarly Press, 2011], 37–58).

demography remains unknown. In species-diverse MDF, the lean periods between mast fruiting years strictly limit predator population sizes and host-specific seed predation among tree species that fruit at these times. Dipterocarp seeds are predated by several species of weevil, none of which are host-specific.

Seed predation by weevils may nevertheless be a major cause of reduced fecundity of *Dipterocarpus* and other dipterocarp species that flower annually in the seasonal tropics. It destroys the out-of-mast fruiting in the everwet tropics, including that of species such as *D. crinitus* that frequently flower out of season. Insect predation fluctuates from year to year and increases during mast years. But the *proportion* of seed destroyed by beetles, like that destroyed by vertebrates, is reduced in mast fruiting years, as the beetle populations cannot increase prior to the event. The predation of seeds and seedlings may cause local extinctions,[9] but it is unlikely to produce more

Fig. 12.7 Large-scale forest destruction by herbivores and disease. *Left,* aerial photograph of regenerating peat swamp forest in a gap attributed to crown predation of the monodominant emergent *Shorea albida* by a hairy caterpillar (Google Earth); *center,* *Cinnamomum ovalifolium,* leaves, dead but still attached, Horton Plains, Sri Lanka (H. H.); *right,* dead mature *Mesua ferrea* in the Sinharaja ForestGEO plot, Sri Lanka (H. H.).

widespread species extinctions in Asian forests, nor to be a primary cause of density-dependent, species-specific seed mortality. Comparative studies between forests are needed, for which the ForestGEO plots are well suited.[10]

No examples of herbivore-induced, species-specific reduced survivorship or mortality have been discovered in Asian rainforests, though they may be expected to exist. During the 1960s, sharp-edged canopy gaps, sometimes more than 1 km in diameter, appeared in monodominant *Shorea albida* stands in Sarawak and were attributed by villagers to a hairy caterpillar (fig. 12.7).

The longhorn beetles (Cerambycidae), a collecting interest of Alfred Russel Wallace, are extraordinarily diverse in tropical Asia. Host specificity varies among species, though narrow specificity is unlikely. Overall, the host specificity of herbivores is no higher in tropical rainforests than in temperate forests, and the high species diversity of insects in tropical forests is due to the high diversity of their multiple plant hosts. Lepš and colleagues found that the relatedness of host plants explained 56% of variation in herbivorous insect diversity in New Guinea, whereas habitat (including succession) explained only 4%.[11] This implies that host specificity is at a higher systematic level than species.

Large animals. Browsing and grazing mammals and their predators are not abundant in equatorial rainforests, and they increase with seasonality. Abundance is greatest in deciduous forests, where periodic fire supports a partially herbaceous field layer and abundant young sucker shoots. There, elephant, rhinoceros, and ungulates, including gaur, wild cattle, and several species of deer, are (or have been) major browsers of tree juveniles (figs. 12.6 and 12.8).

Mammalian herbivory has also been limited in Asia, in comparison with Africa, by the smaller area of natural savanna and grassland. Today, human-induced fire has created widespread, species-poor savanna. Excessive hunting and cattle grazing and browsing have further reduced indigenous mammal biomass in Indo-Burmese forests, and they may alter forest succession. Exclosure experiments have shown how large animals affect forest structure and function. Large animals prevent an over-abundance of lianas. They are also important consumers of large, often odorous, fruits—and thus are dispersers of their seeds. Such seed dispersers can even promote growth and add to the carbon storage in a forest (fig. 12.5).[12]

The Asian elephant. By far the largest of Asian browsing and grazing mammals, the elephant is justly celebrated for its intelligence and calm dignity—maintained most of the time! In the Hindu-Buddhist tradition, the elephant deity Ganesh is the spiritual manifestation of wisdom with humility (the lofty condition to which we researchers aspire). Elephants are generalists, eating in bulk to gain the ingredients needed to sustain health. Browse is essential, for it supplies more than 70% of the amino acids required in their diets. Bark is high in many nutrients, though grass provides more calcium. Elephants occur from desert fringes to equatorial rainforests, usually below 1,000 m, but their optimal habitat is tall deciduous forest. In

Fig. 12.8 Large browsers and seed dispersers. *Left*, guar browsing *Lantana*, Mudumalai (H. H.); *center*, sambur deer browsing *Lantana*, Mudumalai (H. H.); *right*, orangutan consuming *Dipterocarpus sublamellatus* fruit, Borneo (© Tim Laman).

Peninsular Malaysian rainforests, they have been found to consume as many as 400 plant species, 350 palms and rattans, despite the generally low nutrient content of this foliage. They prefer secondary and riverine forest habitats, where they find many pioneer and successional trees whose foliage is comparatively rich in nitrogen and whose fruits are fleshy and nutritious. Elephants must consume a staggering 1.5–2% of their own body weight in forage daily—or 60 kg fresh weight. A herd of 200 individuals can consume 22,800 t fresh weight of vegetation from 20 km^2 per year. They migrate between the forest types and with the seasons, and their home ranges may thus reach 3,900 km^2.

In South Asian deciduous forests, elephants may consume 112 plant species, but concentrate on 25. These include Leguminosae, and subcanopy trees and shrubs in the Malvaceae, notably *Kydia calycina* and *Helicteres isora* (fig. 12.6). Other preferred foods include palms, bamboos, grasses, and sedges. Elephants apparently move from one preferred forage to another over several years, producing a pattern of cyclical recovery that recalls an erratic, Lotka-Volterra predator-prey cycle: the Caughley cycle. Elephants eat the fruits of many species and are important dispersers of large-seeded trees. Asian elephants do little harm to canopy trees and only occasionally push over smaller trees, often to strip their bark. In evergreen forests especially, their apparently indiscriminate depredations, often pushing over trees for no obvious purpose, can lead to semi-permanent patches of secondary forest or even grassland—which they favor.

The wild boar. Pigs are omnivorous, and are major predators both of seeds, which they must occasionally also inadvertently disperse, and saplings. *Sus scrofa,* the domestic pig of temperate Eurasia, extends throughout the continental Asian tropics, and is only domesticated in the Malesian archipelago. The native Peninsular Malaysian *Sus cristatus* is sedentary, roaming only locally. Where forest remnants are surrounded by oil palm plantations, pigs have a ready subsidy between mast years. This has produced high densities within such forest stands, where the sows destroy many saplings to build their nests. At Pasoh, wild domestic pigs have expanded into the forest, where their nest-making activities have killed saplings within dense populations of preferred tree species, showing that large mammals can cause conspecific density-dependent mortality (CNDD; p. 329).[13] Canopy species, including lianas, are most affected. Soil grubbing and seed predation also increase, which is likely to change MDF structure and dynamics toward

Fig. 12.9 Ants and plants. *Left, Dipterocarpus tuberculatus* seedling leaf with sugar glands and a defending ant, Mai-Ping National Park, Thailand (H. H.); *center,* young shoot of *Macaranga bancana* with symbiotic ants, Singapore (W. L.); *right,* hollow stem of *M. beccariana* with *Crematogaster* ants attending scale insects (T. I.).

those in seasonal evergreen forests. The bearded pig of Borneo, *Sus barbatus,* was celebrated for its migrations (p. 252).

Insects as defenders. Many tree species possess extrafloral nectaries—on leaves and young shoots, or other glands that attract insects, particularly ants, that are predators on other insect herbivores (fig. 12.9). Dipterocarp seedlings possess extrafloral nectaries which secrete a sugary liquid that attracts ants that defend the young leaves. In *Shorea acuminata* and perhaps other species, the stipules, which are cupped and do not fall immediately, similarly secrete and thereby attract aphids, which ants "milk."

Among these symbiotic relationships, those between ants of the genus *Crematogaster* and pioneer species of *Macaranga* are the best known in Asia.[14] All *Macaranga* possess extrafloral nectaries along their leaf margins; these only secrete sugars in the non-myrmecophytic species that lack strict mutualism and attract a diversity of ants (fig. 12.9). The mutualistic myrmecophytic *Macaranga* are populated by single *Crematogaster* ant infraspecific taxa, which obtain their sugars indirectly by milking aphid colonies housed within the host trunk.

Some tree species, including *Macaranga,* attract ants by producing energetically costly, oily food bodies (fig. 12.9). The mutualistic gain is linked with a reduction in other defenses, both chemical and physical (such as leaf toughness), compared with other species. *Crematogaster* ants harvest the food bodies to feed their aphids. The ants are highly host-specific obligate symbionts, each dependent on local populations of a *Macaranga* species.

Pathogens. Current empirical and experimental research in the Neotropics implies that, by preventing tree species' population densities from exceeding low thresholds—as stipulated by the Janzen-Connell hypothesis—species-specific pathogens may provide the recurrent, unoccupied habitat spaces into which great numbers of species can fit. Phenolics and other defenses may promote disease resistance. Pathogen-induced, density-dependent mortality is concentrated among juveniles such as those in the less-rich, semi-evergreen forest at Barro Colorado Island in Panama. Current research on conspecific, negative density dependence (CNDD) by Lisa Comita and colleagues suggests that such mortality is less important at later life history stages, less important among closely related species, and less important in fragmented forests; it also suggests that interspecific variation in conspecific, negative density dependence can make species less likely to coexist. In the same species, accumulation of ectomycorrhizal fungi can contribute to density-dependent survival, while the accumulation of pathogens can contribute to density-dependent mortality.[15]

Mass mortality among mature trees of a single species, such as occurred among temperate elms and the American chestnut, has seldom been observed in tropical forests. One striking exception occurred in ~2001, at the ForestGEO demography plot at Sinharaja (Sri Lanka). There, the most abundant species is the canopy tree *Mesua ferrea* (fig. 9.2), locally dominant on high ridges. In that year more than one hundred mature individuals died in and near the 25 ha plot (fig. 12.7), but no more juveniles than would normally be expected. A sister species (*Mesua thwaitesii*, only slightly less abundant and concentrated in adjacent upper gullies) was unaffected. The cause of mortality is unknown, but a species-specific pathogen is suspected. A similar wave of mortality among mature trees occurred among the upper montane *Cinnamomum ovalifolium* on the nearby Horton Plains in 2007 (fig. 12.7).

Claire Elouard studied dipterocarp trees in Java with symptoms of disease: blotched leaves, defoliated saplings, root rot, and other symptoms in *Shorea* (4 species), *Vateria*, and *Hopea*. [16] She isolated cultures of *Fusarium* (3 species, including *F. oxysporum*, the cause of Panama disease in bananas) and *Nectrium* (*N. radicicola*, the cause of black foot disease in grape). She isolated the cultures, re-inoculated normal saplings, and observed the same symptoms. Pathogens, like pollen or seeds, are dispersed mostly within the local host population, yet occasionally over longer distances to other populations. In this way, mutual extinction is generally avoided. In the largely windless climate of species-rich forests, neither freely dispersing spores nor vectors will often be blown off-course. This could lead to a peak in the between-tree distances of the species, irrespective of the species richness of the forest. Trade winds and cyclones of the seasonal tropics should increase dispersal distances both of fungal spores and their insect vectors. These observations point to a major research priority, until now hardly addressed in Asian forests. Lisa Comita and colleagues[17] have shown that the strength of increases in density-dependent mortality lower the tree species' population density, implying that pathogens may be the leading determinants of consistent rank orders of abundance observed in dipterocarp forests. More recently, she and collaborators found a lower diversity of immunity (R) genes in rare species at Barro Colorado Island, in Panama.[18] These results suggest a genetic basis for disease resistance in these trees and its negative density dependence (NDD), which is further discussed in chapter 14.

Crucial symbionts. Bacteria and fungi are important plant partners. Most obvious are the nitrogen-fixing bacteria in root nodules within the Fabaceae. The species diversity of such trees is low in most tropical forests, and they are much less abundant in the Asian tropics than in the New World. Perhaps this is in part due to higher nitrogen content in soils of the former.[19]

Mycorrhizae are fungi that form close cellular associations with tree roots, allowing translocation of some minerals (notably phosphorus) in solution and increasing water uptake.[20] In exchange, they obtain carbon in the form of sugars synthesized by their photosynthesizing hosts, most of which benefit from the full sun of the canopy at maturity. The fungal hyphae vastly increase the effective absorbing surface area, and thus the volume of exploitable soil. Mycorrhizal hyphae can translocate nutrients between roots. Ectotrophic mycorrhizae (basidiomycete root symbionts whose hyphae do not penetrate living root cells, or EcM) are more host-specific than the vesicular arbuscular mycorrhizae (VAM), which are almost universal among tropical trees (fig. 12.10). Differences in mycorrhizae species, or species associations, may cause interspecific differences in translocation efficiency and even relative ion specificity, and may thus restrict translocation to conspecific individuals, particularly parent trees and juveniles. Ectotrophic mycorrhizae are found on all conifers and many temperate tree species, but among tropical tree families, they are predominant only in caesalpinioid legumes, dipterocarps, oaks, and gymnosperms. The angiosperm families

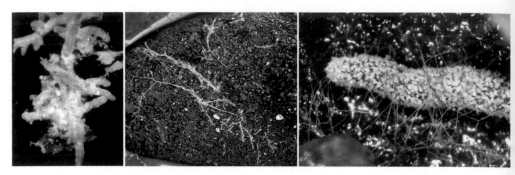

Fig. 12.10 Ectomycorrhizae (ECM) and dipterocarps. *Left*, *Scleroderma*-infected rootlets of *Shorea leprosula*; *center*, *Hopea odorata* roots and fungal mesocarps; *right*, *H. odorata* root tip with fungal mantle and projecting hyphae (all S. Le.).

(but not the conifers) have mast fruiting patterns in everwet tropical Asia and limited seed dispersal. These characteristics may facilitate fungal infection of the seedling roots, including direct infection from mother tree mycorrhizal mycelia. However, EcM do not help dipterocarp seedlings survive.

Ectotrophic mycorrhizae are also considered more efficacious than VAM in water and nutrient uptake. In the tropics worldwide, EcM tree taxa tend toward single-species—or single-family in the case of Dipterocarpaceae in MDF—canopy dominance in seasonal and low-nutrient habitats.

Mycorrhizal infection may also convey resistance to pathogenic fungal infection. This could be achieved indirectly by its production, through co-evolution with the tree, of specific chemical defenses. Species richness is correlated with an extraordinary diversity of EcM mushroom species. There are far more basidiomycetes in the everwet Asian lowlands than anywhere else in the tropics. Kabir Peay and colleagues,[21] using molecular methods of hyphal identification, analyzed EcM communities across the ecotone from udult clay to humult sandy soils within the 52 ha ForestGEO plot at Lambir National Park (Sarawak). Their generic representation and taxonomic diversity resembled that in temperate forests, and the humult soils contained 50% more ectotrophic fungal taxa. Species of ectomycorrhizae co-occur on phylogenetically related hosts, but the array of fungal species that comprise mycorrhizal associations may be host-specific. Not only is their diversity enormous, but variance in host species susceptibility to ectomycorrhizal infection is great. Hence, opportunities for diversification of co-evolutionary relationships, especially where soils vary, might be nearly endless. Recently, Peay and colleagues analyzed mycorrhizae in roots of seedling transplants

of the same species on different soils and found no evidence of host specificity.[22] Soil environment was the primary determinant of EcM diversity
and composition on seedlings. But specificity might well increase as the tree
grows, when interspecific competition increases as leaf and other physiologically significant differences appear. Mycorrhizae could thus be a primary
cause of both the canopy dominance of Dipterocarpaceae and Fagaceae in
tropical lowland and lower montane, as well as broadleaved temperate forests. While fungal pathogens are air-borne above the canopy or dispersed
by flying insects, mycorrhizae, living in soil and bearing carpophores that
arise from it in the windless understory, have limited dispersal opportunities in closed rainforests. Seedling infection by mycorrhizae may be positively density-dependent, sharing mycelia when seedlings are close enough.
Ectotrophic mycorrhizal symbiosis may thus oppose host-specific patho-
genicity by enhancing positive, density-dependent survivorship. Density-
dependent juvenile mortality has not always been evident among dipterocarp seedlings and, where present, may be due to other factors, such as
shading by the parent crown, while enhanced performance observed among
adjacent conspecifics has been attributed to sharing of host-specific mycorrhizae. Ingenious investigation will help us understand this enigma.[23]

Genetic and Species Diversity

Rainforest tree species maintain high genetic diversity within continuous
habitats in their populations, even when restricted to ecological ranges of
extraordinary narrowness and precision This diversity—widespread apomixis apparently notwithstanding—argues that natural selection has played
a dominant role in the dynamics and structure of the forest and in the evolution of its species. That is not to say, though, that chance does not also play
a role, as is obvious in the strikingly limited dispersal of most species' seeds
and the seedlings resulting from them.

Although we may expect that species diversity in tropical rainforests
would be the result of highly specific interdependencies between trees and
other organisms, that expectation has so far only occasionally been supported by research. Among the groups of organisms that directly influence
tree species diversity, the most notable are (1) those that disperse pollen
and seed, thereby enabling high genetic diversity; (2) the pathogenic microorganisms and fungi which, it is becoming clear, are major determinants of
community diversity; (3) the ectotrophic mycorrhizal fungi that appear to

have enabled dipterocarps in the lowlands (and Fagaceae in the mountains) to achieve the high levels of canopy dominance so noticeable in the Asian tropics; and (4) herbivores and their predators. Further research on these organisms will help to resolve the puzzles of the enormous tree diversity in these forests.

Notes

1. Citations seen in *OTFTA*, chap. 5, 361–79.

2. Coley, Phyllis D., and Thomas A. Kursar, "On tropical forests and their pests," *Science* 343 (2014): 35–36.

3. Todesco, Marco, and Quentin Cronk, "The genetic dimension of pest pressure in the tropical rainforest," *Molecular Ecology* 26 (2017): 2407–09; Connell, J. H., "Diversity in tropical rainforests and coral reefs," *Science* 199 (1978): 1302–10; Janzen, Daniel H., "Herbivores and the number of tree species in tropical forests," *American Naturalist* 104 (1970): 501–18; Hülsmann, Lisa, Ryan A. Chisholm, and Florian Hartig, "Is variation in conspecific negative density dependence driving tree diversity patterns at large scales?" *Trends in Ecology and Evolution* 36 (2021): 151–63.

4. Hubbell, Stephen P., *The Unified Neutral Theory of Biodiversity and Biogeography*, Monographs in Population Biology 12 (Princeton, NJ: Princeton University Press, 2001).

5. Smith, James R., Robert Bagchi, Chris J. Kettle, Colin Maycock, Eyen Khoo, and Jaboury Ghazoul, "Predicting the terminal velocity of dipterocarp fruit," *Biotropica* 48 (2016): 154–58.

6. Fleming, Ted H., and W. John Kress, *The Ornaments of Life: Coevolution and Conservation in the Tropics* (Chicago: University of Chicago Press, 2013).

7. Chivers, David J., "The siamang in Malaya: A field study of a primate in a tropical rainforest," *Contributions in Primatology* 4 (1974): 1–333; Chanthorn, Wirong, Thorsten Wiegand, Stephan Getzin, Warren Y. Brockelman, and Anuttara Nathalang, "Spatial patterns of local species richness reveal importance of frugivores for tropical forest diversity," *Journal of Ecology* 106 (2018): 925–35.

8. Wright, Sewell, "The interpretation of population structure by F statistics with special regard to systems of mating," *Evolution* 19 (1965): 395–420.

9. Bagchi, Robert, et al., "Pathogens and insect herbivores drive rainforest plant diversity and composition," *Nature* 506 (2014): 85–88.

10. Basset, Yves, et al., "A cross-continental comparison of assemblages of seed- and fruit-feeding insects in tropical rainforests: Faunal composition and rates of attack," *Journal of Biogeography* 45 (2018): 1395–1407.

11. Lepš, Jan, Vojtěch Novotný, and Y. Basset, "Habitat and successional status of plants in relation to the communities of chewing herbivores in Papua-New Guinea," *Journal of Ecology* 89 (2001): 186–99.

12. Sekar, Nitin, and Raman Sukumar, "Waiting for Gajah: An elephant mutualist's contingency plan for an endangered megafaunal disperser," *Journal of Ecology* 101 (2013):

1379–88; Corlett, Richard, "Frugivory and seed dispersal by vertebrates in the oriental (Indomalayan) region," *Biological Review* 73 (1998): 413–48; Chanthorn, et al., "Spatial patterns of local species richness"; Luskin, Matthew Scott, Kalan Ickes, Tze Leong Yao, and Stuart J. Davies, "Wildlife differentially affect tree and liana regeneration in a tropical forest: An 18-year study of experimental terrestrial defaunation versus artificially abundant herbivores," *Journal of Applied Ecology* 56 (2019):1379–88.

13. Luskin, Matthew Scott, Daniel J. Johnson, Kalan Ickes, Tze Leong Yao, and Stuart J. Davies, "Wildlife disturbances as a source of conspecific negative density-dependent mortality in tropical trees," *Proceedings of the Royal Society B* 288 (2021): 20210001.

14. Heil, M., B. Fiala, K. E. Lensenmair, G. Zotz, and P. Menke, "Food body production in *Macaranga triloba* (Euphorbiaceae): A plant investment in anti-herbivore defense via symbiotic ant partners." *Journal of Ecology* 85 (1997): 847–61; Nomura, M., T. Itioka, and T. Itino, "Variations in abiotic defense within myrmecophyte and non-myrmecophyte species of *Macaranga* in a Bornean dipterocarp forest," *Ecological Research* 15 (2000): 1–11; Itioka, Takao, M. Nomura, Y. Inui, T. Itino, and T. Inoue, "Difference in intensity of ant defense among three species of *Macaranga* myrmecophytes in a Southeast Asian dipterocarp forest," *Biotropica* 32 (2000): 318–26.

15. Bagchi et al., "Pathogens and insect herbivores"; Chen, Lei, et al., "Forest tree neighborhoods are structured more by negative conspecific density dependence than by interactions among closely related species," *Ecography* 41 (2018): 1114–23; Stump, Simon MacCracken, and Liza S. Comita, "Interspecific variation in conspecific negative density dependence can make species less likely to coexist," *Ecology Letters* 21 (2018): 1541–51; Zhu, Yan, S. A. Queenborough, Richard Condit, Stephen P. Hubbell, Keping Ma, and Lisa S. Comita, "Density-dependent survival varies with species life-history strategy in a tropical forest," *Ecology Letters* 21 (2018): 506–15; Krishnadas, Meghna, Robert Bagchi, Sachin Sridhara, and Liza S. Comita, "Weaker plant-enemy interactions decrease tree seedling diversity with edge-effects in a fragmented tropical forest," *Nature Communications* 9 (2018): 4523 | DOI: 10.1038/s41467-018-06997-2; Comita, Liza S., and Simon M. Stump, "Natural enemies and the maintenance of tropical tree diversity: Recent insights and implications for the future of biodiversity in a changing world," *Annals of the Missouri Botanical Garden* 105 (2020): 377–92.

16. Elouard, Claire, "Notes on some *Fusarium* and *Cylindrocarpon* on Dipterocarpaceae of Indonesia," *Biotropia* 3 (1989): 25–40.

17. Comita, L. S., H. C. Muller-Landau, S. Aguilar, and S. P. Hubbell, "Asymmetric density dependence shapes species abundances in a tropical tree community," *Science* 329 (2010): 330–32.

18. Marden, J. H., et al., "Ecological genomics of tropical trees: How local population size and allelic diversity of resistance genes relate to immune responses, co-susceptibility to pathogens, and negative density dependence," *Molecular Ecology* 26 (2017): 2498–2513.

19. Brearley, Francis Q., "Ectomycorrhizal associations of the Dipterocarpaceae," *Biotropica* 44 (2012): 637–48; Liu, Xubing, et al., "Partitioning of soil phosphorus among arbuscular and ectomycorrhizal trees in tropical and subtropical forests," *Ecology Letters* 21 (2018): 713–23; Corrales, Adriana, Terry W. Henkel, and Matthew E. Smith, "Ectomycor-

rhizal associations in the tropics—biogeography, diversity patterns and ecosystem roles," *New Phytologist* 220 (2018): 1076–1109.

20. Brearly, Francis Q., et al., "Testing the importance of a common ectomycorrhizal network for dipterocarp seedling growth and survival in tropical forests of Borneo," *Plant Ecology and Diversity* 9 (2016): 563–76.

21. Peay, Kabir G., P. G. Kennedy, S. J. Davies, S. Tan, and T. D. Bruns, "Potential link between plant and fungal distribution in a dipterocarp rainforest community and phylogenetic structure of tropical ectomycorrhizal fungi across a plant soil ecotone," *New Phytologist* 185 (2009): 529–42.

22. Peay, Kabir G., et al., "Lack of host specificity leads to independent assortment of dipterocarps and ectomycorrhizal fungi across a soil fertility gradient," *Ecology Letters* 18 (2015): 807–16; Segnitz, Richard Max, Sabrina E. Russo, Stuart J. Davies, and Kabir G. Peay, "Ectomycorrhizal fungi drive positive phylogenetic plant–soil feedbacks in a regionally dominant tropical plant family," *Ecology* 101 (2020): e03090; Weemstra, Monique, et al., "Lithological constraints on resource economies shape the mycorrhizal composition of a Bornean rainforest," *New Phytologist* 228 (2020): 253–68.

23. Simard, Suzanne W., Kevin J. Beiler, Marcus A. Bingham, Julie R. Deslippe, Leanne J. Philip, and Francois P. Teste, "Mycorrhizal networks: Mechanisms, ecology and modeling," *Fungal Biology Reviews* 26 (2012): 39–60; Gorzelak, Monika A., Amanda K. Asay, Brian J. Pickles, and Suzanne W. Simard, "Inter-plant communication through mycorrhizal networks mediates complex adaptive behavior in plant communities," *AoB PLANTS* 7 (2015): plv050; doi:10.1093/aobpla/plv050; Liang, Minxia, et al., "Soil fungal networks maintain local dominance of ectomycorrhizal trees," *Nature Communications* 11 (2020): 2636.

With a decade of exploration and research in the magnificent dipterocarp forests of northwest Borneo just behind me, the sounds of their cascading mountain rivers still rushing through my ears, I entered the cloistered life of a university professor. But hardly had I placed my books on the shelf than a request arrived from the Smithsonian Institution in Washington to explore the forests of Sri Lanka with a view to prepare an account of its dipterocarps for a new flora. How could I resist such a temptation?

Within weeks, I was careening down narrow, winding country roads in a massive American pick-up truck (affectionately named 'the Alligator'), in search of the remaining rainforests. I entered their lofty canopy and was amazed that they were dominated by species of Shorea, but of a group that I had never seen before. And among and beneath them were other unfamiliar dipterocarps, with large wingless fruit: Vateria, Stemonoporus. I saw that others were familiar. And then came a flash of recall, strengthening as even the pattern of ridge and slope, clay mull and leached humult soil with associate flora, came back: This is Borneo! But Borneo was 2,500 km to the east across the ocean. That could not be! And yet intriguingly, both similarity and difference were there. — P. A.

13

PHYLOGEOGRAPHY

Plants, sessile with limited means of dispersal, have evolved a great variety of reproductive systems and dispersal strategies. Trees, generally longer-lived than animals, often survive for a century; few exceed a millennium (p. 102). They depend on particular habitats, with limited opportunity for physically direct, interspecific competition. Plants, through seeds and pollen, are dispersed differently—and often more *locally*—than animals. Persistence of the required habitats of a species is therefore vitally important. Historical contingencies leave enduring imprints on plant distributions and influence patterns of diversity.[1]

Interpreting the Past

Various types of evidence help us trace the histories of plant groups.

Fossils. Fossils reveal plant species in place and time. The deeper the past, the more the fossils differ from extant plants, and the more difficult it is to connect them with modern taxa. Fortunately, distinctive pollen grains resist decomposition, particularly in anaerobic or acid-reacting conditions, and evolutionary history has been illuminated by pollen and macrofossils in sedimentary rocks. Pollen deposited in swamps or lakes is generally derived from adjacent sites and can illuminate their floristic composition and community structure, especially in younger deposits from species similar to present ones.

Fossils confirm the minimum age of a species, but rarely the maximum. The record can never be comprehensive, because the conditions favoring fossilization are extremely rare. Certain flowering plant families with thin-walled pollen decompose rapidly and are missing from the record, or produce pollen like gymnosperms.

Phylogeny. Plant species origins and the history of migration routes can be inferred by analysis of current distributions in relation to their phylogenies, derived from molecular analysis of their genetic similarities and combined with information from the fossil record—the developing science of molecular phylogeography (fig. 13.1). These data can provide valuable information at all levels of an evolutionary tree, but missing taxa are likely nearer its base.[2]

Migration routes of plants are constrained by sea barriers, belts of unfavorable climate, and mountain ranges that block movements on continents or, alternatively, serve as corridors linking ecologically similar regions. The Andes do this, as do the eastern Himalayas southeastward into the Malesian "ring of fire" (fig. 3.5). Geological formations with a dominant soil-cover may similarly impede or promote plant migration. However, molecular evidence suggests that many plant species—up to 20% of the flora of Africa and the Neotropics—may have arrived there from across the Atlantic, by rare, long-distance dispersal *after* the drifting-apart of those continents.

Early arrivals provide the vegetational structure within which late-comers must compete, survive, and diversify. Vegetation is somewhat unique to a region, due to the accidents of its biogeographic history.

The *youngest* identifiable taxa from the fossil record are species and subspecies, influenced in their distributions by more recent events permitting migration or producing extinctions. A few taxa are truly ancient; the *Nypa* palm of brackish waterways, a single species now confined to Asian coasts, is recorded from all tropical continents as far back as ~70 Ma. Past events have influenced all taxa at least up to family rank, constrained by the ecological conditions in which they have evolved. Patterns of migration and diversification driven by recent geological and climatic events are the clearest and most interpretable.

Origins. Flowering plants originated in the late Jurassic period (~150 Ma). They diversified toward the end of the Cretaceous period (70–65 Ma). The prodigious diversity of the Asian tropical flora once pointed to Asia as the seat of angiosperm origins.[3] However both the fossil record and contemporary distributions make this unlikely, and its early tropical angiosperm flora was less diverse than the American. Remains of the Magnoliids, basal to other flowering plants, occurred widely in eastern North America,

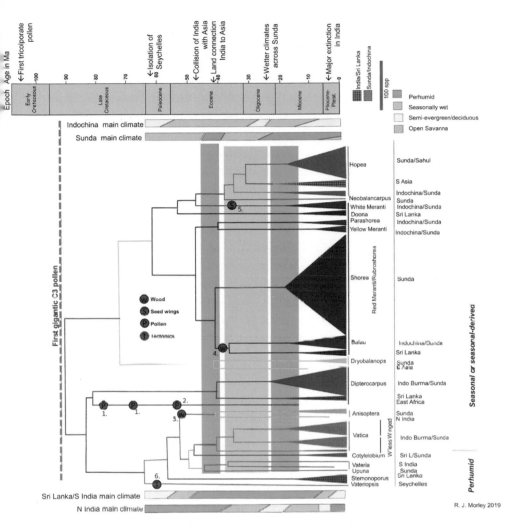

Fig. 13.1 A phylogeny of the Dipterocarpaceae subfamily Dipterocarpoideae, with key fossils located on the tree (see chap. 13, n. 7, and R. Morley and P. Ashton, fig. 6.3, in *OTFTA*, 2014); palaeoclimates and chronostratigraphy of India, Indochina, and Sunda added (see Robert J. Morley, "Assembly and division of the South and South-East Asian flora in relation to tectonics and climate change," *Journal of Tropical Ecology* 34 [2018]: 209–34; and F. Gradstein, J. Ogg, and A. Smith, *A Geologic Time Scale* [Cambridge: Cambridge University Press, 2004]). Intervals with everwet climates occurring contemporaneously in N. India and Sunda are highlighted by darker green bands (R. Morley). Fossils: 1 *Dipterocarpus*-like pollen, Sudan; 2 *Dipterocarpus*-like pollen, India; 3 *Anisoptera*-like twigs, England; 4 *Shorea* wood, Myanmar; 5 *Shorea* fruit, China; 6 separation of Seychelles from India (B. M.).

Northwest Europe, and West Africa. The eudicots (p. 104), basal to the dicotyledons, originated and diversified during the Cretaceous and were concentrated in the tropical climates of western Gondwana—South America and Africa before they divided (~100 Ma). By then, the angiospermous tropical rainforest was well developed.

Evidence from the fossil record at the end of the Cretaceous (72–66 Ma), including findings from molecular phylogenetics, indicates a peak of diversification into contemporary families. The Indian fragment of Gondwana was isolated at temperate latitudes in the southern part of the Tethys Sea 125–113 Ma, when global climates were more tropical. Until the Indian plate collided with Laurasia ~50 Ma, the forests of the tropical Far East contained few flowering plant families. Those had presumably arrived earlier by a northern migration, from North America by way of western Eurasia, when tropical climates extended to high latitudes, and sea barriers there were narrower. The tropical Asian angiosperm flora has therefore overwhelmingly arisen from immigrants.

Many pan-tropical families are archaic and could have existed before the breakup of Gondwana; others are relatively advanced. Until 25 Ma, there were still intermittent connections between North and South America, Europe, and East Asia (fig. 3.1). These bridges were tropical as far north as ~50° N, but Africa was still to the south, and Europe was separated from Asia by the Turgai Strait, running south from the Arctic Ocean.

Tropical Asian Immigrations

Although obscured by the influences of present climates and geology, the ancient tracks by which flowering plants originally moved into tropical Asia can still be inferred. Plants immigrated in waves at different periods. Creating a phylogeography by correlating molecular with biogeographic and other evidence is difficult, not least because the earliest fossils rarely coincide with a taxon's origin. Here, we interpret phylogenies in the context of current geographical ranges and known fossils, seeing five distinct geographical patterns (fig. 13.2). The three main opportunities for the spread of West Gondwanan land plants eastward into the Asian tropics were probably intermittent, owing to changes in climate and land connections.

Pan-tropical distributions. The first distribution was via North America into Europe and southwestern Laurasian Asia, especially when they were connected (~90–60 Ma). This invasion of tropical plants into Asia has left

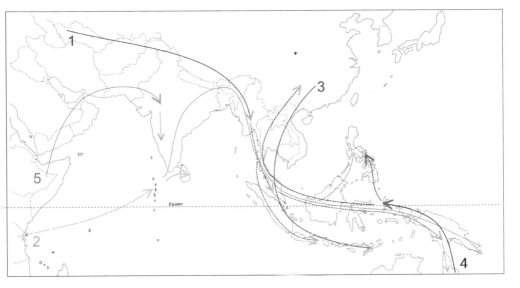

Fig. 13.2 Immigration tracks of flowering plants into tropical Asia, mapped onto current continental configurations. Numbers indicate their sequence over geological time (I. B., after P. A.).

no discernable pattern, such as plant distributions coinciding along a geographical frontier. Pollen related to more advanced families appeared 66–55 Ma, including Sapotaceae, Myrtaceae, Olacaceae, and *Nypa* (Arecaceae). Their route eastward from Europe were enabled by tropical temperatures that prevailed to high latitudes at that time, but rainfall would have been seasonal at best. The southern Laurasian coastline was north of the Tropic of Cancer, and the Turgai Straits presented a migration barrier.

In modern tropical Asia, there are about 125 recognized angiosperm families, including trees, comparable to 127 families in Neotropical West Gondwana. Of the modern Asian families, ~94 are pan-tropical, and almost all include temperate species. Few angiosperm families occur in the Cretaceous or Paleocene fossil floras of tropical East Asia, but these include advanced families like the Asterids, Sapotaceae, and Aquifoliaceae, reflecting early diversification. Climate was periodically dry, and the northern gymnosperms and Laurales indicate high mountains. This first wave of immigrants was only made possible by close land connections between North America and Asia by way of Europe, and by the northerly tropical climates. Climates along this west-to-east tropical corridor were seasonal and dry. Winds, circulating within Hadley cells at the same latitudes as at present

(p. 4), originated off the northern continent until the Indian collision up-lifted coastal uplands.

At ~45 Ma, the global climate was exceptionally warm, and the wave of immigration continued. Land connected the Malay Peninsula with western Borneo and eastern Java—at that time the wettest region. Although North America and Europe still were connected, the Turgai Strait separated Europe from Asia until the Indian collision (p. 62). The very rich Eocene flora preserved in northwest European sediments—the "London Clay"—included many tree families with pan-tropical and warm temperate distributions, which reached tropical eastern Laurasia by the Atlantic track. Lauraceae, well-represented in Far Eastern warm temperate forests and the lowland rainforests throughout Asia, probably followed a similar route, but their pollen is difficult to identify.

The late Cretaceous-Paleocene East Asian fossil record is poor, deserving greater study. Evidence for a northern tropical eastward or perhaps westward migration is weak, though Calamoid palms and some basal Annonaceae may have taken this route. Many characteristic elements of the present flora were absent, including dipterocarps, durians and their allies, Fagaceae, Euphorbiaceae, Rhizophoraceae, and Lythraceae (Sonneratiaceae).[4]

The Indian Noah's Ark. South Asia broke free from the warm temperate Gondwanan supercontinent in the mid-Cretaceous period (~119 Ma) at an early stage of eudicot evolution; its flora resembled the Australasian of that time. The "ark" drifted north into the equatorial regions, along the Madagascar and East Africa coast. During that slow transit, direct migration was initially possible to South Asia from Madagascar, and the SW Sri Lankan endemic *Hortonia* of the southern family Monimiaceae is good evidence of that (fig. 13.3). The tropical African flora immigrated later, when the Indian plate was more distant, though likely connected by an island arc, and it had entered wet equatorial climates (p. 62). The South Asian land connection was likely made with the southern Laurasian coast ~45 Ma, at ~27°N and therefore well within the seasonal tropics (fig. 13.4). That connection likely began before its first contact with the southern Laurasian coast 56–34 Ma, ending as the northern tropical climate dried. Genera still co-occurring in everwet SW Sri Lanka and Sundaland must therefore have migrated eastward across the expanding Bay of Bengal, including the currently extant *Cotylelobium*, *Trichadenia*, and *Axinandra* (p. 284). This was the only opportunity for the migration of taxa confined to that climate that still survive in everwet southwest Sri Lanka and Malesia, including some dipterocarps and

Fig. 13.3 Gondwanan elements in South Asia and the Mascarenes. *Left, Vateriopis seychellarum,* (Dipterocarpaceae, from the Seychelles; P. A.); *center, Hortonia angustifolia* (Monimiaceae, endemic to Sri Lanka; N. G.); *right, Schumacheria castaneifolia,* endemic from Sri Lanka (Dilleniaceae, H. H.).

durians. A northern, wet, tropical overland route to the east emerged later with the rise of the Himalayas, probably ~25 Ma when the global super-hot period ended, and the southwest monsoon began; the new route only supported, at best, taxa of seasonal wet evergreen forests. Everwet climates in Malesia, and seasonally wet ones in Indo-Burma, only spread widely over the Sunda lands 23–20 Ma, when the Indonesian ocean throughflow stopped, and the high Himalayan and Tibetan Plateau started its rise, eventually triggering the southwest Indian summer monsoon ~10 Ma (fig. 3.2). The absence of so many Sunda genera in South Asia suggests that the progenitors of most plants migrated along the northern seasonal track, before Sunda became widely everwet.

Morley[5] argued that the Asian lowland rainforest flora arrived principally via the Indian route. The palynological record indicates a peak in arrivals of flowering plant families from about 40–33 Ma, following the collision of the Indian continental plate. The Asian tropical tree flora still bears the imprint of its mostly African origins, thereby separating both from the Neotropical, but it is poor in endemic, basal flowering plant families—in marked contrast with the East Asian subtropical forest tree flora.[6] But some 20 lowland rainforest families are now recorded from the Paleocene epoch of the Indian plate before the collision, and more may be expected as research continues.

The Dipterocarpaceae occur exclusively throughout the tropics as three subfamilies (p. 104), although the exclusively Neotropical one, with only a single species, is now considered a different family.[7] The Dipterocarpoideae, with 13 genera and ~475 species (fig. 13.1), now occurs only in Asia and

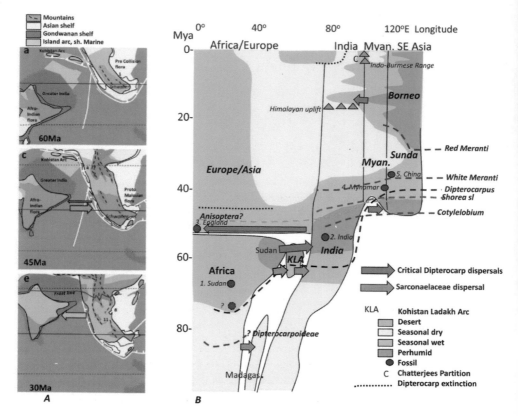

Fig. 13.4 Plate position, climate, and plant migration. *Left*, Conditions were optimal for plant migration across India and into Southeast Asia ~45 Ma. Dark green arrows indicate everwet (perhumid) climate plant dispersals (see Morley 2018, cited in caption for fig. 13.1, and Peter Ashton, Robert. J. Morley, J. Heckenhauer, and V. Prasad, "The magnificent dipterocarps: Précis for an epitaph?" *Kew Bulletin* 76 [2021]: 87–125]; light green arrow shows seasonal climate faunal dispersals. *Right*, whimsical sketch of climate, dispersal paths, and time from Africa to Southeast Asia, showing likely dispersal paths for Dipterocarpoideae based on fossils, climate, and plate tectonics reconstructions (see Morley 2018, cited in caption for fig. 13.1, and www.Scotese.com); 1 *Dipterocarpus*-like pollen from Sudan; 2 Diptercarpoid pollen from Kachchh, India; 3 early Eocene *Anisoptera*-like twigs from London Clay; 4 *Shorea* wood from Myanmar; and 5 *Shorea* fruit from southern China (B. M.).

the Seychelles (fig. 13.3). All dipterocarpoid genera (except the monotypic *Upuna* of Borneo) are either still in South Asia or represented by Cenozoic fossils there. The basal taxa are concentrated in South Asia, and the Neotropical and African subfamilies are basal to the Asian. The main genera of the tribe Shoreae originated early, in South Asia before or soon after the collision. One clade, sect. *Doona*, with nine species, survives only in everwet Sri Lanka.

Lack of dormancy, salt water intolerance, and poor wind dispersal (heavy-winged and gyrating, p. 260) of seeds made ocean dispersal rare. In these characteristics, they resemble Fagaceae, but Fagaceae, unlike dipterocarps, failed to migrate across the Gangetic Plain south into Peninsular India, leaving the dipterocarps that were replacing the oaks to dominate the lower montane forests of everwet southwest Sri Lanka.[8] The absence of basal dipterocarpoids in Madagascar implies that dipterocarps migrated to the Indian plate after separating from Madagascar ~80 Ma. As India moved north into the wet tropics, island chains probably persisted, linking India and Madagascar with continental Africa until the late Cretaceous period and allowing the dispersal of some dipterocarps.

Among taxa with African fossil records are the Malvaceae tribe Durioneae, and the Calamoid palm *Eugeissona*. The borassoid palms came from the rainforests of Africa through Madagascar and the Mascarenes to tropical Asia. Some *Terminalia* species probably immigrated via the Indian rather than the likely semi-arid Laurasian track that opened from Africa in the Miocene epoch.

An eastern Laurasian connection. The mountains of warm temperate and tropical East Asia, fringing the South China Sea from South China and Taiwan into Vietnam, were a corridor of warm temperate climate south into the tropics. They are extraordinarily rich in conifers. The northern tropical lowland frontier was as at least as far north as southern Fujian (27° N) during the Miocene warm period. Dipterocarp fossils indicate that seasonal wet dipterocarp forest prevailed.[9]

The phytogeographic connection between temperate eastern North America and northeast Asia has been well documented since Asa Gray drew attention to it more than a century ago. Opportunity for migratory exchange persisted *via* a warm temperate Bering Strait until 15 Ma. The East Asian temperate flora is currently far richer than the American, partly because of land continuity within humid, warm temperate, and tropical East Asia, which allowed north/south migration during Pleistocene climate change. The richness of the East Asian warm temperate flora in archaic families is due in part to an ancient mild climate. It includes many Far Eastern families, several of which are magnoliids or eudicots, basal in the angiosperm phylogeny. Many of the eudicots are confined to temperate forests in Asia, but

Fig. 13.5 Northern elements in the East Asian lower montane flora. *Left, Maingaya malayana* (Hamamelidaceae, Malaya; D. L.); *center, Magnolia lanuginosa* (Magnoliaceae, Bhutan; H. H.); *right, Schima wallichii* from 1,700 m on Bali (Theaceae; P. A.).

about half penetrate the tropics. There they concentrate in lower montane forests and are also represented in the wet lowlands (fig. 13.5).

The Fagaceae occur in both Asian and American tropics, not in Africa, and they are only abundantly represented in everwet Malesian lowland rainforests. The evergreen genera, all tropical or warm temperate, are the older. Although their current geography and fossil records imply a northern and possibly eastern Laurasian origin, the recent discovery of a beautifully preserved *Castanopsis*-like fossil from late Cretaceous lake deposits in southern Argentina, alongside modern Australasian taxa, brings their origin into question.[10] Immigration from North America is suspected. There is no fossil record of a southern easterly migration, and the deeply and densely toothed leaf margin, inexplicably, more resembles the north temperate deciduous and similar chestnut genus *Castanea*. The Magnoliaceae originated in the tropics in both the New World and Asia in the Eocene epoch, with temperate taxa diverging only later. The Hamamelidaceae and Juglandaceae are represented in both Laurasia and Gondwana (fig. 13.5), though neither they nor the Fagaceae crossed the Gangetic Plain into South Asia. As with Fagaceae, the molecular phylogeny of Hamamelidaceae implies an East Asian origin, with several tropical immigrations from a warm temperate base.[11]

Australasian arrivals. A wet tropical Australasia arrived, and its flora first moved west, after the everwet corridor between East and South Asia was severed (p. 61). It is poorly represented in South Asia and Indo-Burma, except for the speciose genus *Syzygium* (Myrtaceae). Remarkably, Australasian elements had already appeared in Malesia while Australia remained far away, likely still in a warm temperate climate; southern *Podocarpus* pollen originally appeared in West Malesia ~40 Ma, and *Casuarina* (Casuari-

Fig. 13.6 The Australasian elements. Left, *Gymnostoma nobile* fruits and foliage, Gifford Arboretum, Miami (Casuarinaceae); *center*, *Baeckia frutescens* (Myrtaceae, Malaya) and *right*, *Helicia attenuata* (Proteaceae, Malaya; all D. L.).

naceae), *Austrobuxus* (Picrodendraceae), and *Dacrydium* (Podocarpaceae) ~20 Ma later.

Advance fragments of the Australasian plate collided with the Sunda continental shelf, including most lands west of Borneo, ~20 Ma, although the celebrated biogeographic frontier Wallace's Line, isolating Sunda Borneo from Australasian lands to its east, was much wider until late Miocene times, when neighboring Sulawesi first started to appear as island arcs.[12] Everwet taxa such as the lower montane podocarp *Phyllocladus* appeared only after ~5 Ma. Since Australasia arrived close enough for substantial plant dispersal only recently, its geographic imprint is more recognizable.

Everywhere in the tropical Far East, Australasian elements in the tree flora are best represented in montane forests, where Cunoniaceae, Podocarpaceae, Araucariaceae (*Agathis*), and Myrtaceae frequently dominate the canopy. *Leptospermum* (fig. 10.10), *Baeckia* (fig. 13.6), and *Xanthomyrtus* (Myrtaceae) are occasionally found at low altitudes there. The success of Australasian and Indian-Laurasian elements in the Far Eastern and Malesian flora reflects the climate and soils of their lands of origin; Indian-Laurasian taxa migrated east to dominate New Guinea's lowland forest, whereas Australian elements dominated the subalpine thickets as far west as Kinabalu.[13]

The Saharo-Sindian element. Many tree families occur in the wet Asian and New World tropics, but not in Africa. That reflects the drying of Africa ~23–25 Ma and its floristic impoverishment before dry land routes were established from Africa across the Middle East to the Asian tropics. Trees of Salvadoraceae and Moringaceae are distributed in sub-Saharan Africa and in arid South Asia. Many genera and some species in several families are shared, comprising what has long been recognized as the Saharo-Sindian element in the South Asian flora (fig. 13.2). Particularly well represented are Fabaceae (*Acacia*), Combretaceae (*Anogeissus*), Rhamnaceae (*Zizyphus*),[14] Capparaceae, Zygophyllaceae, and Hernandiaceae—also represented in lowland tropical rainforests. These, along with seasonal or wet tropical elements, including the vine genus *Artabotrys* (Annonaceae), are considered to have migrated *via* Arabia, at a concurrent time of greater coastal climatic and vegetation heterogeneity.[15]

Subsequent diversification in tropical Asia. In all five immigration patterns, very few species survive in their places of origin as well as colonized regions. Current Far Eastern rainforest tree species originated in the lower Miocene epoch (23 Ma), when everwet climates began to become widespread, and their major diversification continued at least into the Pleistocene epoch.

Whereas temperate deciduous taxa, such as *Ulmus*, deciduous *Prunus*, *Fraxinus*, *Rhus*, *Alnus*, and *Betula* reach similarly low latitudes in the mountains of both Old and New Worlds, tropical lower montane evergreen Hamamelidaceae and *Nyssa* do not occur in the Neotropics, while evergreen *Quercus* only reaches northern Colombia. This distribution implies a northern, warm temperate origin.

Similarly, among Australasian immigrants, only Proteaceae (fig. 13.6) and Monimiaceae have tropical lowland representatives in Asia. Apparently, the flowering plants that migrated to tropical and temperate Asian climates diversified very early. Early-evolving familial traits—many of which may appear to have lost adaptive significance (Gould and Lewontin's "spandrels of San Marco")[16]—continue a fundamental ecological conservatism. Limits to ecological diversification were set early, and their constraints on migration have been insufficiently appreciated.

Diversification through Speciation

Forest species composition may change across barriers to dispersal such as ocean straits, high mountain ranges, or major river valleys and floodplains. Such barriers may also establish *after* the spread of a species, subsequently promoting diversification. Alternatively, small founder populations may have diversified (p. 331) from seed dispersal across a confining barrier. Changes in forest structure also facilitate diversification. The stature and well-developed emergent canopy of forests on zonal, yellow-red soils in the

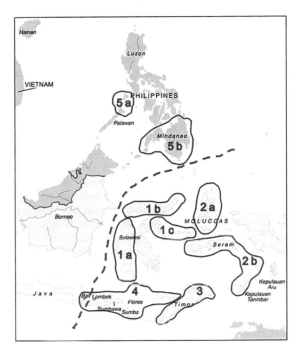

Fig. 13.7 Plant geographical regions in Wallacea (see Bernard Michaux, "Biogeology of Wallacea: Geotectonic models, areas of endemism and natural biogeographical units," *Biological Journal of the Linnean Society* 101 [2010]: 193–212). Some, at least, are correlated with differences in rainfall seasonality and may therefore have an ecological rather than historical explanation. Dashed blue line is the Wallace Line, as modified by Ernst Mayr (I. B., after B. Michaux).

everwet Far East has enabled the diversification both of canopy dipterocarps and subcanopy species (chapter 7).[17]

The Australasian frontier: Wallace's Line. The biogeographic division delineated by Alfred Russel Wallace 150 years ago is still recognized by zoogeographers as Wallace's Line (fig. 13.7). Its central track, between Borneo and Sulawesi, marks the eastern edge of the Sunda continental shelf. The southeast arm of Sulawesi is geologically part of southeast Sunda Borneo; these land masses were united in Eocene times and separated by 47 Ma, and the isolation of Sunda from Australasian lands remained wide until Sulawesi started to re-arise in the upper Miocene.[18] That bridge may have been an initial toehold for the eastward spread of the Laurasian tropical flora once Australasia had approached.

Surprisingly little of the indigenous Australasian flora has spread into the Far Eastern lowland tropics from its continental shelf, the Sahul Shelf. Nonetheless, Australasian Gondwana penetrated the everwet tropical zone for 20 million years, shortly after the expansion of everwet climate through the expanding Sunda lowlands. Most of the lowland wet tropical Asian plant families and half of the species spread eastward across the line. The migration of Dipterocarpaceae and Fagaceae was constrained by their limited means of seed dispersal and lack of dormancy. There are eight Fagaceae in Sulawesi (all endemic) and 141 in Borneo; *Quercus* is absent from the Philippines and east of the line. The seven Sulawesi dipterocarps (in three genera) hardly compare with the 269 species in Borneo.

Wallace's Line has been more of an ecological constraint on the dispersal of animals than on that of plants. Whereas the tropical Asian flora has invaded the lowlands of eastern Malesia, the Australasian flora has mostly spread westward along the chain of upper montane forests leading into continental Southeast Asia, in several cases also descending to lowland *kerangas* in Borneo, which shares low-nutrient substrates with much of the Australian continent.

New Guinea and Australia share the Sahul continental shelf and were connected by land as recently as 12 Ka. The rainforests of Queensland—well within the tropics—are about as seasonal as southeast New Guinea. The relative poverty of the Queensland rainforest tree flora is due to its arid Pleistocene history.

The Philippines as a stepping stone. The Philippines were close to the main eastward route of the Malesian lowland rainforest flora since arriving as an oceanic volcanic island arc from the northeast ~15 Ma. Of 356 Sunda genera unknown east of Wallace's Line, 61% occur in the Philippines, including Palawan. There is also a strong Australasian wet forest presence at all altitudes there, but the less diverse Philippine MDF is primarily West Malesian.

The isolated rainforests of South Asia. Following the rise of the high Himalayas 12–10 Ma, which initiated the southwest monsoon, arid episodes have caused extinctions and rainforest retreat in South Asia, particularly in the Western Ghats.[19] Southwest Sri Lankan rainforests became isolated from the peninsula. The South Asian rainforest is much less diverse than that of the Far East. Remarkably, apart from *Michelia* (Magnoliaceae) and *Ilex* (potentially of Gondwanan origin), none of the northern lower montane families currently appear in the Ghats rainforests. Fagaceae, which dominate the canopy of both tropical lower montane and warm temperate Himalayan forests, are absent. In the lowland rainforests there are absences too, notably the canopy and emergent legume genera. The forest is devoid of true emergents, even in wind-protected valleys.

For its size, Sri Lanka is the richest area of South Asia, with 1,052 genera and 3,210 species of flowering plants (table 13.1). Ninety percent of these

Table 13.1 Tree species endemicity within the Pasoh 50 ha and Lambir 52 ha ForestGEO plots. Percentage of total within plot in parentheses.

Locality	Number of species	Widespread Sunda lands	Widespread Borneo (Lambir) or Peninsular Malaysia (Pasoh)	Local: Riau Pocket (Lambir), southern Peninsular Malaysia (Pasoh)
Lambir Hills, Sarawak, Borneo	780	530 (68%)	138 (18%)	112 (14%)
Pasoh Forest, Negeri Sembilan, Peninsular Malaysia	463	379 (82%)	60 (13%)	24 (5%)

species are confined to its small area of everwet forest, where an extraordinary 95% are endemic to it or to other South Asian rainforests. Sri Lanka possesses a full complement of dipterocarp genera, but few are endemic.

Chatterjee's Partition. The pockets of moist, semi-evergreen forest in valleys of the higher hills of the Eastern Ghats north to Odisha (Orissa) contain a flora with greater affinity to that of the Western Ghats, which are 1,200 km to the west-southwest, than to that of the Khasi hills 600 km to the northeast. None of the Himalayan tree families absent from the Western Ghats are present in the Eastern Ghats, though most are present in the Khasi Hills, isolated from the Himalayas by the Brahmaputra valley but less so from the Burma frontier range arching southeast from the Himalaya. The geographic boundary between the evergreen forests of Peninsular India and Indo-Burma is therefore a partition made by the deciduous forests of the Gangetic Plain and lower valley. Conversely, the deciduous forests of India and Burma are separated by a gap of only tens of kilometers: an evergreen forest barrier between the dry valleys that run west into India and east into Burma along their political frontier. Seasonal evergreen dipterocarp forests run down the high ridges, which curve south from the Himalaya, down the frontier into the Chittagong Hill Tracts of Bangladesh, down the Burma coast, and to the Bay of Bengal as the Andaman Islands. This range was lifted to current levels 7–3 Ma, consistent with Mammalian fossil evidence for a contemporary increase in grazing taxa and the spread of deciduous forests and savanna.[20]

This boundary was described by D. Chatterjee in 1940,[21] recognizing the relationship between eastern Himalayan and northern Burmese evergreen floras. Chatterjee's Partition is as distinct, in relation to the woody and especially the deciduous flora, as is Wallace's Line, but it has not received comparable attention. It has been both a continuing *barrier* to deciduous flora migration, and a periodic migration *track* of evergreen forest flora from and to northern monsoon India via the semi-evergreen forests of the Himalayan foothills and down the Indo-Burmese partition ranges and the Isthmus of Kra (and alternatively the Andamans and Sumatra), to Sundaland. This lack of research attention may result partly from the Partition's lack of a distinct zoographical signal—though more study of insects would be rewarding. What better way to distinguish these two deciduous forest regions than by their glittering denizens, the blue Indian and the emerald Indo-Burmese peacocks, their ranges separated by the dense forests along the Indian-Burmese frontier?

The lowland tropical margin as a plant geographic frontier. The northern tropical margin is clear from its winter frosts.[22] In the floristic ecotone from tropical lower montane to warm temperate forest on slopes and ridges, there is only a small change in generic composition, but a major one in species—greatest in the canopy. In lowland South China, there is also a quite narrow northern margin to forests of almost exclusively tropical species composition. Many genera are shared between tropical and warm temperate evergreen forests. More palynological evidence and documentation of rare killing frosts in this region would help resolve biogeographic questions. A few families, including Betulaceae, Fagaceae, and the archaic Hamamelidaceae and Magnoliaceae contribute to the tropical flora, while several tropical families, including Theaceae, Rutaceae, and Sapindaceae have moved into temperate climates in East Asia. Despite the greater seasonal temperature extremes in the dry tropics, the floristic margin is less clear than in the wet tropics. Although the tropical margin presently hugs the Tropic of Cancer near the Vietnam-Guangxi frontier, dipterocarp fossils from late Miocene times indicate that it was then as far north as the southern Qin Ling Shan foothills.

The regional rainforests and their origins. The lands of West Malesia, the Sunda continental shelf (Java and southernmost Sumatra excepted), have a uniform, warm everwet climate (p. 52). It coincides with the range of MDF, peat swamp, and other associated forests.

Varying climates. Although the Sunda lowlands have existed since ~23 Ma, the extent of everwet climate has varied during that time. Cycles of seasonality and low sea levels, associated with changes in Earth's orbital eccentricity, began ~5 Ma and continue to the present. They led to the Pleistocene northern ice ages, correlated with drops in sea levels since 900 Ka.

During these cyclical changes, the current rainforest flora of Sunda diversified.[23] The most extensive, hyper-diverse lowland forests in the Old World are circumscribed by the Sundaland everwet climate and by Wallace's Line. There are 117 endemic genera and an impressive ~15,000 endemic species in Sundaland and the Philippines, or 60% of the estimated total for Malesia. But there are no endemic tree families. Instead, many families are vastly richer in species, generating the highest community diversity in Asia (chapter 14).

Morley has argued that the Far Eastern tropics were seasonal from the late Eocene to the end of the Oligocene epoch,[24] that is, *after* the families, which rafted to Laurasia on Gondwana India, had arrived. However, everwet taxa apparently migrated eastward over the nascent Bay of Bengal (p. 284) before initial contact, when northern India had already become seasonal. They survived in everwet lowland refugia in the Far East when the regions between South and East tropical Asia had become seasonally dry. The Sunda region was mostly mountainous until the Miocene epoch. Extensive everwet lowlands were a prerequisite for the diversification of the current tree flora—rich compared to that of the eastern Andean foothills of the Amazon, which shares the same climate. The modern floras of the mast fruiting evergreen forests of Sundaland can therefore be no more than 23 million years old, but they are ancient in comparison with those of cool temperate and boreal forests, which were reduced during the Pleistocene ice ages.

Migratory linkages. The survival of genera known exclusively from MDF and everwet climates in both Sri Lanka and Malesia implies that there must once have been a migration route between them. Some genera are clearly ancient: the dipterocarp *Cotylelobium,* in a basal group of the Asian subfamily with two species in Sri Lanka and three in the Far East, and *Axinandra* (Crypteroniaceae) with one species in Sri Lanka and three in the Far East. No everwet overland corridor for migration of exclusively everwet regional elements ever existed, because the relentless northeastward rafting of South Asia carried the migratory route far north of everwet equatorial climates. The last time the seasonal evergreen dipterocarp forest corridor was bridged is unknown, but since we are currently in a warm period, it is likely to have been pre-Pleistocene, or more than 2 Ma. Many South Asian dipterocarps appear to have sister species in Far Eastern seasonal evergreen forest, though this awaits molecular genetic confirmation.

The Kangar-Pattani Line. The Isthmus of Kra has long been known to mark a major turnover in plant and animal species (fig. 13.8). It marks the

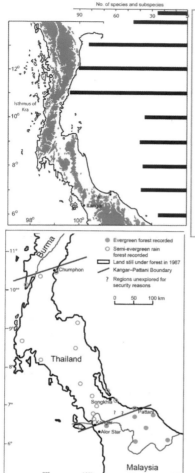

Fig. 13.8 The Kangar-Pattani Line. *Top left*, the mountain ranges above 500 m down the southern Isthmus of Kra; bars indicate the range limits of resident birds. *Bottom left*, boundaries and the distribution of forest types down the isthmus (both I. B., after D. S. Woodruff; see David Woodruff, "Neogene marine transgressions, palaeogeography and biogeographical transitions in the Thai-Malay Peninsula," *Journal of Biogeography* 30 [2003]: 551–68.). *Top right*, ranges of *Dipterocarpus* species down the isthmus (I. B., after C. F. Symington; see chap. 13, n. 25).

northern limits of the emergent, light hardwood red and yellow meranti *Shorea* species that dominate the MDF canopy in western Malesia and the Philippines.[25] The Kangar–Pattani Line coincides with the northern limit of the everwet climate and the onset of a reliable annual dry season of more than two months (p. 52). Even lower montane species change at the line; of 64 Fagaceae in Peninsular Malaysia, only four species are found on both sides. Although long regarded as ancient, mid-Pleistocene fossils of the exclusively MDF leguminous genus *Koompassia* in NW Thailand confirm periodic northern extension.[26]

Chatterjee's Partition, the tropical margin, the continental coast of the

South China Sea and the Kangar–Pattani Line together define a distinct *Indo-Burmese floristic province* (table 13.1). Its forests range from semi-arid, deciduous thorn to seasonal evergreen lowland forests and lower montane forests, yet they share a unique floristic imprint. Altogether, these habitats are estimated to contain ~13,500 vascular plant species, half endemic.

Land continuity across the Sunda Shelf. Current sea levels are probably within 5 m of the highest they have been since the Miocene, but occasionally they have fallen to as much as 120 m below the present level (p. 65; fig. 4.7). The decline in sea level from ~2 m higher at ~10 Ka produced the extensive coastal peat swamps of eastern Sumatra, western Peninsular Malaysia, and Borneo. These are the youngest natural rainforests of the region, and they are among the youngest on Earth.

During the northern ice ages when sea levels fell, Sumatra, Peninsular Malaysia, Borneo, and the Philippine island of Palawan became a single landmass almost as extensive as modern northern South America, drained from the northeast by a mighty river, the Proto-Mekong (p. 67). Sea levels dropped to their lowest at the climax of the last northern ice age, though they rarely fell by more than 30–40 m at other times. The low sea levels produced drier climates, affecting Southeast Asia less than South Asia. On Java, there probably were open, semi-evergreen forests with an abundance of savanna ungulates and large predators, including "Java man," *Homo erectus.* The current impact of ENSO-associated droughts and consequent human-initiated fires in southeast Borneo has promoted speculation that fire has long affected these areas.

A corridor of savanna across the equator? A large grazing and browsing fossil mammal fauna, dependent on seasonal grass swards, existed in Java from 2.4 Ma. The fauna immigrated across the equator from seasonal Indo-Burma. Before mankind, true savannas (open grassland with scattered trees), likely never existed in the Asian tropics except in floodplains. Ancestral fire-setting by our hominin ancestors, known in seasonal Asia since 1.8 Ma (p. 164), could have established a grassland corridor through the plains to the east and north of the Barisan and Javan "ring of fire" during drier continental periods, accompanied by mammals of the savanna, but not long enough for any but a few common tree species to migrate. Perhaps rainfall was reduced at times without strong seasonality, and everwet but drier conditions prevailed, supporting MDF of typical composition but with the addition of a few drought-tolerant coastal species. Or, perhaps there were times during the Pleistocene epoch, when MDF and its everwet lowland

Fig. 13.9 The Riau Pocket. *Left*, The expanded Riau Pocket plant geographical province. Three localities beyond which Riau Pocket dipterocarps occur, in the southern Borneo foothills, indicated (I. B.; see R. Morley 1999, cited in chap. 13, n. 5). *Right*, the northwest Borneo refugium near Gunung Santubong, Sarawak (© Tim Laman).

climate retreated to patches, some large, as in Borneo north of the equator, or smaller, as in Peninsular Malaysia east of the main range and Perak in the northwest.

The Riau Pocket. Assuming that drier climates prevailed periodically, how might the rich endemic MDF flora, now strictly confined to everwet climates, have survived? Shared species distributions define a floristic province that includes northwest Borneo, eastern coastal Peninsular Malaysia, and its west coast from the mouth of the Perak River north to Thailand (fig. 13.9). This plant geographic region was originally recognized and named the Riau Pocket.[27] It is exceptionally rich in endemic species. Northwest Borneo, the largest area, is particularly diverse.[28] The Riau Pocket regions were ecologically separated by the wide alluvium of the Proto-Mekong and its tributaries, either by flooding or by being covered in peat, as now.

Northwest Borneo is easily the richest part of the Riau Pocket, but then it is also the largest area. The whole Sunda region is unified by its everwet climate, but some of the coastal lowlands are drought-prone, partly due to ENSO. Occasional aseasonal drought itself enhances the raw humus accumulation, forming the habitat of so many Riau Pocket endemics. There is also endemism in the Riau Pocket flora on other, non-humic soil types, especially in northern Borneo but also in southeast Johor and coastal Perak (fig. 13.9). Perhaps these floristic provinces of the Riau Pocket were refuges for an everwet Pleistocene flora in general.

Lowland floristic hotspots. Lowland hotspots of exceptional floristic richness are correlated with high rainfall and habitat diversity (table 13.1), as well as with climatic stability over geological time.

Sunda centers of tree richness. In Borneo, Slik and colleagues identified the southeast as the region richest in tree *genera* (see n. 28 above). Southern Borneo was the most directly connected by upland habitats to Java and Sumatra, and thence to Peninsular Malaysia. Its generic richness reflects its connection to the central core of the Sunda flora, and also the higher survival of seasonal forest trees. Based on herbarium collections, Raes showed that the center of *species* diversity in Borneo is the northwest lowlands,[29] encompassing the richest region of the Riau Pocket (fig 13.9), and it has been identified as globally the richest in tree species. The 50 ha ForestGEO plot at Pasoh (Peninsular Malaysia) has almost the same number of families and genera as the 52 ha plot at Lambir (northwest Borneo; table 13.2), but it has far fewer species. However, the Lambir plot has more habitat diversity.

Peninsular Malaysia's and Sumatra's everwet eastern lowland MDFs are strongly connected. Nevertheless, before its almost complete conversion to commodity plantations, Sumatra's tree flora was poor, with low endemicity. Its frequent and recently prolonged separation by sea from other Sunda landmasses, especially from Borneo since ~3 Ma, helps explain these patterns, as may the massive Toba eruption ~700 Ka.

Other centers of richness. The Sri Lankan, Philippine, and New Guinean hotspots correspond with the most constantly wet—though not necessarily the highest overall—rainfall in their regions. The Peninsular Malaysian and Bornean hotspots at the center of the Riau Pocket are areas of exceptional habitat diversity (notably in soils), as well as likely Pleistocene refugia. The northeast Indo-Sino-Burmese regional *tropical* hotspot correlates with a seasonal rainfall climate made continuously moist by winter dew, and by less climate variability than in India, owing to its proximity to the Pacific Ocean and its distance from the Indian monsoon. It is also a region of exceptional habitat diversity, owing to rugged topography and abundant karst islands, as well as topographic variation in rainfall seasonality under the northeast trade winds, and temperature and soil variation at the margin of the ecological tropics. Abutting it to the north is the richest tree flora in the temperate world, sharing more than half its genera with its tropical neighbor.

Lowland centers of endemism. Hotspots of species richness are often, but not always, centers of endemism. Generally, the proportion of endemic species in a locality correlates with duration of constant climatic conditions.

The ancient and geologically uniform landscape of everwet southwest

Table 13.2 Hotspots of tree species richness and endemism in tropical Asia. Figures are approximations.

Region	Approximate number of tree species	Percentage endemism
SW Sri Lankan lowlands	850	70
Montane Sri Lanka	500	70
Western Ghats	800	60
Lower montane northern Indo-Burma, Meghalaya	1,000	65
Lowland northern Indo-Burma, South China	1,500	60
Indo-Burmese southern seasonal evergreen forest	2,500	60
Andaman Islands	1,000	30
Lowland Peninsular Malaysia	3,000	30
NW Borneo	4,000	35 (to Borneo)
NE Borneo	4,000	35 (to Borneo)
Borneo: Kinabalu	2,300	25 (to the mountain)
Sulawesi	600	40
Lowland New Guinea	3,000	65
Montane New Guinea	3,000	75

Source: P. P. van Dijk, A. W. Tordoff, J. Fellowe, M. Lau, and J. Ma, "Indo-Burma," in *Hotspots Revisited: Earth's Biologically Richest and Most Endangered Terrestrial Ecoregions*, ed. R. Mittermeier et al. (Mexico City: Cemex, 2004), 323–32.

Sri Lanka is a major repository of tree species endemism (70%), although endemic genera are few. In a flora of species mostly wide-ranging within this small region, the endemic dipterocarp *Stemonoporus* contains more than 20 species. Confined to lowland and lower montane forest, most *Stemonoporus* species have small distributions, but a few species are relatively widespread, such as the montane *S. gardneri* (occurring throughout the main massif, p. 197) and the lowland riparian *S. wightii*. *Stemonoporus*—by reason of some characteristic of their breeding system or population genetics yet to be ascertained—has diversified rapidly.

Endemism in the large island of Borneo is concentrated in the northern half, consistent with palynological evidence that it has been a refuge of everwet climate since ~23 Ma. Most Bornean endemic dipterocarps are confined to Brunei and the Malaysian states of Borneo (fig. 13.10). At Pasoh,

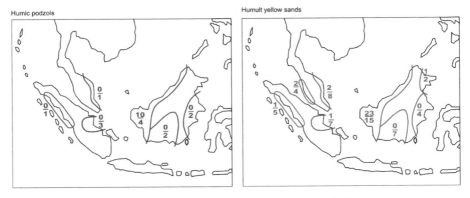

Fig. 13.10 The number of dipterocarp species endemic (top number) and widespread (bottom number) in each of the lowland regions of Sundaland in which podzols (*left*) and humult yellow sandy soils (*right*) prevail (I. B., after P. A.; see Peter Ashton, "Biosystematics of tropical forest plants: A problem of rare species," in *Plant Biosystematics*, ed. W. F. Grant [Toronto: Academic Press, 1984], 497–518).

few of the dipterocarps are peninsular endemics, with low, whole-flora endemicity. At Lambir, 54 (61%) of the dipterocarps are endemic to Borneo, and more than 55% of the flora as a whole (table 13.1).

High endemism in NW Borneo (particularly between the wide lower Kapuas and Lupar river valleys) and unique co-occurrences of species here and in east coastal Peninsular Malaysia are explained by the origin of the Lupar river by mid-Miocene capture of the headwaters of an earlier, eastward-flowing, Proto-Kayan drainage, with headwaters in the uplands between those two landmasses. Central Sarawak and Brunei, the lowlands east of the Lupar to SW Sabah, is now recognized as floristically the globally richest region for its area, on account of its edaphic diversity and climatic continuity through the Pleistocene epoch.[30] Regional endemism of the northwest Borneo Riau Pocket flora is complemented by a second center in Sabah. This northeast Bornean floristic province includes Mount Kinabalu and also several islands of ophiolitic extrusion and ultramafic soils. Like the northwest Borneo province, this region has likely served as a refuge. Its clay loam soils, though rich in provincial endemics, link the widespread element of its flora to the general Sunda MDF flora on clay-bearing substrates. In contrast, the endemics of the NW Borneo Riau Pocket flora predominate on a more fragmented terrain of humult, mostly sandy soils typical of the Riau Pocket elsewhere, although clay loam endemics are also well represented.

Endemism is promoted by several factors. Limited seed dispersal and ecological specialization influence tropical tree community composition within and between local habitats. Some rivers and associated habitats are barriers to the migration of tree species, such as dipterocarps, with their poorly dispersed seeds. Finally, fragmentation can constrain the geographical ranges of individual species.

Point-endemism (with very small species ranges) in the everwet lowlands of the main Sunda land masses is often correlated with limestone and ultramafic habitats.

Montane diversity and endemism. Mountain floras occupy ecological islands of varying age and connectivity.[31] During the northern ice ages, tropical high mountain forests of the seasonal Asian tropics were reduced by frost to refuges at lower altitudes and in valleys. Today, endemism varies between upper and lower montane forests. Lower montane forests include several genera endemic to them (chapter 10), but most of these genera have few species. The upper montane flora has many genera rich in local endemic species, including sister species of lowland taxa. Global comparison of floristic data indicates that diversification of current tropical montane floras, which are richest on higher mountain systems, stems from ~6 Ma but has been enhanced by more recent uplift and climate change since the Pleistocene epoch.[32]

Point-endemism is highest in the least seasonal montane climates on the most extensive and isolated massifs. The ancient mountains of Sri Lanka have been intermittently uplifted since the Indian collision with Laurasia, more than 35 Ma. These mountains shelter 51 endemic tree species, and an additional 12 also occur in the Western Ghats.

In the everwet tropics, connectivity between subalpine and upper montane forests increased as forest zones declined in elevation during cooler periods, but lower temperatures *narrowed* the lower montane forest zone, because its lower ecotone, broadly correlated with the cloud base as well as the temperature, changed little (p. 211). The summits of granite and sedimentary peninsular Malaysian mountains arose intermittently prior to the origin of flowering plants, and ceased before ~65 Ma. In contrast, the current Bornean mountain chains—probably humid through the Pleistocene—have arisen since 2 Ma. This may explain the lower point-endemicity on Borneo mountains (Kinabalu excepted), which at the same time provides a conservative time scale for the rate of speciation among montane trees. Similarly, the montane tree floras of the Sumatran and Javan "ring of fire" volcanoes are quite low in tree endemics. Although Gunung Kerinci in Sumatra's Barisan range reaches 3,805 m, it has few endemic trees and other plants, attributable to its youth and uniform soils.

Montane endemism is markedly higher in plants with shorter life cycles—more in shrubs than trees. *Rhododendron* sect. *Vireya,* despite its easily dispersed, dust-like seed, is particularly rich in endemics. In Borneo, *Rhododendron* endemicity matches patterns among lowland trees. Kinabalu (table 13.1) has exceptionally high woody plant endemicity (p. 209). In a massive study of species pairs from Kinabalu,[33] 16 endemic species were derived from others at lower elevations on the mountain, and 8 species came from a long distance. The mean divergence times of 20 endemics was less than 6 Ma, the age of the mountain.

Age and speciation of rainforest tree species in Asia. Areas of high species richness and endemicity may be attributable to high rates of recent evolutionary diversification from favorable recent conditions, but they may equally be due to climatically or geomorphologically induced past extinction events that eliminated earlier rich floras. The tree flora of the MDF on the clay and sandy clay loams of the everwet Sunda lowlands has the lowest regional level of endemicity of all the upland forest types. This widespread flora may include progenitors that fanned out into the emerging Sunda uplands and lowlands when they became widespread 20–15 Ma. Thus, these are the oldest rainforests in the Far East.

The only Asian tropical tree taxon whose history of diversification is fully documented in the fossil record is the small mangrove genus *Sonneratia* of the Indian and Western Pacific Ocean coasts. Along with upland *Duabanga,* ancestral *Sonneratia* diversified from the Lythraceae (with affinity to *Lagerstroemia*). The two genera may have diverged ~22 Ma; their modern species, ~10 Ma. Many surviving, widespread species have been the progenitors of others currently in more isolated habitats, with higher local endemism. Point-endemism is rare in the Borneo lowlands, more frequent in the north. Examples like *Hopea vaccinifolia,* known only from the *kerangas* on a Pleistocene, raised marine beach on the eastern Brunei-Sarawak frontier, implies recent origin, though it could have survived from an extinct habitat. The exceptional *regional* richness of the northern Bornean tree flora, unmatched elsewhere in the Old World with ~4,500 species, is due to the lowland soil diversity and to the persistence of an everwet climate.

Rainforest tree species have originated by many mechanisms. Rates of speciation might have increased in cooler climates by extinctions and reduction of competition following Pleistocene climate fluctuation or by ever more complex biotic interactions.[34] Speciation has occurred in the Asian tropics by dispersal of founders across barriers and by vicariant diversification of a metapopulation that was subsequently divided by a barrier. Speciation has responded to strong ecological pressure, even in the absence of physical barriers; sympatric speciation remains controversial. *Syzygium* is particularly rich in sister species inhabiting different habitats within a shared landscape (p. 314). *Leptospermum javanicum* and *L. recurvum* (Myrtaceae, p. 206) similarly evolved on high Sabah mountains. Cryptic species pairs, identical in the herbarium but differing in reproductive size and reproductive phenology, co-occur at Pasoh, suggesting differentiation of populations within a common landscape.[35] Different breeding systems explain some patterns of variation among families (chapter 11). Rainforest tree species may also originate by polyploidy or other processes that prevent subsequent introgressive hybridization in nascent populations (p. 230). Apomixis, especially by means of adventive embryony, may promote the rapid spread of fit genotypes where competition is comparatively relaxed. Many mechanisms help explain the evolution of this enormous diversity.

Notes

1. Content of this chapter is from *OTFTA*: chap. 6, 381–447; review in Lohman, David J., et al., "Biogeography of the Indo-Australian Archipelago," *Annual Review of Ecology, Evolution and Systematics* 42 (2011): 205–26.

2. Eiserhardt, Wolf L., Thomas L. P. Couvreur, and William J. Baker, "Plant phylogeny as a window on the evolution of hyperdiversity in the tropical rainforest biome," *New Phytologist* 214 (2017): 1408–22.

3. Takhtajan, A., *Flowering Plants: Origin and Dispersal*, trans. C. Jeffrey (Edinburgh, UK: Oliver & Boyd, 1969).

4. Couvreur, T. L. P., et al., "Early evolutionary history of the flowering family Annonaceae: Steady diversification and boreo-tropical geodispersal," *Journal of Biogeography* 38 (2011), 664–80; Wolfe, Jack A., "Some aspects of plant geography of the Northern Hemisphere during the late Cretaceous and Tertiary," *Annals Missouri Botanical Garden* 62 (1975): 264–79.

5. Morley, Robert J., *Origin and Evolution of Tropical Rainforests* (Chichester, UK: Wiley, 1999).

6. Slik, J. W. Ferry, Janet Franklin, et al. (~184 co-authors), "Phylogenetic classification of the world's tropical forests," *Proceedings of the National Academy of Sciences U.S.* 115 (2018): 1837–42.

7. Heckenhauer, Jacqueline, et al., "Phylogenetic analyses of plastid DNA suggest

a different interpretation of morphological evolution than those used as the basis for previous classifications of Dipterocarpaceae (Malvales)," *Botanical Journal of the Linnean Society* 185 (2017): 1–26.

8. Ediriweera, Sisira, et al., "Changes in tree structure, composition, and diversity of a mixed-dipterocarp rainforest over a 40-year period," *Forest Ecology and Management* 458 (2020): 117764.

9. Shi, G., F. M. B. Jacques, and H. Li, "Winged fruits of *Shorea* (Dipterocarpaceae) from the Miocene of Southeast China: Evidence for the northward extension of dipterocarps during the mid-Miocene climatic optimum," *Review of Palaeobotany and Palynology* 200 (2014):97–107.

10. Wilf, Peter, Kevin C. Nixon, Maria A. Gandolfo, and N. R. Cúneo, "Eocene Fagaceae from Patagonia and Gondwanan legacy in Asian rainforests," *Science* 364 (2019): eaaw5139.

11. Xiang, Xiaguo, Kunli Xiang. Rosa Del C. Ortiz, and Wei Wang, "Integrating palaeontological and molecular data uncovers multiple ancient and recent dispersals in the pantropical Hamamelidaceae," *Journal of Biogeography* 46 (2019): 2622–31.

12. Mansyursyah, A., S. Nugraha, and R. Hall, "Late Cenozoic palaeogeography of Sulawesi, Indonesia," *Palaeogeography, Palaeoclimatology, Palaeoecology* 490 (2018): 191–209.

13. Brambach, Fabian, Christoph Leuschne, Aiyen Tjoa, and Heike Culmsee, "Predominant colonization of Malesian mountains by Australian tree lineages," *Journal of Biogeography* 47 (2020): 355–70.

14. Hauenschild, Frank, Adrien Favre, Ingo Michalak, and Alexandra N. Muellner-Riehl, "The influence of the Gondwanan breakup on the biogeographic history of the ziziphoids (Rhamnaceae)," *Journal of Biogeography* 45 (2018): 2669–77.

15. Chen, Junhao, Daniel C. Thomas, and Richard M. K. Saunders, "Geographic range and habitat reconstructions shed light on palaeotropical intercontinental disjunction and regional diversification patterns in *Artabotrys* (Annonaceae)," *Journal of Biogeography* 46 (2019): https://doi.org/10.1111/jbi.13703.

16. Gould, Stephen J. & Richard Lewontin, The Spandrels of San Marco. The Panglossian paradigm: a critique of the adaptationist paradigm," *Proceedings of the Royal Society B* 205 (1979): 581–598.

17. Wong, Khoon Meng, and L. Neo, "Species richness, lineages, geography, and the forest matrix: Borneo's 'Middle Sarawak' phenomenon," *Gardens' Bulletin Singapore* 71 (2019, Suppl. 2): 463–96.

18. Merrill, E. D., *An Enumeration of Philippines Flowering Plants* (Manila: Bureau of Printing, 1926).

19. Bose, Ruksan, Brahmasamudra Ranganna Ramesh, Raphaël Pélissier, and François Munoz, "Phylogenetic diversity in the Western Ghats biodiversity hotspot reflects environmental filtering and past niche diversification of trees," *Journal of Biogeography* 46 (2018): 145–47.

20. Patnaik, R., "Diet and habitat change among Siwalik herbivorous mammals in response to Neogene and Quaternary climate changes: An appraisal in the light of new data," *Quaternary International* 371 (2015): 232–43.

21. Chatterjee, D., "Studies in the endemic flora of India and Burma," *Journal of the Royal Asiatic Society of Bengal (Sci.)* 5 (1939): 19–67.

22. Ashton, Peter S., and Zhou, H. 2020. The tropical-subtropical evergreen forest transition in East Asia: An exploration. *Plant Diversity* 42 (2020): 255–80.

23. Raes, Niels, et al., "Historical distribution of Sundaland's Dipterocarp rainforests at Quaternary glacial maxima," *Proceedings of the National Academy of Sciences U.S.* 111 (2014): 16790–795.

24. Morley, Robert J., "Palynological evidence for Tertiary plant dispersals in the SE Asian region in relation to plate tectonics and climate," in *Biogeography and Geological Evolution*, ed. R. Hall and J. D. Holloway (Leiden: Backhuys, 1998), 211–34; Morley, *Origin and Evolution of Tropical Rainforests.*

25. Symington, C. F. (revised by P. S. Ashton and S. Appanah), *Forester's Manual of Dipterocarps*, Malayan Forest Records 16 (Kepong, Malaysia: Forest Research Institute of Malaysia & Malayan Nature Society, 1943 and 2004); Parnell, John, "The biogeography of the Isthmus of Kra region: A review," *Nordic Journal of Botany* 31 (2013): 1–15.

26. Philippe, Marc, Nareerat Boonchai, David K. Ferguson, Hui Jia, and Wickanet Songtham, "Giant trees from the Middle Pleistocene of Northern Thailand," *Quaternary Science Reviews* 65 (2013): 1–4.

27. Corner, E. J. H., "The evolution of tropical forest," in *Evolution as a Process*, ed. J. S. Huxley, A. C. Hardy, and E. B. Ford (London: Allen and Unwin, 1954), 34–46.

28. Slik, J. W. Ferry, et al., "Soils on exposed Sunda shelf shaped biogeographic patterns in the equatorial forests of Southeast Asia," *Proceedings of the National Academy of Sciences U.S.* 108 (2011): 12343–347; Slik, J. W. Ferry, et al., "An estimate of the number of tropical tree species," *Proceedings of the National Academy of Sciences U.S.* 112 (2015); K. Wong and Neo, "Species richness, lineages, geography, and the forest matrix; Barthlott, W. et al., "Geographic patterns of vascular plant diversity at continental to global scales," *Erdkunde* 61 (2007): 305–15; Kier, G., et al., "Global patterns of plant diversity and floristic knowledge," *Journal of Biogeography* 32 (2005): 1107–16.

29. Raes, Niels, *Borneo: A Quantitative Analysis of Botanical Richness, Endemicity and Floristic Regions Based on Herbarium Records* (Leiden: National Herbarium of the Netherlands, 2009).

30. Morley, Robert J., Harsanti P. Morley, and Tony Swiecicki, "Mio-Pliocene Palaeogeography, uplands and river systems of the Sunda region based on mapping within a framework of VIM depositional cycles," *Indonesian Petroleum Association, Proceedings*, May 2016: IPA16–506-G.

31. Morley, Robert J., "The complex history of mountain-building and the establishment of mountain biota in Southeast Asia and Eastern Indonesia," in *Mountains, Climate and Biodiversity*, ed. Carina Hoorn, Allison Perrigo, and Alexandre Antonelli (London: Wiley-Blackwell, 2018), 475–93.

32. Muellner-Riehl, Alexandra N., et al., "Origins of global mountain plant biodiversity: Testing the 'mountain-geobiodiversity hypothesis,'" *Journal of Biogeography* 46 (2019): 2826–38.

33. Merckx, Vincent S. F. T., et al. (50 co-authors), "Evolution of endemism on a young tropical mountain," *Nature* 523 (2015): 347–50.

34. Igea, Javier, and Andrew J. Tanentzap, "Angiosperm speciation cools down in the tropics," *Ecology Letters* 23 (2020): 692–700.

35. Thomas, Sean, *Interspecific Allometry in Malayan Forest Trees*, PhD thesis, Harvard University, 1993.

In 1982, Ian Baillie and I presented a paper at a symposium on rainforest ecology at Leeds University. Based on evidence from plots of dipterocarp forests of central Sarawak, we revealed correlations between soils and the tree flora, concluding that the commoner species are soil specialists. Steven Hubbell, then from the University of Iowa and unknown to me, described his analysis of tree populations in one huge, 1,000 × 500 m plot on Barro Colorado Island, in Panama. It showed no floristic correlation with habitat—only the inevitable clumping of individuals to be expected from limited seed dispersal. He concluded that these species were essentially dispersed by chance and were ecologically complementary. I was shocked and defensive, but I introduced myself and suggested we discuss our differences over a pint in a nearby pub. We did and agreed to replicate his plot in Malaysian dipterocarp forest. Enlisting the support of Salleh Mohd Nor and the Forest Research Institute, we jointly applied to the National Science Foundation for funding. Thus, the Center for Tropical Forest Science at the Smithsonian (later ForestGEO) was born. — P. A.

14

FOREST AND TREE DIVERSITY
Why Does It Vary, and How Is It Maintained?

Tropical rainforests contain more than half of the world's terrestrial biodiversity and probably most of world's genetic diversity. Most living plant biomass consists of trees that provide the structure of forest ecosystems, providing habitats for most other organisms.[1] Estimates of tree species richness and diversity therefore provide a simple and practical "proxy" measure of overall biodiversity. We distinguish between species *richness* and *diversity* (p. 14). Richness, as understood here, is therefore but one aspect of diversity, or α–diversity (p. 14). Genetic diversity is more difficult to measure. We therefore generally resort to *species richness* as a reasonable approximation.

To understand how biodiversity is maintained, we must know how it originated in the first place and how it continues to evolve, much of which has been discussed in earlier chapters. Understanding the factors influencing species diversity in ecosystems, with tropical ones as the most spectacular examples, is one of the deepest challenges in ecology and evolution.[2] We here emphasize local variation in diversity.

Diversity through Ecological Equivalence or by Specialization?

Whether niche-breadth varies with species diversity is the question central to understanding how species diversity has arisen and is maintained. Naturalists assume that each animal species occupies a unique role in an ecosystem, but plants–especially large trees–may not. Dispersal may be constrained, reducing physical competition among individuals of any pair of species. Most seeds drop close to the parent crown, so most competition is at first intraspecific; but interspecific competition increases as the tree grows and occupies a larger area.

Mating is similarly restricted by limits on pollen dispersal, particularly in the most species-rich communities, such as MDF. Many tree species could be ecologically equivalent, surviving in a mixture where limited dispersal restricts the distance between seeds.

Stephen Hubbell, from data at the Barro Colorado Island (Panama) and Pasoh (Malaysia) ForestGEO 50-ha plots, showed that distributions of tree populations within such samples of species-rich rainforests are consistent with the ecological equivalence of tree species, and that constraints on their seed dispersal are sufficient to account for the observed spatial structure of these forests (p. 306).[3] Hubbell's *Unified Neutral Theory of Biodiversity and Biogeography* extended MacArthur and Wilson's *Theory of Island Biogeography*,[4] applying it to communities, landscapes, and regions *within* continental land masses. It assumes ecological complementarity and uniformity of distribution over the landscape at every scale. But how far does the relative role of ecological equivalence—"dispersal assembly" over "niche assembly"—extend beyond the 50 ha plots to larger scales? We argue that this hypothesis cannot be universally applied, even though limited seed dispersal undoubtedly constrains competition between individuals of different species in hyper-diverse communities on uniform terrain. The roles of natural selection and niche specificity, on the one hand, and random factors (including history and dispersal constraints), on the other, are theoretically and practically important. How predictable, for example, is species composition within a specific habitat, irrespective of its history? Hubbell's theory provides us with a rigorous null hypothesis against which to test conclusions reached from empirical observations. However, his neutral model fits other hypotheses of niche differentiation as well as those supporting coexistence by fitness equivalence. Ultimately, these questions can be resolved with rigorous experimentation made feasible by the ForestGEO plots (p. 429), with hypotheses derived from natural history. We will now summarize conclusions of previous chapters in a dynamic evolutionary context, using tree species richness as the central focus. What patterns can be discerned in the variability of tree species richness and community diversity among the forests of tropical Asia? What historical, physical, or biological environmental factors are correlated with these patterns?

Species diversity can be estimated by many measurements. Some are highly sensitive to the size of the sample, and some assume random distributions of individuals that are clumped within populations. The measure most commonly used for plant communities is Fisher's α. It may underestimate species diversity where populations are highly clumped, but it is a valid comparison among different tropical evergreen forests within a region, since seed dispersal distances vary more between genera than between species, and regional forests share most of their genera, with constant numbers in the evergreen forests of tropical Asia.

To summarize, in Hubbell's words: "The argument between niche-assembly and dispersal-assembly perspectives is long-standing. It has persisted so long precisely because each perspective has strong elements of truth and because reconciling them is non-trivial."[5]

Patterns of Tree Species Richness and Community Diversity

Various factors help explain tree species richness as well as community diversity. patterns. Are they consistent with the neutral theory, or with niche specificity?

The regional scale. Patterns may vary at the regional scale not only from the influence of climate, but also from different histories of extinction and opportunities for immigration. In tropical Asia, the most recent period of catastrophic climate change, the Pleistocene epoch, has been particularly influential (p. 64).

The theory of island biogeography (p. 308) predicts that, for every order-of-magnitude increase in island area, species number should double, if immigration rates are equal. It is a "neutral" theory, predicting that species are equally likely to occupy any space. It predicts that restricted seed dispersal leads to distance-correlated changes in the composition of plant communities that occupy common habitats, and that these changes are measurable at regional and local scales. Plant diversity within a community is also dependent on the conditions for growth and production. High tropical biodiversity is in fact concentrated in a relatively small area of the wet tropics, in the lowlands below ~700 m. Species diversity also varies greatly among different tropical forest ecosystems, even within the most biodiverse equatorial regions. The maximum tree species diversity within a homogeneous climate and soil environment is similar on all three tropical continents, independent of the different forest areas under a given rainfall regime in each. Nevertheless, other patterns caused by historical geography, varying dispersal opportunity, and regional catastrophe histories affect diversity in each region.

Precipitation. Worldwide in the tropics (and even at the local scale), high lowland tree species diversity correlates with rainfall aseasonality. This correlation itself implies specialization, and it is therefore contrary to neutral theory at the regional scale—though not necessarily at the local scale.

Evergreen forest communities are generally more species-rich than deciduous ones (table 14.1). They differ both in the Far East and in South Asia, where lowland deciduous forests in monsoon climates are a quarter as diverse as those of the lowland evergreen rainforests in everwet regions. Species survival is reduced by drought, fire, and cyclonic storms.

Age. Species richness may vary with the geological age of community habitats. In Asia, expectations from island biogeography theory appear to be contradicted, for example, when the number of tree species per area in the rainforest of everwet southwest Sri Lanka (less than 1,000 species in ~10,000 km^2) is greater than that number in the Western Ghats seasonal evergreen forests (~800 species in ~40,000 km^2). Assuming island biogeography is correct, these areas either had different immigration or extinction rates—perhaps as historical extinction events from which they are still recovering, not yet reaching equilibrium.

In everwet West Malesia, as in southwest Sri Lanka, climate and island biogeography—the two fundamental influences on species diversity—can be teased apart. The total tree flora of Sumatra is poor compared with that of much smaller Peninsular Malaysia (table 14.2). Neither is the tree flora of Borneo richer than that of Peninsular Malaysia in the proportion predicted by island biogeographic theory, despite its greater soil diversity. The estimated combined tree flora of lowland Sabah, Sarawak, and Brunei in everwet northern Borneo (~5,500 species in ~75,000 km^2) is as great as—or greater than—the tree flora of the everwet lowlands of upper equatorial Amazon (~150,000 km^2). These data are inconsistent with predictions of the theory of island biogeography (and therefore neutral theory).

In summary, the numbers of tree species known from each of these regions contravene simple island biogeographic expectations, but they do so for reasons consistent with another aspect of neutral theory: tropical tree species disperse and evolve slowly, delaying predicted equilibria.

Neutral theory also predicts that, owing to the effect of limited seed dispersal, the ranges of species that have yet to exploit the full extent of their available habitat vary geographically at random. Condit found that species

Table 14.1 Composition of the tree floras from the ForestGEO and selected other plots in the Asian tropics SEDF = southern seasonal evergreen dipterocarp forest.

Location	Area (ha.)	Min. diam. (cm)	Trees/ha	Basal area/ha (m²)	No. species	Fischer's α	Forest type	Altitude (m)
Pasoh	50	1	1788	31	814	100	MDF	100–125
Lambir	52	1	2664	43	1192	153	MDF	200–235
Bt. Timah	2	1	2473	35	347	60	MDF	69–124
Palanan	16	1	3379	40	336	46	MDF	100–150
Sinharaja	25	1	5016	46	205	23	MDF	430–570
Khao Ban Thad	20	1	4110	34	596	84	S SEDF	150–300
Bidoup (Vietnam)	25	1	4913	…	450	55	Semi-evergr.	1516–1597
Huai Kha Khaeng	50	1	1476	31	291	31	Semi-evergr.	549–638
Thai tall decid.	16	1	…	…	…	10–23	Tall decid.	300–650
Mudumalai	50	1	1250	25	71	9	Tall decid.	550–650
Halmahera	0.5	10	742	…	76	…	E. Mal. SEDF	630
Wanang (PNG)	1	5	5395	…	580	70	TRF	440–600
Xishuangbanna	20	1	4792	…	468	42	Northern SEDF	600–700
Nanjeng Shan	3	1	3582	36	118	16	N. coast SEDF	150–200
Doi Inthanon	15	1	4910	40	192	20	Lower montane	1663–1737
Gutianshan	24	1	5827	…	159	18	Warm temp. evergr.	446–714

Source: OTFTA, table 3.7, p. 196.

Table 14.2 The areas and number of tree species on the three major Sundaland land masses

Land mass	Area (km^2)	Number of Tree Species
Peninsular Malaysia	131,565	3,100
Sumatra	524,100	~2,500
Borneo	757,050	~6,000

Source: OTFTA, table 7.1, p. 452.

turnover rates were too low to match neutral theory predictions at regional scales in the Neotropics but were consistent at scales of less than 50 km.[6] Similar results have been documented over regionally uniform terrain in West Africa and have also been reported from Pasoh.[7]

Remarkably, each of the large lowland plots of the ForestGEO network captures, within 25 ha, approximately one-half of its known *local* tree flora, and fully a quarter of its entire *regional* flora. Tropical trees have high capacities to disperse over time, despite short-term dispersal constraints. Around 12 Ka, species reoccupied continuous habitats within large, previously seasonal areas in Peninsular Malaysia and Borneo. Rarely, some dipterocarps have hopped islands from Sunda to New Guinea.

Phylogenies of species-rich rainforest tree genera indicate that most speciation has been quite recent, between ~10–3 Ma, and has subsequently declined. Speciation spurts during "evolutionary episodes," when areas of new habitat become available. Afterward, rates may slow as niche space saturates under uniform climate. Species occupying the continuous loam habitat in southern Borneo dispersed up to 500 km from refugia during the past 12 Ka, at rates of ~50 m/y. Either dispersal constraints are less than those assumed by neutral theory, or late Pleistocene climate change was minimal, with shorter reoccupation distances.

Habitat islands. Some habitats are naturally more fragmented within a region. In Borneo, white sand podzols are confined to plateaus, ridges, and raised Pleistocene beaches, forming "islands" rarely greater in area than 50 km^2. Yellow humult sandy soils occur on exposed geological formations, rarely exceeding 1,000 km^2. The most species-rich Sunda floras of habitat islands, *kerangas* and the MDF flora on yellow sand, occur in northwest Borneo.

The Riau Pocket. Northwest Borneo has the richest tree flora in the Old

World, ~4,500 species. The richest individual communities there are on the yellow sandy humult soils. These harbor most species of the plant geographic province known as the Riau Pocket (p. 296). Its soils are principally responsible for the high species diversity. Yet they are discontinuous through the region—ecological islands isolated by river systems. The Miocene sandstone of the Lambir Hills, an ecological island of less than 500 km², contains ~2,500 tree species, half the estimated total for all northern Borneo. Four-fifths as many dipterocarp species are confined to the ecological islands of deep yellow humult sands of the Lambir and Andulau hills (each ~500 km² and separated by 60 km of peat swamp) as are restricted to the clay loams that continue unobstructed for ~50,000 km² across northeast Borneo. The extraordinarily slow-growing, emergent heavy hardwood, *Shorea geniculata,* whose heavy, wingless, golf-ball-sized fruits have no known means of disper-

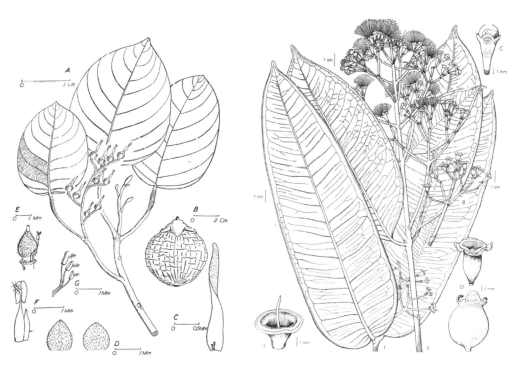

Fig. 14.1 *Shorea* versus *Syzygium* in Borneo. *Left, Shorea geniculata,* with wingless, golf-ball-sized fruit and small flowers; *right, Syzygium stipitatum,* with showy shaving brush flowers and fleshy fruit (P.A.; see Peter Ashton, "Dipterocarpaceae," and "Myrtaceae," in *Tree Flora of Sabah and Sarawak,* ed. E. Soepadmo, L. G. Saw, and R. C. K. Chung [Kepong: Forest Research Institute of Malaysia, 2004 and 2011], 5:63–388, and 7:87–330).

Table 14.3 Degree of endemism of Bornean *Syzygium* (Myrtaceae), with vertebrate dispersed fruit, and *Shorea* (Dipterocarpaceae), with fruit dispersed by gyration or not at all, compared. Numbers of species, with percentages of genus total in parentheses.

Genus	Occurring beyond Borneo W. of Wallace's Line	Widespread in Borneo only	Borneo N. of the Equator	Local Bornean habitat island(s)
Syzygium (*N* = 181)				
Totals and percentages	72 (40)	31 (17)	29 (16)	49 (27)
Montane alone	0	0	8 (4)	15 (8)
Shorea (*N* = 296)				
Totals and percentages	106 (36)	44 (15)	109 (37)	37 (12)
Montane alone	2 (1)	0	11 (4)	3 (1)

Source: OTFTA, table 7.4, p. 457.

sal, occurs in some of these habitat islands from west Sarawak to Brunei, 500 km along the coast (fig. 14.1). Despite these dispersal limitations, *S. geniculata* persists on ecological islands of less than 500 ha.

Syzygium versus Shorea. The two most species-rich genera of trees in Borneo, *Shorea* (Dipterocarpaceae) and *Syzygium* (Myrtaceae), vary morphologically within species and habitats throughout northwest Borneo (table 14.3). They are dispersed differently. *Syzygium* species are as well dispersed as any in the forest's mature phase (p. 105). *Syzygium* is bee-pollinated, but some species attract butterflies and birds. Its seeds are dispersed by birds, primates, and small, arboreal vertebrates, such as squirrels. Many species arrived in Africa by dispersal from Asia across the Arabian Sea. Fleshy-fruited Myrtaceae have diversified on a much larger scale than genera with small-winged seeds and more limited dispersal in windless climates.[8]

This contrasts with the Bornean *Shorea*, pollinated mainly by thrips—no more capable of dispersing pollen over distance than the wind itself—and other small insects (p. 240). Their fruits are either winged and gyrate to the ground, only occasionally being caught in squalls, or are wingless and dispersed by water or gravity. These species should be more habitat-specific, with more fragmented distributions and higher local endemicity. However, endemism and intraspecific morphological variation among the two genera are almost identical. *Syzygium* contains more local endemics than *Shorea* (table 14.3). These endemics are habitat specialists, and sometimes are sister species to others in adjacent habitats. Biological limitations of *Syzygium* may counteract its apparent dispersal advantages over *Shorea*.

At the community scale. Within a regional forest formation, communities with different species compositions establish under different influences.

Influence of the regional flora. following neutral theory predictions, if communities contain a random mix of species constrained only by variation among species in their evolutionary age and means of dispersal, species diversity *within* communities should be correlated with the size of their regional flora—the number of species available for immigration. This is only partly the case (table 14.1). The 25 ha ForestGEO plot in Sinharaja Forest (Sri Lanka) contains only 208 tree species, barely a quarter of the number at Pasoh. Similarly, Fisher's α among plots in Indo-Burmese tall deciduous forests ranges from 10.5–23, higher than its value for the tall deciduous forest at Mudumalai in South India (table 14.1).

In these cases, differences in species diversity do reflect differences in the richness of their regional tree flora. But the decline in community species richness from everwet Pasoh to Khao Chong, in Peninsular Thailand, with its short dry season, parallels the decline in community richness from everwet Sri Lanka to the somewhat seasonal southern Western Ghats, unrelated to the sizes of their tree floras.

Communities with higher species richness than those in northwest Borneo are in South Kalimantan and the Tigapuluh mountains in East Sumatra (table 14.4), and also occur with similar richness on loams in East Kalimantan. The Sumatran sample is particularly surprising because of the small lowland Sumatran tree flora (table 14.2). From Brunei and ForestGEO Sunda plots, the greatest number of tree species in a community or uniform habitat is about 750. Remarkably, the range of Fisher's α in 0.4 ha subunits of the physically more uniform Pasoh ForestGEO plot (42–162 for all trees >1 cm DBH) falls *within* the range in the Lambir plot (14–191, fig. 14.2); these two forests (table 13.2), with only one-quarter of species in common (22% of Lambir species at Pasoh, 32% of Pasoh species at Lambir), have the same richness in shared habitats. The estimate of Fisher's α for MDF in Sumatra was 224, a global record, promoted by high habitat diversity. Sumatran MDF communities are similar in richness to those elsewhere in the Sunda forests. This implies an asymptote in maximum available niche space in MDF on zonal udult soils, but some species richness of individual communities correlates with their differing habitats.

If a ceiling in community richness is reached in a habitat, irrespective of

Table 14.4 Species diversity (Fisher's α), recorded by various workers in Sunda MDF

Site (altitude, m)	Area (ha)	No. of species	No. of individuals	Spp./500 individuals	Fisher's α
Borneo					
MDF					
Andulau, Brunei (50)	0.96	256	621	235	163
Kuala Belalong, Brunei (250)	1.0	231	550	220	150
Wanariset, E. Kalimantan (50)	1.6	239	866	168	109
Lempake, E. Kalimantan (60)	1.6	209	712	171	100
Kerangas					
Mulu, Sarawak (20)	1.0	123	708	109	109
Bukit Sawat, Brunei (15)	0.96	121	734	114	114
Malay Peninsula MDF					
Sungei Menyala, Negeri Sembilan (30)	2.02	232	953	177	97
Pasoh, Negeri Sembilan (90)	50	683	26,450	201	125
Bukit Timah, Singapore (30)	2	182	847	123	71
Sumatran MDF					
Batang Ule (150)	3	502	1885	263	224
Ketambe (50)	1.6	116	861	97	36

Source: OTFTA, table 3.7, p. 196.

area or age, do species only immigrate by competitively excluding an existing member? Why does this apparent asymptote vary with soil and climate (fig. 14.3)? Long-term monitoring of immigration and extinction rates at community scale is possible in the large tree demography ForestGEO plots.

Latitude, temperature, and moisture. Does diversity vary between communities with different habitats? Mean winter temperatures decline with increasing latitude, but summer monsoon temperatures vary more with atmospheric humidity. In Indo-Burma, the northern seasonal evergreen forests are poorer in tree species than the southern, but history and isolation may

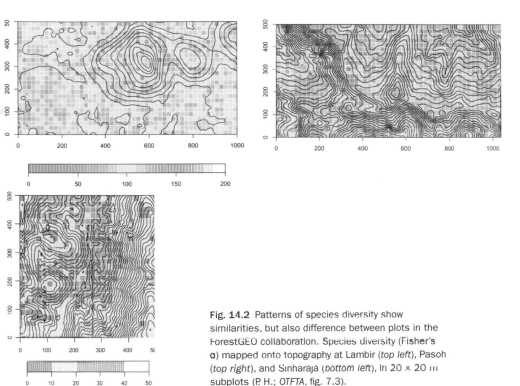

Fig. 14.2 Patterns of species diversity show similarities, but also difference between plots in the ForestGEO collaboration. Species diversity (Fisher's α) mapped onto topography at Lambir (*top left*), Pasoh (*top right*), and Sinharaja (*bottom left*), in 20 × 20 m subplots (P. H.; *OTFTA*, fig. 7.3).

better explain this than temperature. Temperature-derived patterns of richness among the lowland forests of tropical Asia are obscured by other factors. Species richness of evergreen forests also declines with altitude in everwet regions (p. 199). This regional decline in richness relates more to the reduced evergreen species' component, with smaller compensatory increase in deciduous species in the forest formations from MDF through seasonal evergreen and semi-evergreen to tall and short deciduous forest. This correlates more with rainfall seasonality rather than temperature or latitude. It is only broadly reflected in the Amazonian hylaea, where species richness in climates with up to three dry months near Manaus can be greater than near the aseasonal Andean foothills. There, the dry season is more intermittent than under the Asian monsoon climate.[9] Drought may lower production and increase tree mortality in everwet forests, thereby affecting forest structure and composition.

The highest species diversities are in tropical forests with very high pre-

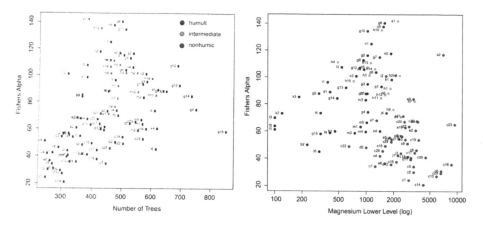

Fig. 14.3 Correlations with species diversity (Fisher's α) in 105 0.6 ha plots (trees >9.6 cm DBH (see *OTFTA*, fig. 7.2, 459). *Left*, with total concentrated HCl extractable magnesium in the mineral soil at 70–80 cm; *right*, with stand density—or the number of trees in the plots (P. H.). Black dots are humult sandy and sandy clay soils, and red dots are udult clay soils (P. H.).

cipitation totals every month, even though forest structure and geological and climatic history may vary.

Most organisms increase in diversity toward the equator. Analyzing seedling competition within ForestGEO plots at different latitudes has revealed a mechanism explaining gradients in tree diversity. Conspecific negative density dependence—which provides more space for other species, as predicted by the Janzen-Connell hypothesis (p. 254)—was reduced with higher latitude. Furthermore, the longer growing season at lower latitudes reduces potential inter-species conflicts, and reductions in climatic stress at low latitudes contribute to this diversity gradient.[10]

Forest structure. Light intensity and sun-fleck frequency within forest canopies increase with height above ground, creating a vertical resource gradient (p. 94). Subcanopy species within series of co-occurring congenerics do indeed differ in the height at which reproduction commences and in light response according to their height. Trees (like fish) continue to grow throughout life in both diameter and crown size, but reach an asymptotic height, as injury increasingly constrains further extension (p. 100). The range of species heights at maturity is also constrained by canopy stratification (p. 118). Height diversity is least in the emergent stratum, although variation in overall mature-phase canopy height is greatest in forests on soils of intermediate fertility, where species diversity also reaches its peak.

Canopy height is reduced by the frequency and severity of water stress, mediated by the soil's capacity to store and release water (p. 83). In MDF, in all three canopy-height guilds, species diversity is greatest on intermediate-capacity soil types. Peak species diversity therefore coincides with forests of most complex structure (and their peaks in heterogeneity of mature stand structure) on udult short lattice clay and udult and humult sandy clay soils. Forest structure tends to covary with both soil fertility and water economy (fig. 7.1).

Mixed dipterocarp forest. Variation in species richness, as Fisher's α, can be mapped as subplots as small as 20 × 20 m (400 m^2) within ForestGEO MDF (fig. 14.2). We include three large plots: Pasoh, in Peninsular Malaysia (p. 94); Lambir, in Sarawak (p. 41); and Sinharaja, in southwest Sri Lanka (p. 74). Pasoh, with gentle topography and relatively uniform sandy clay soils, has the least variation in species diversity. The other two have steep topography and are more comparable. The Lambir plot's ridge consists of a sandstone slab, with slopes of varying nutrients (p. 186). Sinharaja has no such gradient in soil nutrients or physical composition. Soils on ridges are shallow at Sinharaja, but at Lambir, they are sandy, deep, and free-draining—and drought-prone.

Forest dynamics. Species diversity varies over these three landscapes. In all, pioneers vary little in richness, and contribute little to total diversity. Diversity is governed by variation in the later successional and mature-phase species. In both Lambir and Sinharaja, the lower slopes have clay loam soils of moderate nutrient level, with frequent ground movement and canopy disturbance. They are among the most species-diverse habitats, which is attributable to the mostly mesophyll, late succession species. At Pasoh, there is a flat "upper alluvium," where the microphyll heavy hardwood *Shorea maxwelliana* dominates the canopy. At Lambir, stable ridgetop sites with shallow or sandy soils, dominated by microphyll *Dryobalanops aromatica* (fig. 9.4), also support less diverse communities than the unstable lower clay slopes. However, its most diverse subplots are on gently sloping sandstone slopes, with deep, *stable* low-nutrient sandy soils, a more diffuse notophyll-dominated canopy, and mostly small single-tree gaps.

Variation in diversity among Far Eastern MDF correlates with mature-phase forest structure and soil physical characteristics (figs. 7.1 and 14.2; p. 174). In a forest on moist, friable clay loams (fig. 7.1), overall diversity is low, particularly in the dense emergent and diffuse main canopy strata, whereas the subcanopy synusium is rich in shade-tolerant species benefiting from the low but homogeneous light climate. Here, growth rates among the species, from pioneers to shade-tolerant understory, vary the most. The emergent canopy consists of a few, predominantly fast-growing, light red and yellow meranti *Shorea*, which have repressed all but their fastest-growing competitors by overtopping them with their dense, mesophyll foliage. In MDF on drought-prone sandy and other shallow soils, as at Bako (fig. 7.1), in a low-diversity community, species growth rates hardly vary. The forest is short, and the emergent stratum is diffuse, with periodic and species-specific, drought-induced mortality, while canopy foliage is notophyll and microphyll, and is itself diffuse, permitting more and patchier light beneath. It has relatively low richness, but more in main and subcanopy. In contrast, the distinctly 3-layered MDF on intermediate sites, as on deep, yellow humult sandy soils on the Lambir dipslopes (fig. 7.1), has high richness in all three strata. There the emergent canopy is patchy but reaches height comparable to that of trees growing on loam. Growth-rate variation is intermediate. Late succession and mature-phase juveniles prosper in a heterogeneous subcanopy light climate, while even the pioneer flora is richer than in other sites. The canopy gaps are small, and there is little soil surface disturbance, where a diversity of seedlings compete for light.

More canopy disturbance explains why, unlike in Far Eastern MDF, subcanopy species diversity at breezy Sinharaja is higher than that of the canopy. Stands on lower slopes at Sinharaja have variable, mature-phase canopies, including emergent dipterocarps, that vary in gap sizes. Thus, canopy structural and dynamic diversity account for the greater overall tree species diversity of valley forests here. At Lambir, species diversity is higher on the less-fertile and freely draining sandy soils, where single-tree mortality predominates, than it is on clay loams, where larger gaps from windthrow prevail. Richness can decline where surface soil has been disturbed, both in recent windfall gaps and on steep, unstable slopes with clay or sandy soils.

In these examples, the influence of topographic diversity and/or soil-nutrient levels affects *mature-phase* canopy structure sufficiently to override whatever impact there might be on richness from gap succession *within* communities.

Overall species diversity in different communities is influenced by individual structural and dynamic guilds with different proportions of species diversity. One striking example: species diversity at Lambir is similar to but a little lower than that in the 25 ha ForestGEO plot at Yasuni (Ecuadorian Amazon), which has a similar rainfall distribution (fig. 14.4). Soils and to-

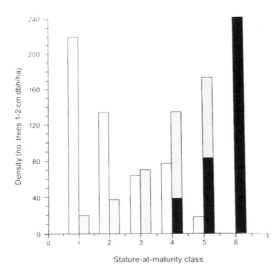

Fig. 14.4 The distribution of species diversity among canopy species in Far Eastern MDF strikingly differentiates them from hyper-diverse Neotropical forests: density of trees, by diameter classes, of families at Yasuni, Ecuador that do not occur in Asia (white columns), and at Lambir, Sarawak (grey, Dipterocarpaceae in black), that do not occur in the New World (I. B., after J. La Frankie; see chap. 14, n. 12). Size at maturity assumed from maximum DBH: 1–5 cm (1), >5–10 (2), >10–20 (3), >20–40 (4), >40–80 (5), >80 (6) (*P* < 0.0001).

pography at Yasuni are more uniform. Lambir's richness at the whole-plot scale is partially attributable to its high β-diversity, correlated with its diversity of soil types. Whereas tree species diversity is concentrated in the main canopy at Lambir, it is greatest in the understory at Yasuni, where the emergent canopy is scattered and species-poor.[11]

The latitudinal gradient in species richness and dynamics from northern Borneo to the forests fronting the Pacific coast of Taiwan recall the transect across coral reefs that inspired Connell (p. 254) to propose his "intermediate disturbance hypothesis."[12] Such disturbance provides different stages of successional recovery and opportunities for different organisms within a local habitat. The peak in species richness and diversity in tropical rainforests can also be reached where canopy disturbance is moderate, from drought and squalls, or modest from lightning or pathogens. In that calm Sunda climate, the role of pioneer diversity has been overtaken by that of structure and leaf physiognomy.

Soil nutrients. Species distributions are largely determined at the regional scale by rainfall seasonality; within a given rainfall regime, however, they are determined by soil-nutrient levels and—perhaps to a lesser extent in wet climates—by frequency and intensity of soil-water stress (p. 170). Similar results are now being reported from the Neotropics.[13] In Sarawak MDF, and probably in Peninsular Malaysia and elsewhere in the everwet or seasonal tropics, the primary influence of soil nutrients makes the difference between fertile udult loams lacking surface raw humus, and less-fertile humult soils on which a densely rooted surface mat of raw humus is present (p. 76). The physical and chemical environments in which seeds germinate and seedlings establish are substantially different on these soils. Regions where both soil types occur consist therefore of "mosaics," generally delineated by underlying lithology and each with characteristic species assemblages. In 105 0.6 ha plots in lowland MDF in Sarawak, nutrients correlated with variation in species composition, particularly on the less fertile soils (p. 318). The soil nutrients phosphorus, magnesium, and calcium were especially important (fig. 14.3). Maximum diversity peaked toward the low end of the nutrient range in all.[14] At Pasoh and Lambir, the values of α at the 400 m^2 scale, 14-191, varied in relationship to both soil nutrients and topography, and therefore to drought-induced canopy heterogeneity (table 13.2).

Such distinct patterns of species richness correlated with the physical habitat conflict with the neutral theory. On the other hand, the peaked distribution of maximum species richness in the forests' mature phase against soil-nutrient concentrations was predicted by David Tilman, who reasoned that diversity will be low in any habitat in which resources (such as nutrients) are limited, because the options for differential resource exploitation limit the available ecological niches.[15] As nutrient availability increases, variations in availability among the essential nutrients also increase, forming a spatial mosaic of nutrient availability with increased opportunities for niche specialization. As nutrient (particularly nitrogen) availability grows, and where other limiting factors (notably water stress) are low or constant, conditions will increasingly favor those species that are most competitive in capturing light. Tilman argued that increased maximum growth rates of a few canopy species will suppress species richness overall. In that, Tilman and Connell agree. Tilman's hypothesis is consistent with Russo's observations that species on low-nutrient soils compete through survivorship rates, whereas those on high-nutrient soils do so through growth rates (p. 186). In the Asian tropics, at least, nutrients are more tightly correlated with species distributions on the low and intermediate soil-nutrient levels than on the more fertile loams. The extraordinary number of tree species within a habitat means that Tilman's resource-ratio hypothesis cannot be the sole explanation for how such species richness is maintained. Species richness is therefore greatest where intermediate levels of nutrient availability and disturbance coincide. The level of disturbance determines a forest's structural complexity and the relative representation of its structural guilds; nutrient levels—when below a certain threshold—influence both the species composition and diversity of those guilds. In the Lambir ForestGEO plot, the comparatively low-nutrient, deep yellow sandy humult soils have higher species diversity, but diversity is reduced when samples include recent canopy gaps, especially those caused by landslides (fig. 14.2). The correlation between species diversity and soil nutrients does not, therefore, exactly parallel the correlation between species diversity and degree of canopy disturbance. Both independently lead to peaks at intermediate levels, but those peaks only occasionally coincide within a single habitat. The influence of each factor may therefore be identified separately. Both Connell and Tilman are right!

Ranks in abundance. Neutral theory also predicts no consistency in the rank order of abundance of species populations within a community other than at short distances, where seed dispersal limitations concentrate the leading species' progeny. Species composition is surprisingly consistent among communities sharing the same soil in Bornean MDF, irrespective of distance between sites (p. 318).

Dominance. Some species are common, and others are rare. The more common species are more consistent in their rank order of abundance, even in hyper-diverse MDF. Dominance of one species over others in mixture necessarily reduces the available space for other species occupying the same spatial guild, thereby reducing overall species diversity. Dominance by a dense-crowned species may reduce species richness by shading. A partial cause of the extraordinary diversity in coexisting species of upland tropical wet evergreen forests everywhere is their lack of individual species that achieve sole dominance in canopy or understory.

The consistent canopy dominance of dipterocarps in many Asian forests is unmatched by other families elsewhere in the tropics. In Asian lowland forests, legumes are among the tallest trees, but are less numerically dominant than dipterocarps. Such family dominance is confined to lowland yellow/red soils and everwet climates. Individual, drought-adapted dipterocarp species do achieve local dominance; sal (*Shorea robusta*) is the regional dominant on siliceous soils of northeast India (p. 184).

The tendency toward dipterocarp dominance in the everwet lowlands is likely enhanced by their unique reproductive biology (p. 236). Several dipterocarp species are monodominant in the emergent canopy (fig. 14.5), including *Shorea robusta*; *Shorea curtisii*, in the Peninsular Malaysian hill and coastal hill dipterocarp forests on the freely draining, granite-derived soils;

Fig. 14.5 Some dipterocarps frequently occur as monodominant stands. *Left, Shorea trapezifolia* in flower at The Peak Sanctuary, Southwest Sri Lanka; *right, Dryobalanops rappa* in the Anduki Forest Reserve, Brunei (all P. A.).

and *Dryobalanops aromatica,* formerly in extensive stands on sandy leached soils on the east coastal plains of Peninsular Malaysia and Sumatra.

Seven species of *kapur* (*Dryobalanops*) occur in northwest and northeast Borneo, where each occupies a distinct habitat with no overlap at maturity. Each establishes gregarious "drifts" of dominance in the canopy (fig. 14.5). Most are microphyll emergents. *Dryobalanops* species are pollinated by honeybees and flower more frequently than in the ENSO-related mass flowering years (p. 232). They establish higher densities of juveniles than other light hardwood dipterocarps. Fully explaining this extraordinary plasticity among *Dryobalanops* seedlings remains a challenge.

Single species dominate individual stands across nutrient gradients within the everwet tropics. However, dominance at *community* scale is most frequent toward the two ends of the nutrient gradient. Canopies of *Dryobalanops lanceolata,* or a small group of light hardwood *Shorea* species on well-watered fertile soils, dominate in stands with depressed richness in all vertical guilds, as Tilman predicted. Single-species dominance is also more common with increasing seasonality.

Single understory species may outcompete all others and, in the process, suppress the regeneration of canopy species. In the Asian lowland tropics, the preeminent examples are the bamboos and bertam, the stemless palm *Eugeissona tristis.* Bertam forms continuous populations along ridges in the

Peninsular Malaysian granite Main Range and the coastal hills, and is associated with seraya (*Shorea curtisii*), whose shade- and drought-tolerant seedlings survive beneath its fronds, where these are not dense. The dynamic relationship between bertam and seraya has yet to be fully deciphered. In seasonal Asia, dominance of genera or single species of bamboo in evergreen seasonal, semi-evergreen, and tall deciduous forests is widespread. The intermediate disturbance hypothesis (p. 321) predicts single-species dominance where disturbance is frequent or intense and species diversity is low.

A species may be regionally rare, or rare at the landscape scale, because its habitat is scattered or it occupies only a small area, but such species may be common and even dominant within that rare habitat. The converse attribute unique to speciose tropical lowland forest is that *most* species persist in low population densities within their optimal communities (fig. 14.6). At low densities, such populations may disperse in clumps due to seed dispersal limitations (fig. 14.7).

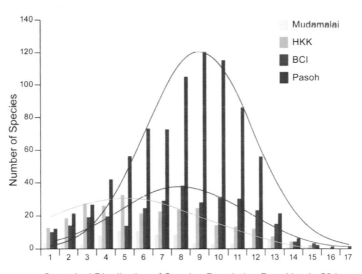

Canonical Distribution of Species Population Densities in 50 ha

Fig. 14.6 The number of species according to their population densities in ForestGEO plots, arranged by classes of doubling abundances (canonical distributions). Plots in climates of increasing rainfall seasonality with trees >1 cm DBH ha⁻¹: yellow, Mudumalai (353 trees); green, Huai Kha Khaeng (1,450 trees); red, Barro Colorado Island, Panama (4,600 trees); blue, Pasoh (6,700 trees). The peaks of species numbers move toward greater rarity in closed-canopy forests of decreasing seasonality, correlating with declining stand density (P. H.; see N. Manokaran et al., cited in fig. 12.3 caption; see also *OTFTA* fig. 7.8).

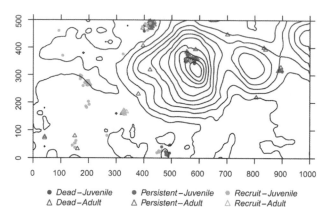

Fig. 14.7 Clumped population distribution over 20 years of *Koilodepas longifolium* (Euphorbiaceae), an understory species with explosively dispersed seeds, in the 50 ha plot at Pasoh, West Malaysia (P. A.).

● *Dead–Juvenile*　　● *Persistent–Juvenile*　　● *Recruit–Juvenile*
△ *Dead–Adult*　　　△ *Persistent–Adult*　　　△ *Recruit–Adult*

Some rainforest species are widespread throughout Sundaland, but at low density. These are niche-specific (notably with regard to soils). Wild mangoes and Malvaceae (with some durians) are always rare. Their seeds are dispersed by large mammals and birds, and seedlings are seldom clumped. All canopy durians are chiefly pollinated by the wide-ranging bat *Eonycteris spelaea* (p. 242); wild mangoes, by the equally wide-ranging giant honeybee (and other bee species, p. 248).

The relative abundances of *common* species within a specific habitat are more consistent than neutral theory predicts. Their abundance is due in varying degrees to competitive assets of the species themselves and to the prevailing disturbance regimes in the landscapes and communities they inhabit. Species with exceptionally low population densities in the humid tropics apparently avoid extinction at very low population levels. Such species in hyper-diverse rainforests die at high densities (p. 265).[16] This may prove to be a key mediator of species richness in hyper-diverse communities. It implies that the great majority of rare species are—in every respect but pathogen specificity—ecologically complementary within their guilds. Some may be holdovers from periods when the climate was drier or more seasonal, and they were more abundant. These species, where rare, may become ecologically complementary to more abundant species when in very low numbers, and thus conform to the predictions of neutral theory. They may gradually accumulate in equable climates and be a major cause of species richness in regional refugia.

Congeneric species series. Most of the variation in species diversity among similar forests is due to co-occurring species within a small number of genera in several different families. In each forest, these variably speciose genera come from the same small generic pool, and the number of species substantially determines the overall species diversity of a community.

The main reason for variation in species richness among ForestGEO plots in tropical Asia is the size of their congeneric species-series. The Pasoh ForestGEO plot in Peninsular Malaysia has far fewer species compared to Lambir in Borneo, yet the numbers of genera and families are similar (table 13.2). Areas with more species in higher-level taxa—such as northern Borneo—have had more speciation since ~8 Ma; much of it is a response to soil diversity. Northern Borneo's climatic stability has allowed higher taxa to persist and fostered speciation within them. High local endemism may thus reflect high rates of survival, rather than high rates of speciation.

Furthermore, the genera that vary in co-occurring species remain the same across major biogeographic boundaries such as Chatterjee's Partition and Wallace's Line, where overall community richness changes. Evergreen forests have more speciose genera than deciduous forests because speciose deciduous genera are rare or absent from evergreen forests. *Dalbergia* and *Lagerstroemia* are the only speciose deciduous genera in Asian tropicsl forests, and both contribute to the richness of semi-evergreen forest. Deciduous forests are less rich because they have few speciose genera, and also because they have few species overall, especially in Indo-Burma. Semi-evergreen and tall deciduous forests gain some species from each, compensating for their general poverty of species in the larger evergreen genera. Fedorov suggested that these species series, so characteristic of diverse tropical rainforests, are ecologically equivalent and therefore adaptively neutral.[17] However, if niches are partitioned among them, they should be subject to the most stringent competition. Therefore, this pattern can only be explained by something intrinsic to the genera themselves. Evidence for competition leading to niche differentiation among species co-occurring within a community is thus a critical test of the role of neutral factors in structuring species α-diversity. The fate of hybrids and introgressants among seedlings of *Shorea* species in a Singapore forest is such a test (p. 229).

Certain genera neither form co-occurring species series nor are rich in endemics, including those, often in low densities (p. 261) whose seeds are dispersed by primates and other large mammals (such as *Durio, Mangifera,* and *Nephelium,* among which up to four may co-occur), and genera whose flowers are bat-pollinated. Givnish suggested that the dense subcanopy of evergreen forest supports more sedentary frugivorous birds, resulting in short seed dispersal distances and therefore higher local endemism.[18] Many congeneric species series are indeed in predominantly subcanopy, bird-dispersed genera—including Myristicaceae and Ebenaceae, and the predominantly pioneer *Macaranga*—but with no greater local endemism. Then there are the rich congeneric species series of mostly emergent *Shorea* and *Dipterocarpus* (p. 238).

Species differences in seed dispersal and pollination do not exclude other ecological equivalencies, which may affect their interspecific competitiveness at another stage of their life cycle. Co-occurring figs bear fruit of different crop sizes, fruit dimensions, and colors, attracting different though overlapping guilds of frugivores, and establish at different sites within the forest canopy (p. 245).

There are 52 distinct *Syzygium* taxa within the ForestGEO Lambir plot, and 45 within the edaphically and topographically more homogeneous Pasoh plot. It is unlikely that each species has a significantly distinct set of life-history characteristics. Six *Shorea* (sect. *Mutica*) species occur at Pasoh, 14 at Lambir, and 11 in the Semengoh forest in West Sarawak, but only four in each are abundant and flower sequentially. The rest are intruders from adjacent soils in which they maintain viable populations; most species with low numbers in one habitat are more abundant in another. The proportion of those species that exist at low numbers in the absence of preferred habitats and may survive by chance needs study.

The coexistence of so many species-rich subcanopy and canopy *genera* is baffling. One possible explanation is that the genera differ in arthropod predators as well as pathogens (p. 268);[19] thus, species population sizes are limited. Such species within each genus survive with species in other co-occurring genera under reduced competition, with greater ecological complementarity in other respects.

Competition and spatial patterns. Common coexisting congeneric species may respond differently to light, to canopy gap size, and to position within gaps. Nevertheless, competition for space among individuals may be sufficiently frequent to differentiate their population habitats over many generations. Detecting such changes and proof of their cause is central to defining the relative roles of chance and determinism in mediating the structure of hyper-diverse plant communities—but it is fiendishly difficult to achieve!

Most species surviving at low densities don't compete for space. General competition among individuals for spatially variable resources (light, water, nutrients) is more likely to form complex local mosaics *within* communities.

Differentials in survivorship and performance *together* can mediate the varying outcome of multispecies competition in a mosaic of soils of differing fertility, thereby leading to changes in the rank order of abundance, at least of commoner species on different soils (p. 174).[20]

The *relative* strength of neutral versus deterministic influences on competition in hyper-diverse forests is central to our understanding of how species diversity is maintained. The enigma remains: absence of habitat-correlated spatial patterning must imply random (and therefore ecologically neutral) interspecific competition, until other evidence is found to the contrary.

Community reproductive phenology. Extended seasons of flowering and fruiting with little year-to-year variation in phenology should offer the best opportunities for niche specialization among pollinators and seed dispersers. These conditions should promote niche partitioning, through sequential flowering, of tree species sharing the same pollinator (p. 238). However, the supra-annual canopy mass flowering and intervening famine years of MDF in the aseasonal wet tropics of Asia provide the opposite. Phenological constraints on tree species diversification in MDF are overridden by other supporting forces. All the same, it is in this climate that communities supporting the diversity of dipterocarps appear to be in part due to sequential flowering times sustained by competition (p. 238)

Pathogens and herbivores. The great diversity of plant chemical defenses has promoted a comparable diversity of strategies among herbivores to circumvent them (p. 268). This interaction is a fundamental cause for the evolution of species diversity. Janzen and Connell (p. 254) hypothesized that tree species diversity should be enhanced by species-specific seed predators and tree pathogens killing conspecifics in high-density populations, thus freeing up space for individuals of other species to invade. These invading species could in principle differ solely in their degree of susceptibility to the same enemies. However, the general level of predator specificity appears no greater in biodiverse tropical forests than in species-poor temperate forests;[21] host-herbivore co-evolution cannot explain the greater species richness of tropical trees, but it could provide one explanation for their vastly greater generic representation. Pathogens may still provide opportunities for otherwise ecologically complementary species in different higher taxa to coexist.[22]

Juvenile (particularly seedling) mortality is correlated with the spatial density of a juvenile population. Species-specific pathogens can spread more easily when the hosts are close to one another (p. 254). Further, individuals in species with low population densities may be susceptible to pathogens, but they survive through escape (p. 254). This phenomenon, if indeed universal in speciose rainforests, could by itself explain how so many apparently ecologically similar species can coexist. It would do so by providing a further axis in niche space—in that respect contravening neutral theory.

Symbiosis and species diversity. Ectotrophic mycorrhizae (p. 269) are notably characteristic of the two tree families that dominate Asian rainforests—Dipterocarpaceae in the lowlands and Fagaceae in lower montane forests. Mycorrhizae form complex associations both with particular soils and with the particular dipterocarp species that grow in them (p. 270).

Has the ectotrophic association in Asia with Dipterocarpoideae promoted family dominance and hyper-speciation in some genera?

Hybridization between tree species grown together in arboreta is commonplace, so it is surprising that morphologically intermediate individuals of co-occurring congeneric species are so rare in nature. Hybridization among *Shorea* sp. at Singapore suggests that competition may select against the hybrids (p. 229).

Putting It All Together

Influences on species diversity in the tropics are various and complex in their interactions. All the major hypotheses that explain higher species diversity in the tropics operate in the Asian tropics, including the neutral theory of Hubbell, the density-dependent mortality hypothesis of Connell and Janzen, the intermediate disturbance hypothesis of Connell, and the resource-ratio hypothesis of Tilman. The strongest force promoting species richness has been that of natural selection, operating on gradients in climate, soils and their nutrient content, topography, water availability, interactions with other organisms, and time. These factors influence forest structure and complexity, further affecting species diversity.

Species age and turnover. The oldest tree species endemic to rainforest regions can be inferred from the durations of their climates. For everwet western Malesia, this is about 23 million years. Some widespread species with limited seed dispersal may indeed have arisen that long ago, but most endemic species are more recent. Low regional endemism within widespread habitats, such as the MDF of lowland Sumatra, implies that such forests are younger, reestablishing after the return of everwet conditions ~10 Ka. High endemism in species-diverse habitat-islands, such as Sarawak's Lambir

Hills, and the sharing of endemics between isolated islands imply longer periods of climatic stability—and opportunities for both immigration and speciation.

We can also infer the age of species from their geological and geomorphological histories, geographical distributions, and phylogenetic evidence. On this basis, most species are at least half a million years old, and many more than three million years. They have had time to colonize their appropriate habitats even when these are scattered over a region, and they have had opportunities for genetic exchange in continuous habitats.

Rates of tree speciation, especially in widespread and relatively continuous habitats, may be slowing. Extinction rates may be exceptionally slow in the everwet climate of Sundaland, offsetting low speciation rates in setting uniform maxima in regional species richness and community species diversity. Low speciation rates explain the rarity of point-endemics. Co-evolution of trees and the mobile organisms with which some are specifically interdependent (p. 223) results from the evolution of specific chemical signals by tree taxa.[23] Tropical biodiversity could be explained by faster rates of genetic processes in the tropics, thanks to the influence of continuous high temperatures. Whether this is so remains to be confirmed, but the rates of speciation inferred from the evidence reviewed here are unexceptional.

Community drift. The neutral theory predicts community species richness will eventually peak within any uniform habitat in a region sharing a common history, with a drift in species composition over distance; this is consistent with evidence, especially over Sundaland. Constraints of limited seed dispersal combined with random immigration and extinction from local to regional scales will produce drift. Compositional drift would be enhanced by speciation.

Local extinction and immigration events do occur among tropical tree populations, according to the neutral theory. Community composition and diversity within stands and small areas of uniform habitat (especially among species of lower density) may fluctuate at random. Community diversity within larger areas may remain constant, especially among more common species. Neutral forces mediated by seed dispersal alone can therefore help determine community composition and diversity. But the patterns of diversity predicted by neutral theory may also be consistent with those arising from niche differentiation.[24] Competitive success and niche breadth cannot be defined along single environmental gradients, but they occur through combinations of life-history traits. Furthermore, competitive attributes vary at developmental, interspecific, intergeneric, and interfamilial levels. Phylogenetic relationships therefore mediate the outcome of competition and the biological diversity that results.[25]

Gamma diversity. The regional gamma diversity in all the Asian tropics is probably much greater than in Africa and the Neotropics—despite the insufficient study of taxonomic diversity, even among trees, in tropical America. This may result from the complex geological and climatic history of Asia and partly from the dramatic effects of plate tectonics. The resulting landscape of large land masses, mountains, rivers, peninsulas, and islands along a longitudinal distance of 9,100 km is of a variety not matched by other tropical regions. Whether it is twice as great on an area basis (p. 11) remains to be determined.

Notes

1. This chapter condenses and updates *OTFTA*, chap. 7, 449–98.

2. Some recent articles representing a range of approaches to the problem of diversity: Terborgh, John. W., "Toward a trophic theory of species diversity," *Proceedings of the National Academy of Sciences U.S.* 112 (2015): 11415–422; Comita, L. S., et al., "Testing predictions of the Janzen-Connell hypothesis: A meta-analysis of experimental evidence for distance- and density-dependent seed and seedling survival," *Journal of Ecology* 102 (2014): 845–56; Ricklefs, Robert E., and F. He, "Regional effects influence local tree species diversity," *Proceedings of the National Academy of Sciences U.S.* 113 (2016): 674–79; Ricklefs, Robert E., and Susanne S. Renner, "Global correlations in tropical tree species richness and abundance reject neutrality," *Science* 335 (2012): 464–67; Kohyama, Takashi S., Matthew D. Potts, Tetsuo I. Kohyama, Abd Rahman Kassim, and Peter S. Ashton, "Demographic properties shape tree size distribution in a Malaysian Rainforest," *American Naturalist* 185 (2015): 367–79; LaManna, Joseph A., et al., "Plant diversity increases with the strength of negative density dependence at the global scale," *Science* 356 (2017): 1389–92; Zhu, Yan, Lisa. S. Comita, Stephen. P. Hubbell, and Keping Ma, "Conspecific and phylogenetic density-dependent survival differs across life stages in a tropical forest," *Journal of Ecology* 103 (2015): 957–66; Umaña, Natalia M., et al., "The role of functional uniqueness and spatial aggregation in explaining rarity in trees," *Global Ecology and Biogeography* 26 (2017): 777–86.

3. Hubbell, Stephen P., *The Unified Neutral Theory of Biodiversity and Biogeography*, Monographs in Population Biology 12 (Princeton, NJ: Princeton University Press, 2001).

4. MacArthur, R. H., and E. O. Wilson, *The theory of island biogeography*, Monographs in Population Biology 1 (Princeton, NJ: Princeton University Press, 1967).

5. Hubbell, *Unified Neutral Theory*, 25–26.

6. Condit, R., et al., "Beta diversity in tropical forest trees," *Science* 295 (2002): 666–69.

7. Okuda, T., H. Nor Azman, N. Manokaran, L. Q. Saw, H. M. S. Amir, and P. S.

Ashton, "Local variation of canopy structure in relation to soils and topography and the implications for tree species diversity in a rainforest of Peninsular Malaysia," in *Tropical Forest Diversity and Dynamism: Findings from a Large-Scale Plot Network*, ed. E. C. Losos and E. G. Leigh Jr. (Chicago: University of Chicago Press, 2004), 221–39.

8. Biffin, E., E. Lucas, L. A. Craven, I. Ribeiro da Costa, M. G. Harrington, and M. D. Crisp, "Evolution of exceptional species richness among lineages of fleshy-fruited Myrtaceae," *Annals of Botany* 106 (2010): 79–93.

9. Duque, Alvaro, et al., "Insights into regional patterns of Amazonian forest structure, diversity, and dominance from three large terra-firme forest dynamics plots," *Biodiversity Conservation* 26 (2017): 669–86.

10. J. LaManna et al., "Plant diversity increases with the strength of negative density dependence"; Usinowicz, Jacob, et al., "Temporal coexistence mechanisms contribute to the latitudinal gradient in forest diversity," *Nature* 550 (2017): 105–8; Chu, Chengjin, et al., "Direct and indirect effects of climate on richness drive the latitudinal diversity gradient in forest trees," *Ecology Letters* 22 (2019): 245–55.

11. Ashton, Peter S., "A contribution of rainforest research to evolutionary theory," *Annals of the Missouri Botanical Garden* 64 (1977): 694–705.

12. Connell, J. H., "Diversity in tropical rainforests and coral reefs," *Science* 199 (1978): 1302–10; Connell, J. H., J. G. Tracey, and I. J. Webb, "Compensatory recruitment, growth, and mortality as factors maintaining rainforest tree diversity," *Ecological Monographs* 54 (1984): 141–64.

13. La Frankie, J. V., "Lowland tropical rainforests of Asia and America: Parallels, convergence and divergence. In *Pollination Ecology and the Rainforest*, ed. D. J. Roubik, S. Sakai, and A. A. Hamid (New York: Springer Verlag, 2005), 178–90; La Frankie, J. V., et al., "Contrasting structure and composition of the understory in species-rich tropical rainforests," *Ecology* 87 (2006): 2298–2305.

14. Duinvoorden, J. F., J.-C. Svenning and S. J. Wright, "Beta diversity in tropical forests," *Science* 295 (2006): 636–37; Tuomistu, Hanna, K. Ruokalainan, and M. Yu-Halla, "Dispersal, environment, and floristic variation of western Amazonian forests. *Science* 299 (2003): 241–44.

15. Tilman, David, *Resource Competition and Community Structure*. Princeton Monographs in Population Biology 7 (Princeton, NJ: Princeton University Press, 1982).

16. Comita, L. S., H. C. Muller-Landau, S. Aguilar, and S. P. Hubbell, "Asymmetric density dependence shapes species abundances in a tropical tree community," *Science* 329 (2010): 330–32; Yenni, G., P. B. Adler, and S. K. M. Ernest, "Do persistent rare species experience stronger negative frequency dependence than common species?" *Global Ecology & Biogeography* 26 (2017): 513–23.

17. Federov, A. A., "The structure of the tropical rainforest and speciation in the humid tropics," *Journal of Ecology* 54 (1966): 1–11.

18. Givnish, Thomas J., "On the causes of gradients in tropical tree diversity," *Journal of Ecology* 87 (1999): 193–210.

19. Novotny, Vogitech, et al., "Why are there so many species of herbivorous insects in tropical forests?," *Science* 313 (2007): 1115–18; Novotny, Vogitech, et al., "Low beta diver-

sity of herbivorous insects in tropical forests," *Nature* 448 (2007): 692–97; Segar, Simon T., et al., "Variably hungry caterpillars: predictive models and foliar chemistry suggest how to eat a rainforest," *Proceedings of the Royal Society. B* 284 (2017): 20171803.

20. Russo, Sabrina E., S. J. Davies, D. A. King, and S. Tan, "Soil-related performance variation and distribution of tree species in a Bornean rainforest," *Journal of Ecology* 93 (2005): 879–89; Russo, Sabrina E., P. Brown, S. Tan, and S. J. Davies, "Interspecific demographic trade-offs and soil-related habitat associations of tree species along resource gradients," *Journal of Ecology* 96 (2008): 192–203.

21. Novotny et al., "Low beta diversity of herbivorous insects."

22. Roslin, Tomas, et al., "Higher predation risk for insect prey at low latitudes and elevations," *Science* 356 (2017): 742–44; Freckleton, R. P., and O. T. Lewis, "Pathogens, density dependence and the coexistence of tropical trees," *Proceedings of the Royal Society. B* 273 (2006): 2909–16; Terborgh, John W. "Enemies maintain hyperdiverse tropical forests," *American Naturalist* 179 (2012): 303–14.

23. Coley, Phyllis D., "Herbivory and defensive characteristics of tree species in a lowland tropical forest," *Ecological Monographs* 53 (1983): 209–33; Coley, Phyllis D., and Thomas A. Kursar, "On tropical forests and their pests," *Science* 343 (2014): 35–36.

24. Chave, J., "Neutral theory and community ecology," *Ecology Letters* 7 (2004): 241–53; Chave, J., "Spatial variation in tree species composition across tropical forests," in *Tropical Forest Community Ecology*, ed. W. P. Carson and S. A. Schnitzer (Chichester, UK: Wiley Blackwell, 2008), 11–30.

25. Wills, Christopher, et al., "Persistence of neighborhood demographic influences over long phylogenetic distances may help drive post-speciation adaptation in tropical forests," *PLoS ONE* 11 (2016): e0156913. doi:10.1371/journal.pone.0156913; Y. Zhu et al., "Conspecific and phylogenetic density-dependent survival"; Zhu, Yan, S. A. Queenborough, Richard Condit, Stephen P. Hubbell, Keping Ma, and Lisa S. Comita, "Density-dependent survival varies with species life-history strategy in a tropical forest," *Ecology Letters* 21 (2018): 506–15.

With my former student, colleague, and friend, I Goeste Made Tantra, I visited the sacred spring at Tampak Siring, on a mountain slope in Bali and at the base of what still remains of the Hutan Larang, a sacred forest (fig. 15.1). I bathed in the pool into which the spring gushes and meditated at the temple. I gained spiritual strength from immersion in these sacred waters and forests, as I also have in Java, India, Sri Lanka, Japan, and yes, Ireland, with the same experience gained by the forest farmers millenia ago. Sadly, the revered hill of the ancient Bretonic Celts near my home, Glastonbury—"hill of the sacred oak enclosure," is bald now. — P. A.

15

PEOPLE IN THE FOREST

In Asia, distinctive climatic patterns influence forest landscapes. These have affected human habitation and patterns of land use and wider aspects of social organization. Although our ancestors burned and thinned deciduous seasonal forests for two million years, until the advent of agriculture, the Asian tropics were almost completely covered by closed forest, except in floodplains where major rivers issue from the Himalayas, and on high mountain summits.[1]

Early People of the Asian Tropics

Hominids (early humans and their close relatives) have lived in tropical Asia for at least two million years (stone axes in Peninsular Malaysia have been dated to 1.8 Ma; Java man at 0.7–1.0 Ma; and modern humans have been here for at least 50,000 years). Humans produced a remarkable assemblage of cultures closely attuned to the environments where they lived. Today, they continue as people of the forest, variously described as tribals, or Adivasi (in India), *orang asli* (i.e., "traditional" in Malaysia and Indonesia), and Montagnards, or hill tribes (in Indochina and Thailand). Such groups in the Asian tropics speak ~1,900 languages, out of a global total of 7,100, with 840 in Papua New Guinea and 707 in Indonesia alone. The tropics are generally home to the greatest linguistic diversity, but the Asian tropics contain the greatest diversity of language and culture anywhere in the world, each language full of knowledge of the biota, the medicinal and food uses of

Fig. 15.1 In Bali, the Tampak Siring water temple has been a place for self-purification for at least a millennium (Ma. A.).

plants, and the ecology of the forest. Intrepid researchers have investigated a small portion of this cultural diversity: cultural anthropologists, linguists, agricultural ecologists, and others, under the umbrella of political ecology.[2]

Making a living: swidden. Forest inhabitants discovered means to exploit the resources surrounding them. They developed the technology (such as blow guns) to hunt and kill animals. By gathering, they discovered and used plants that were nutritionally and medicinally valuable (e.g., some that provided essential starch). Over the past 3 millenia, they gathered valuable resins, latexes, and medicinal plants for trade with a wider world.

People used fire, first to drive game, and later to clear land for agriculture, particularly in deciduous forests. Regions with two or more dry months each year have supported irrigation-based civilizations since ancient times; irrigated, wet rice cultivation had appeared about 7 Ka in the plains of warm temperate China, where it spurred increases in agricultural productivity and cultural complexity. Our first records of wet rice cultivation in the moist tropics come from northeast Indo-Burma, at 6.5 Ka. The techniques appear to have become widespread in monsoon tropical Asia by 6.0 Ka.

In the wet tropics, humans discovered that fire could be used to open small forest areas to grow crops for 2–3 years in cleared gaps of usually less than a hectare, particularly rice (fig. 15.2). Ash contained nutrients that were absorbed into the soil. Then the land was left as fallow, perhaps with a few fruit trees added, for ~15 years (or 5 years in the more nutrient-rich seasonal forests) until nutrients were restored, and the land could be prepared for another crop cycle. *Swidden* is derived from an old English word, "to singe," and described similar cropping cycles developed in Europe—also known as shifting agriculture, or slash and burn. In tropical Asia, the details of swiddening varied among groups. Swiddening was typically practiced by individual households in a village (or longhouse), and households controlled their plots. Traditionally, at extremely low population densities, swiddening

Fig. 15.2 Swidden, or shifting, cultivation. *Top*, Iban longhouse (Rumah Antawan) on upper Rejang River, Sarawak. Swidden plots of different ages are seen above the house. *Bottom left*, swidden plot cultivated by the Temoin tribe in Ulu Kenaboi, West Malaysia; cassava (*Manihot esculenta*, of South American origin) is the important crop; *bottom right*, two Temoin men, stopping by former village and swidden plots in Ulu Kenaboi, with abundant durian and other fruit trees (all, D. L., 1975).

mimicked the natural regeneration cycles in forests.[3] It also was combined with hunting and gathering in a household economy. The long-term effect of swidden on the ecology of tropical forests is controversial. There is certainly wide variation in these effects, given the surrounding forest types and different swidden techniques. Traditional swidden-based subsistence systems in Southeast Asia were certainly less damaging to soils and hydrology than more intensive forms of cultivation, particularly since the latter commonly involve the construction of roads and infrastructure, but swidden under short-fallow times stores *less* carbon and *decreases* soil quality compared to long-fallow times.

Sacred forests. These cultures established beliefs of the sacredness of forests in which they lived. They preserved cosmologies that described their origins in the forest. Certain animals and trees were sacred (particularly dangerous animals, such as tigers, and odd plants, such as strangler figs), and dusk was a precarious time to visit the forest (there were many taboos among their beliefs). Some cultures, particularly in South Asia, preserved forest areas as sacred.[4]

Early Civilizations in the Asian Tropics

Between 6,000 and 4,500 BP, at the northwest edge of the Asian tropics, in the Indus River valley, a remarkably uniform "nation" of small city-states formed. This civilization was very different from contemporary Mesopotamia to the north and west, but it depended on the same temperate cereals, rather than on rice. Their sophisticated flood control and urban drainage systems, intercity communication and water storage tanks all imply a long familiarity with control of water, if not crop irrigation. The domestic ox was certainly known to them.

Gods of the later Hindu pantheon were venerated by this early civilization, but it is still unclear exactly who these people were. The Indus Valley civilization endured for a millennium and a half. It collapsed shortly before—or perhaps because of—the migration of Indo-European, "Aryan" pastoral peoples from Central Asia. The Indo-European peoples brought horses, which they had domesticated by 6.4 Ka, and horse-drawn chariots. They cultivated rice, the dominant cereal in the Gangetic plain to the present day. They spread slowly southward and eastward into the lowlands of modern northern India and Pakistan. By 2.8 Ka, forest clearance was widespread, cities arose, and plains were irrigated. Irrigation spread with the expansion of Buddhism, arriving in Lanka (now Sri Lanka) ~2.4 Ka. The first cities on the Deccan Plateau date from 2.2 Ka.

The Indo-Europeans venerated mountains and forests, the springs that issue from forested mountains, certain tree species, and mother-goddesses of abundance and fertility. Much of their world view as pastoralists from the dry steppes and mountains of west-central Asia was incorporated into modern Hinduism as well as Buddhism.[5]

Millennia before the spread of the Abrahamic religions, variants of this basic worldview had extended over much of Eurasia, from eastern Indonesia as far north and west as the Scottish Highlands and Ireland. These variants survive particularly in the wetter regions of South Asia and Bali, and in the tradition of the sacred groves, many of which are still strictly protected (figs. 15.1, 15.3). Such beliefs and practices probably spread along ancient trade routes, especially in East Asia.

Fig. 15.3 Temple, sacred forest, and spring at Alagar Kovil, near Madurai in South India. *Top left*, temple and Alagar Hill in background; *top right*, shrine of sacred jambolan tree (*Syzygium cumini*) in forest; *bottom left*, Alagar Bund, sacred springs and pilgrimage site; *bottom right*, bathers in sacred springs (all D. L.).

Fig. 15.4 Sacred fig trees. *Left*, offerings at the foot of a strangler fig at Tampak Siring, Bali (P. A.); *center, F. virens* tree enclosed within temple at Vajreshwari, Maharashtra, India (D. L.); *right*, peepal (*F. religiosa*) worshipped at Vajreshwari temple (D. L.).

Within defined areas, often catchments flowing from inland mountains into a larger river or the seacoast, communities were established. Their deities oversaw the farming in local landscapes. Such deities arose locally and were often influenced by the Indo-European nature goddesses in South Asia or by Hindu gods whose influence moved eastward. They were worshiped to assure success in farming and to prevent harm from danger in the adjacent forest. Villagers also venerated certain sacred trees, such as the strangler fig. Its extraordinary form (branches adorned with ghoulish descending roots) and its life-habit (as a killer parasite of noble forest giants) commanded a certain awe. Forest people also knew that strangler figs presided over the most fertile soils, and that their fruit was the famine food of forest wildlife (fig. 15.4; p. 246).

Between 1.5–1.2 Ka, Indian traders and *sadhus* spread their worldview eastward throughout Indo-Burma and southeast to Sumatra, especially through the fertile volcanic and seasonally dry lowlands of East Java to Bali, and on to the Lesser Sunda Islands. Ancient understandings of the interdependence between man and nature, pervasive in humid climates north and east to Tibet, China, Korea, and Japan, were assimilated. Later, as Buddhism spread, it was incorporated into this worldview or, like the Shinto tradition of Japan, prospered alongside it. Indigenous people throughout the Asian tropics retained their own ecological and spiritual world views, sometimes influenced by contact with these early civilizations.

Thus, people living within or adjacent to forests accumulated a sophisticated ecological knowledge, and they developed agricultural systems of considerable technological complexity. Their religious traditions suffused the agricultural and forest landscapes, contributing to their sustainability. That knowledge was often demeaned by colonial and then government officials, who looked to technical advice from "foreign experts" (such as the authors), who provided narrow technological solutions and ignored broader and more ecologically based advice known to the residents.[6]

Irrigation: A fundamental influence. Monsoon lands—with a shorter dry season, or perennial streams descending from the mountains—supplied water for agriculture. The ancient technological innovation of terraced rice (padi) irrigation apparently approached its present form in northern Indo-Burma and South China around 2.5 Ka (fig. 15.5). It spread west and eventually south to the Sunda Islands. Farmers gradually enriched the nutrient supply in their fields by mulching, which also helped control weeds. In wetter seasonal climates, up to three crops of rice could be harvested annually, without cleaning or intervening restorative crops, if the water supply were controlled. Terraced rice cultivation reached its peak in the seasonal mountain landscape of northern Luzon, where people with irrigation technologies of great subtlety settled several thousand years ago. They still exist in tropical and warm temperate Asia, from Luzon and Indo-China northeast as far as Honshu (fig. 15.5). Such crops are seasonally dependent on water captured from fog by the ridge forests high above. In Indo-Burma, as north-

Fig. 15.5 Rice cultivation. *Left*, deeply terraced canyons of padi in the Ifugao Mountains of the Northern Philippines (Wikimedia: Magnus Manske); *right*, rice terraces and other crops near Bukit Tinggi, West Sumatra, 1975 (D. L.).

ern Han populations expanded, hill tribes such as the Hmong were pushed south and brought with them the warm-temperate traditions of terraced but non-irrigated vegetable cultivation for market. Consequently, particularly in northern Laos and Vietnam, little lower montane forest remains.

The diversion of water from rivers and its distribution among peasant farmers required sophisticated technology and communal cooperation. This became the technology of dynasties, the so-called hydraulic societies. Such sophisticated, intensive systems of rice culture, combined with transportation by cart or by water, fostered the most sophisticated agrarian civilizations at that time. Of the two potential influences on tropical Asia (China: the Han, 200 BCE–220 CE, and Song, 10th–13th centuries CE; and India: the Chola, 300 BCE–13th century CE), the culture and technology of South Asia was more influential and spread rapidly through trade and settlement into the similar climatic regions of the Far East. The Chola culture influenced the major civilizations of tropical Asia: Champa (Vietnam, 2nd–18th centuries), Srivijaya (Sumatra, 7th–13th centuries), Khmer (Cambodia, 9th–15th centuries) and many lesser ones. The great 9th-century monuments of Java, Prambanan (Hindu), and Borobudur (Buddhist) resemble South Indian Pallava art of the same period (fig. 15.6). Modern Malay languages, including Bahasa Malaysia and Indonesian, are spoken by some 300 million people in the region (more than French, and globally the fifth largest). They contain many Sanskrit borrow words, such as kepala (head), guru (teacher), desa (countryside) and angkasa (sky). These were likely absorbed by speakers in the Indian-influenced civilizations, as the Srivijaya.

In the seasonal Asian tropics, irrigation spread wherever there was

Fig. 15.6 Cultural influences of South India in Southeast Asia. *Left,* Mahaballapuram, descent of the Ganges (Pallava, 7th century); *left center,* Borobudur entrance detail (Java, 8th century); *right center,* Prambanan (Java, 8th century); and *right,* Prambanan, bodhi tree detail (all D. L., 1975).

Principal habitat	Animal
Summer: alpine pastures; winter, temperate forests	Yak, *Poephagus grunniens*
As above, but more often lower	*Zho* (yak x domestic cattle)
Temperate forests and meadows	Domestic cattle
Tropical lower montane forests	*Mitun* (domestic cattle x gaur)
Tropical lowlands and lower montane forests; not domesticated	Gaur, wild Indian ox (*Bos gaurus*)

Fig. 15.7 Cattle and grazing. A variety of animals, domesticated ancient varieties and crosses (indicated by x) with the wild gaur, were used in the Asian tropics. Top, table of cattle and origins in the Bhutan Himalayas; *bottom left*, the mitun (H. H.); *bottom right*, hay ricks of a smallholder in a dry district near Udaipur, Rajasthan (P. A.).

enough water. Forest on irrigable land had been almost entirely converted to human use long before the arrival of global commodification and industrialization. In continental Asia, where water supply was inconsistent, and forests survived on shallow soils, cattle were kept for milk; these had long ago competed successfully with indigenous grazing and browsing mammals (fig. 15.7). Such forests also provided fuelwood and famine food for humans in drier regions. Forests have always been of utmost value to rural communities, and still are today. Their use was often locally and informally regulated within increasingly deforested agricultural landscapes. Forested uplands unsuitable for permanent agriculture became the perquisite of rulers, traditionally invested with divine powers. These were their hunting grounds, but they were duty-bound to protect the forest for their subjects as well. Rulers of traditional societies expected fealty from smallholders and assistance in times of war, in return for upholding certain land rights. Grazing access for commoners was limited in years of average weather, but available for the collection of fodder and famine foods in times of hardship.

Perhaps these traditional systems of resource-use regulation in precolonial Asia were not sustainable. As human populations slowly expanded in Asia, there was a need for resource-productive landscapes. This led to the development of a variety of systems of land "ownership" or control that were well-adapted to local conditions.

The Mughal influence in South Asia. Islamic influence in India began in the 7th century, but taxation on a large scale was not imposed until the Mughal Empire (16–17th centuries), when taxes were collected as rents by *zamindars. Zamindar* is a word that derives from the Persian, approximately meaning "land owner." *Zamindars* were not technically proprietors of land, but they effectively controlled the uses of land, sometimes as enormous territories, and were also given the judicial authority of a rural magistrate over disputes. The *zamindari* system thus gave a hereditary aristocracy practically unlimited powers over the lives of peasants, reinforcing the rigid caste system throughout much of rural India. This despised feudalistic system endured into the colonial era and was unfortunately co-opted as a practical measure by the British Raj. Forests under *zamindari* authority were often degraded, which was challenging for the young Indian Forest Service.

Agriculture in the Everwet Far East

Water was always difficult to control in everwet regions prone to flash floods, especially where sandy soils over siliceous rocks prevented the construction of stable bunds and banks. Heavy, continuous rainfall prevents accumulation of humus and leaches nutrients, while absence of a clearly defined dry season reduces the reliability and volume of rice harvests. Nevertheless, settled activity in New Guinea stretches back 26 Ka, some 25 thousand years after humans arrived. Evidence of an increase in fires appears at 18 Ka in the northern Sumatran mountains, suggesting human modification of the landscape.

Until the colonial era, settlement and political influence in Malesia lay along the coasts, where extensive sedimentation from the rain-washed hills supported mangrove and productive fisheries, and where the sea facilitated communications and trade. Sultanates developed around settlements in the upper tidewaters of the major rivers and exerted control over both the rich coastal fisheries and the hinterlands, while retaining access to plentiful fresh water. Well-placed communities dominated the regional trade in natural resources, including rice. Control of the forested hinterlands was less pervasive, and the upland peoples were fewer and more widely dispersed.

Fig. 15.8 The traditional land use in everwet and moist seasonal southwest Sri Lanka. More productive annual crops occupy the floodplain and lower slope at the base; the predominantly woody home garden is on the mid-slopes, merging with those of neighbors to form a continuous forest band. Above are steeper slopes used for swiddening if the paddy harvest fails, while aboriginal forest along the ridge provides fuel-wood and non-timber forest products (M. A.; see Mark S. Ashton, S. Gamage, I. A. U. N. and C. V. S. Gunatilleke, "Restoration of a Sri Lankan rainforest: Using Caribbean pine as a nurse for establishing late successional tree species," *Journal of Applied Ecology* 34 [1997]: 915–25).

A common tradition and history of land use. Throughout the well-watered regions of South Asia and the Far East, wherever land tenure is secure, rural inhabitants live in villages and towns strategically placed on the lower slopes of the hills (fig 15.8). Around their houses they cultivate miniature groves, or "home gardens," each containing a rich mixture of useful herbs, vege-tables, medicinals, cosmetics, and fiber sources, as well as fruit and timber trees. The wetter the climate, the more diverse are the species and the richer the mixture. These groves still incorporate mostly native species, but seeds and cuttings were exchanged throughout tropical Asia. Fruit trees native to the Western Ghats and northeast India moved to Java and the Far East, such

Fig. 15.9 Forest understory trees with edible fruits that spread to moist tropical forests throughout Asia early in history, and more recently as fruit crops, assembled and photographed in Selangor, Malaysia. *Top left*, mata kuching = *Dimocarpus didyma; top center*, rambai = *Baccaurea motleyana; top right*, carambola = *Averrhoa carambola; middle right*, salak = *Salacca zalacca; bottom left*, rambutan = *Nephelium lappaceum; bottom center*, duku = *Lansium domesticum;* mangosteen = *Garcinia mangostana* (D. L.).

as jack (*Artocarpus heterophyllus*), mango (*Mangifera indica*), jambu (*Syzygium aqueum*), the *Averrhoa* starfruits, ornamental flowering trees, probably the useful sugar palm (*Borassus flabellifer*), and even the riverbank tree *Terminalia arjuna,* said to reduce the hardness of water and combat heart disease (fig. 15.9). Some of these South Asian species appeared in frescos at the 9th-century Buddhist shrine at Borobudur in East Java. Trade was dynamic in both directions; *Syzygium jambos, S. samarangense,* mangosteen (*Garcinia mangostana*), rambutan (*Nephelium lappaceum*), bananas (*Musa* spp.), durian (*Durio zibethinus)* and coconut (*Cocos nucifera*) arrived in southern India and Sri Lanka from the opposite direction. In the Far East, forests were modified by selectively promoting useful species and by trees growing from swidden (p. 338) plots. Such early spread may increase the alpha diversity of many local forest plots, as has been shown in Amazonia.[7]

The upper slopes—especially above 1,200 m, where few lowland crops bear fruit—were generally left under forest (New Guinea excepted) and venerated as the homes of benevolent deities and the reliable source of the all-important water supply (fig. 15.10). Nearer to habitations, some forest was cut for swidden, the frequency depending on food availability and family size.

In other places, the ebb and flow of armed conflict led to the abandonment of land. In the West Himalayan foothills, Banj oak (*Quercus oblongata*) forests rose from the deserted farms of the formerly powerful Kumaon nation. In Pegu (Burma), much of the finest teak grew around abandoned farms of Chinese settlers evicted in the 18th century. Still, great tracts of aboriginal forest survived in South Asia on lands that were formerly often under the protection of rulers, who hunted in them with their courts and tolerated the forest tribes.

Upland peoples. It is unclear when domestic cattle entered the Asian tropics, but they were certainly present in India after the Indo-European immigration. Cattle do not prosper in the lowland wet evergreen forests, but graziers have long left cattle to roam in the deciduous forests and the lower montane evergreen forests of the Himalayas and Indo-Burma, where

Fig. 15.10 Veneration of mountain forests in Thailand, Doi Inthanon, Thailand's highest mountain. *Left*, Naphamethinidon Chedi, built in 1987; *right*, more modest temple at the summit (all D. L., 2005).

remnants of an annual transhumance cycle of extraordinary complexity continue (fig. 15.7).

Elsewhere in the everwet tropics and large hill tracts of the drier tropics, where ridges are high enough to attract rain, the climate or topography was unsuitable for irrigation. Tribes survived by swidden (p. 348) in such areas.

In India, the lowlands of the Carnatic coast and the foothills abound in sacred groves (fig. 15.3), while the upper slopes and hills remained until recently under primary forest cover, except for the expansion of upper montane grasslands under cattle grazing. Spices, such as pepper and cardamom, grown in agroforestry home garden systems were exported from the Malabar Coast for millennia.

Land use and inequality. Lowland irrigation produced a fundamental division within societies between people who occupied the plains and lower slopes amenable to irrigation, and those in the hills, who relied on the rains directly. Up to 1,200 m, most slopes are amenable to swidden. In everwet climates, only the less steep lower slopes were cultivated continuously (unless the soils were exceptionally stable and fertile, such as those near the volcanic Barisan range of Sumatra). On the one hand, the ox or buffalo and plough combined with naturally high irrigation-based productivity and good communications to support political consolidation and power in the lowlands, which also fostered specialization of work and social hierarchy. On the other hand, the people of the high hills were restricted to the dibble stick, extensive farming practices, and diffuse populations. Isolation probably encouraged the separation of societies and the diversification of languages. Political power among these indigenous groups was limited and local, lying ultimately with the "foreigners" of the plains, who controlled the trade in minor forest products and other natural resources.

The beginning of agriculture in the everwet tropics of Asia is unknown, although taro (*Colocasia esculenta*) was probably cultivated in New Guinea a remarkable 10 Ka. Swidden (p. 348) was sustainably practiced by communal traditions of considerable sophistication. In some locales it continues— provided that human population densities are stable (fig. 15.2). Even swidden cycles could only be sustained over time on the more fertile substrates and the deeper clay loam soils of lower slopes and alluvium. Swiddeners in everwet regions learned that burning, essential for release of nutrients from felled woody biomass, also consumed the greater part of the soil organic matter in less fertile humult soils: the raw humus that takes generations to regenerate. Most of the Sunda Hills and much of the Philippines and New

Guinea remained under primary forest until European colonization. Irrigated agriculture reached into the less fertile areas relatively recently, and generally only under one of three conditions: (1) when local populations flourished (e.g., when Javanese and other Hindu-Buddhist satellite nation-states expanded, such as Sri Vijaya in Sumatra, Samarinda in West and East Kalimantan, and at Ligor in peninsular Thailand/northern Peninsular Malaysia)]; (2) when spill-over populations moved into marginal areas; or, (3) when alien states invaded, and inhabitants retreated onto land they had previously avoided.

Upland swidden cultivators developed deep knowledge of their demanding and variable terrain and, in times of stable population numbers (often reduced by periodic warfare or malaria), sustained productivity by mutually accepting traditional rules of constraint. Traditional swiddening tribes did not always live in sustainable harmony with the land upon which they depended. Some, like the Iban of Sarawak, in northeast Borneo, were often expansionist, moving onto and taking over the lands of other tribes after their own land was invaded with perennial weeds, especially the deep-rooted lalang (*Imperata cylindrica*) grass. Yet, the Iban of West Borneo were more settled in their swidden practice. The Bajau (seafaring peoples of south Philippines and northeast Borneo, who spread over much of Indonesia) habitually settled near the coast, cut nearby forests to cultivate, then moved on in search of new land rather than adopting permanent swidden or other agriculture. The Hmong of Laos and Thailand intensively exploited hill land until it was exhausted, then moved on, often invading the lands of more sedentary tribes. The lalang grasslands of plateaus on the Annamite range (Indochina) are testimony to the failure of sustainability in past hill tribe cultivation.

Village governance and the state. Before the arrival of European colonialism, the political concept of the nation-state did not exist. The disposition of land, especially uncultivated forest land, was broadly under the authority of the local ruler. It was determined by local communities, with hunting often more strictly under the ruler's control.

A History of Global Trade

The Asian tropics and temperate Eurasia have traded since antiquity. Asian spices and perfumes were used in Palestine in biblical times, such as cinnamon and agarwood (*Aquilaria* sp., Thymeliaceae). Later, during the Roman

Empire, Asian tropical rainforests were sources of raw materials—including timber, rattan, and birds' nests—for China. High-value medicinals, such as the antidote bezoar stone, still extracted from the guts of infected monkeys for export, were exchanged with China for ceramics along with other products of its advanced economy. As early as the 14th century, Arab traders discovered the spices and aromatics of South Asian and Far Eastern rainforests, and lively trade ensued in nutmeg, pepper, cinnamon, camphor (the aromatic resin of *Dryobalanops*), sandalwood, and agarwood. This trade incited European interest in the Far East, and it is explored in the next chapter.

Prior to colonization, the population of tropical Asia remained low. By 1600, the population of Southeast Asia was 22 million with an average density of $5/km^2$. With population concentrated in a few areas, as in Java, the density in the forested areas was much less. Population in South Asia was much greater, 115 million, or $33/km^2$, but much lower in extensive forested areas).[8] With colonization and independence, it grew much more rapidly.

Notes

1. This chapter is a condensation of *OTFTA*, chap. 8, 501–10.

2. Dove, Michael R., Percy E. Sajise, and Amity A, Doolittle, eds., *Beyond the Sacred Forest: Complicating Conservation in Southeast Asia* (Durham, NC: Duke University Press, 2011); Orr, Yancey, J. Stephen Lansing, and Michael R. Dove, "Environmental anthropology: Systemic perspectives," *Annual Review of Anthropology* 44 (2015): 153–68; Bryant, Raymond L., Jonathan Rigg, and Philip Stott, "Introduction: Forest transformations and political ecology in Southeast Asia," *Global Ecology and Biogeography Letters* (issue title: "The Political Ecology of Southeast Asian Forests: Transdisciplinary Discourses") 3, no. 4/6 (1993): 101–111; Hill, Christopher V., *South Asia: An Environmental History* (Santa Barbara, CA: ABC-CLIO, 2008); Boomgard, Peter, *Southeast Asia: An Environmental History* (Santa Barbara, CA: ABC-CLIO, 2007).

3. Dove, Michael, "Theories of swidden agriculture, and the political economy of ignorance," *Agroforestry Systems* 1 (1983):85–99; Dove, Michael R., "Linnaeus' study of Swedish swidden cultivation: Pioneering ethnographic work on the 'economy of nature,'" *Ambio* 44 (2015): 239–48; Dressler, Wolfram, "Disentangling Tagbanua lifeways, swidden and conservation on Palawan Island," *Human Ecology Review* 12 (2005): 21–29; Lawrence, Deborah, David R. Peart, and Mark Leighton, "The impact of shifting cultivation on a rainforest landscape in West Kalimantan: Spatial and temporal dynamics," *Landscape Ecology* 13 (1998): 135–48; van Vliet, Nathalie, et al., "Trends, drivers and impacts of changes in swidden cultivation in tropical forest-agriculture frontiers: A global assessment," *Global Environmental Change* 22 (2012): 418–29; Mertz, Ole, et al., "Swidden

change in Southeast Asia: Understanding causes and consequences," *Human Ecology* 37 (2009): 259–64.

4. Wadley, Reed L., and Carol J. Pierce Colfer, "Sacred forest, hunting, and conservation in West Kalimantan, Indonesia," *Human Ecology* 32 (2004): 313–38; Grim, John A., ed., *Indigenous Traditions and Ecology: The Interbeing of Cosmology and Community* (Cambridge, MA: Harvard University Press, 2001).

5. Chapple, Christopher Key, and Mary Evelyn Tucker, eds., *Hinduism and Ecology: The Intersection of Earth, Sky, and Water* (Cambridge, MA: Harvard University Press, 2000); Tucker, Mary Evelyn, and Duncan Ryūken Williams, eds., *Buddhism and Ecology: The Interconnection of Dharma and Deeds* (Cambridge, MA: Harvard University Press, 1998).

6. Brosius, J. Peter, "Local knowledges, global claims: On the significance of indigenous ecologies in Sarawak, East Malaysia," in *Indigenous Traditions and Ecology*, ed. John A. Grim (Cambridge, MA: Harvard University Press, 2001), 125–57; Ellen, Roy, ed., *Modern Crises and Traditional Strategies: Local Ecological Knowledge in Island Southeast Asia* (New York: Bergahn Books, 2007); Wadley and Pierce Colfer, "Sacred forest, hunting, and conservation."

7. Levis, C. et al., "Persistent effects of pre-Columbian plant domestication on Amazonian forest composition," *Science* 355 (2017): 925–931.

8. Reid, Anthony, "Low population growth and its causes in pre-colonial Southeast Asia," in *Death and Disease in Southeast Asia: Explorations in Social, Medical, and Demographic History*, ed. Norman Owen (St. Lucia, Australia: University of Queensland Press, 1987), 34–35; https://en.wikipedia.org/wiki/List_of_countries_by_population_in_1600.

In 1985, my friend the forest botanist K. M. Ko-chummen took me to Serting Forest Reserve, in West Malaysia. It had been logged in 1935 and then treated by a silvicultural forerunner of the Malayan Uniform System, celebrated as the only demonstrably sustainable method of managing tropical rainforest for timber. It was now ready to be felled again (following which it was converted to oil palm). The MUS represented the successful culmination of a century of silvicultural field experimentation aimed at sustainable rainforest management. The dense stand of straight-boled canopy merantis was beautiful, surely at least half as much greater in volume than their predecessors in the primeval forest. After fifty years, their sale value would have achieved about 5% per annum, nothing compared with the profit from a single final felling. Then I considered: Was such a continuing revenue the optimal value of this forest resource, or would it have been preferable to conserve it for its service values—of clean reliable water supply, soil conservation, recreation for hiking and local hunting, and yes—biodiversity, its protection funded by the public will? — P. A.

16

FOREST HISTORY: THE EUROPEAN INFLUENCE

The value of tropical Asian spices, dyes, and perfume oils stimulated exploration by major European powers.[1] The Spanish claimed the Philippines in 1521, during Ferdinand Magellan's expedition of circumnavigation (he was killed there), and the Spanish retained control until 1898. The Portuguese, through Vasco de Gama's discovery of a route to the east, arrived in India in 1497. In the following century, Portuguese coastal trading ports (fig. 16.1), wealthy for a time, left a major mark on the cultures of tropical Asia and introduced many trees and shrubs brought from Brazil: papaya (*Carica papaya*), sapodilla (*Manilkara sapota*), avocado (*Persea americana*), custard apple (*Annona* spp.), chili (*Capsicum*), and manioc (*Manihot esculenta*). And their country music still echoes through their former trading stations. Imagine a curry once spiced solely by the indigenous black pepper (*Piper nigrum*)! The Dutch (1596) and British (1601), arrived later and effectively removed the Portuguese monopoly of the spice trade. The Dutch, through the Dutch East India Company (VOC, 1603 on), controlled trade in Indonesia. In 1826, the company was dissolved, and the territories became a formal colony of the Netherlands. The British, through their East India Company (EIC, 1601 on), controlled trade in India and adjacent regions. The company was dissolved (but reestablished in 2004!), and its territories became dependencies within the British Empire in 1874. The French, a global power next to the British, took an interest in its future Indochinese colonies through its Jesuit missions in the 17th century, never established a single trading company, but used military force to establish colonies, the Indochinese Federation, in 1887.

Fig. 16.1 Malacca, an old trading port in West Malaysia, taken over by the Portuguese in 1521, who were later supplanted by the Dutch and then the British. *Left,* A Famosa, the Portuguese fort built in 1511; *right,* the town center, with the *stadthuys* from 1640, on the left, and the Christchurch on the right, built by the Dutch in 1754 as Dutch Reformed, and reconsecrated as the Church of England by the British in 1824 (all D. L.).

Commodities

In contrast to the Portuguese, the later Dutch and British brought neither plants for home gardens nor folk music to enliven a languid evening. Initially, they all attempted (with varying success) to control the lucrative trade in non-timber forest products. Then they focused their commercial interests on land acquisition and commodity crops. Botanical gardens were established, at Calcutta in 1787, Buitenzorg (Bogor) in 1817 (fig. 16.2), Peradeniya in 1821, and Singapore in 1859, for cultivating and improving plants of commercial interest and exchanging them between the continents. These gardens also became centers for taxonomic and ecological studies of the unknown plants in Asian tropical forests. William Roxburgh, a superintendent of the Calcutta garden, was such an avid student of plant diversity that he is today known as the "father of Indian botany" (fig. 16.3).

Coffee and tea were introduced into the hills (fig. 16.4), and later Brazilian rubber and West African oil palm were introduced in the lowlands (fig. 16.5). In the late 19th century, a Euphorbiaceous, late-successional light hardwood from Amazonia, *Hevea brasiliensis* (Brazilian rubber), was found to be far more lucrative than the native *gutta percha* (*Palaquium gutta*), which was felled to extract its latex.

Fig. 16.2 The Indonesian National Botanic Gardens at Bogor (Buitenzorg), founded in the early 19th century by the Dutch. The monument is a memorial to Marianne, first wife of Stamford Raffles, who was the governor of the East Indies while the Dutch were under the sway of Napoleon (P. A.).

Fig. 16.3 Three influential forest botanists in British India. *Left*, William Roxburgh (1751–1816); *center*, the forester Dietrich Brandis (1824–1907); and *right*, Joseph Dalton Hooker (1817–1911; all Wikimedia).

Fig. 16.4 Tea and coffee cultivation. *Bottom left* and *right*, tea cultivation in Munaur, South India (L. E.); *top left*, coffee cultivation in Sirumalai Hills, South India; *top right*, coffee flowers and fruit (D. L.).

The exotic Brazilian rubber was the first lowland commodity crop suitable for wet climates and soils of unexceptional fertility. It not only caught the attention of the colonial administrations but quickly moved into the economies of the forest dwellers throughout the humid tropics of Asia. The swiddeners had been collecting forest latexes and aromatics for centuries (p. 347), which provided income for items they could not make, such as steel axes and knives. Depending on the success of swidden cultivation, they could collect minor forest products as income, allowing them to keep some of their swidden plots in fallow. Rubber could grow as if a tree in the recovering forest, could be tapped during the year, and the latex was easily pro-

Fig. 16.5 Plantation crops of the lowland tropics: rubber and oil palm. *Top left*, oil palm plantation in Selangor, West Malaysia, reserve forest in background; *top center*, individual palm; *right*, fruit cluster, ~30 cm long. *Bottom left*, rubber plantation in Perak, West Malaysia; *bottom center*, tapped trees; *bottom right*, curing rubber sheets, Selangor (all D. L.).

cessed into sheets that could be sold to traders. Most of the rubber produced under colonial rule was from smallholders and not from large plantations. However, colonial administrators struggled to more directly control rubber production through plantations and not from smallholders.[2]

Having wrested control of forests from rulers and constrained the activities of the swiddeners, they cleared wet upland forests previously uncultivated. The lower montane forests on less-steep slopes in wet seasonal climates, with their humus-rich loams, were ideal for woody, tropical commodity crops. In the mid-19th century, coffee (*Coffea arabica*) was introduced, but was largely eliminated by a fungal pathogen. It was replaced by tea (*Camellia sinensis*), a species indigenous to warm temperate East Asia and to the lower montane forests of Indo-Burma (p. 202). Its phenolic metabolites defended tea against pathogens. The lower montane oak-laurel, *shola* and *rata dun* forests of South Asia were thereby widely eliminated on all but the steepest slopes.

Land Ownership and Forestry

The colonial powers first dealt with local rulers to purchase non-timber products for trade in Europe, including spices, dyes, and perfume oils (fig. 16.6), then later to gain access to land for production of commodities. Access to forests was more difficult because the ownership was not often well defined, and they were used and occupied by tribes that collected the valuable goods and traded with coastal rulers—and had no written records of ownership of the lands they used. For instance, ownership by the Temiar of Peninsular Malaysia eventually was documented through their singing of oral traditions.[3]

Thus, a central conflict arose, complicated by the poor documentation of ownership by forest dwellers and the approaches of the different colonial administrations. It continues to animate many of the conservation disputes in tropical Asia—between varied, locally adaptive patterns of control over land and forest versus larger-scale, centralized administration and commercial rental or ownership. The Asian tropics encompass a range of ownership arrangements. Many feature customary *communal control* of land, often in combination with privately held plots. Such arrangements have long been part of the fabric of traditional society, such as in Adivasi areas of India.

The establishment of reserved forests for timber production by colonial governments and the regulation of forest use transferred power from the village to the metropolis. Such a change produced harsh dissonance between people's needs and experiences and the legal-administrative frameworks available to accommodate them. In many forests, they are still not resolved.

Thus, most forest land was controlled and managed by European colonial agencies, with all the ambiguous administrative fall-out that status implies. In India—as it also was in medieval Britain—*forest* remains largely an administrative category: forest land is largely land that is *not* subject to the authority of the Revenue Department. The introduction of large-scale commodity crops for export during the colonial period, largely coffee and tea, was responsible for most forest conversion, while the rural economy remained at subsistence level, rice cultivation ensuring food security.

In the drier Asian climate regions, population pressures led to overexploitation of forest for fuelwood, grazing, hunting, and hardscrabble dry-

Fig. 16.6 Non-timber forest products. *Center*, baskets and cordage from various sources, near Madurai, South India; *left*, cloves and cardamom, Kanyakumari, South India; *right*, sandalwood plantation, Karnataka, South India. Sandalwood trees, on the left, parasitize roots of the casuarina trees, right (all D. L.).

land agriculture. That practice had worked adequately until the 19th century, when rural populations increased beyond the supporting capacity of the land. Beginning in the 1920s in northern India, *panchayat* forests were legislated in some states to ensure the needs of village communities for forest goods. These moves were consolidated nationally in the 1927 Indian Forest Rights Act and the 1931 Van Panchayat Act. This early attempt at community management could be successful only where earlier traditions of continuous forest use were maintained, since the laws enshrining the reserved timber production forests remained in force, deterring alternative uses. Meanwhile, less-restricted state lands and community-managed forests degraded.

South Asia, West Java, and Bali supported the densest rural communities in the seasonal wet tropics of Asia. In the latter regions, fertile, base-rich volcanic soils supported irrigated rice throughout the lowlands. In Sri Lanka, where private land title was long the norm, and where traditional respect for nature is also strongly engrained, a populist conservation movement among rural and urban groups achieved legislation of policies for sustainable use and conservation.

Efforts to Sustain Declining Forests

Western trading companies were interested in a limited number of forest products, among which timber was foremost by the end of the 18th century. Durable, medium-density hardwoods, such as teak, were particularly important in maintaining navies and merchant fleets. The companies managed the resources that yielded goods of value until those resources were depleted. European forestry practice in the tropics, focused on sustaining timber yields, intervened eventually but developed differently under the varying climatic conditions (and hence silvicultural ecology) of the forests in question, and under the direction of different countries. In India and Burma, where it developed first, tropical forests are predominantly deciduous; in Malesia, they are evergreen.

European forestry practice. In India a single species, teak (p. 183), became indispensable to the British navy for decking and ship furniture. By the early 19th century, the market value of teak rose in response to the gradual exhaustion of accessible sources along the west coast of Peninsular India. Forest degradation accelerated soil erosion and the siltation of harbors, and ships were forced to anchor offshore. Due to the discoveries of Hales and Priestley that plants mediate climate and rainfall, the Royal Society promoted forest conservation in England and its colonies. The droughts and massive famines of the late 18th and early 19th centuries were blamed in part on deforestation (fig. 16.7). The young Scottish surgeons of the British East India Company, such as William Roxburgh (fig. 16.3), agitated for forest conservation in India. The first commercial teak plantation was established by Henry Conolly in Malabar. In 1840, he recommended to his EIC superiors that active management of regeneration should be enshrined in law. His recommendations constitute the basic principles of forest management that continue in India to the present day: to obtain "a complete knowledge" of the quantity and quality of timber in a prospective forest; to forestall any kind of depredation being committed on forests, whether government-owned or leased; and to improve the forests by new plantings and by fostering the growth of young trees. The teak forest at Nilambur, successfully established by Conolly and Chathu Menon, was the first commercial timber plantation in the tropics, and Conolly was the first forester in the Western "scientific" tradition. The trees of Conolly and Menon became so magnificent that the plantation has become a conservation area, its trees revered.

By mid-century, commercial deforestation in India had become so serious that the British Association for the Advancement of Science, at its annual conference in 1852, sponsored a session on deforestation in the tropics! A committee recognized the necessity of converting accessible forest land to agriculture to feed and employ a population that was increasing under stable rule, but emphasized that much land was too steep, or its soils too infertile,

Fig. 16.7 Famine in 19th-century India, near Madras, South India, in 1877 (from the *Illustrated London News* and Wikimedia).

for continuous agriculture, and that it should therefore be maintained under forest management. Recommendations from this meeting led to establishing the Indian Forest Service, an early action of the British government that had assumed the governing of India from the humiliated and disbanded charter company following the great rebellion of 1857.

In 1855, Governor-General Dalhousie issued the Charter of the Indian Forests. In 1864, Dietrich Brandis (fig. 16.3) was appointed inspector-general of the new Indian Forest Department, which became the Imperial Forest Service (IFS) in 1866. He was responsible for the establishment of an institution of preeminent reputation and *esprit de corps*. No British university had the capacity to train professional foresters for another 40 years, and then only because of the growing demand from India (The US Forest Service was effectively founded in 1881, partly in consultation with Indian foresters). The British Forestry Commission was founded only in 1922. France and Germany, then having the most distinguished record for "scientific"— that is, investigative and systematic—forestry, were the sources of professional foresters.

The future development of scientific forestry, in India and in the tropics, was fortunate in the appointment of Brandis, the father of Indian forestry. In his first posting in Burma, he immediately came up against the leading challenge in tropical forest management, which in various forms continues to the present day: the chronic disharmony between commercial and conservation interests. Commercial timber interests favored high profit from a single crop and disregarded damage to the forest's regenerative capacity. In contrast, the forester or conservator, representing the owner—namely, the nation—endeavored to sustain the value of the forest by regeneration following harvest. British commercial interests had negotiated favorable logging concessions regarding the magnificent teak from local rulers that were disastrous for the forest. Brandis's task was to correct that, and he quickly instituted practices fundamental to sustainable forestry in the tropics.

Brandis took early steps toward community involvement in forest management. In contrast to India, where forests were in danger of degradation by cattle browsing and fuelwood collection near villages, in Burma the upland people widely practiced a form of swidden, *taungya*. Brandis observed that teak regenerated in disturbed semi-evergreen forest near villages. Brandis advocated demarcating remaining good stands as strict forest reserves for teak, within which *taungya* should be banned. He also advocated reservation and sustainable management of scrub forests in the Irrawaddy plains and elsewhere. Communities depended on such forests for fuel, browse, and other products, and they were susceptible to degradation. One of Brandis's conservators persuaded Karen upland swiddeners to plant teak before they abandoned their fields to woody fallow. This approach gradually became accepted. It has been adopted worldwide in a hybrid system known as *taungya* forestry: forest villagers plant crops for several years in reserved timber production forests in return for their labor in clearing and planting seedlings of timber species. When the seedlings grow large enough to shade out the crops, new areas are cultivated.

The equitable sharing of forest resources. Brandis wisely saw opportunity in gaining the participation of forest villagers in the management of teak rather than excluding them, which, in so many parts of India, was advocated by foresters trained in the tidy management traditions of temperate Europe. Nevertheless, collaborative forest management with forest-dependent groups increasingly eluded colonial foresters and remains an elusive goal to this day. After all, the disciplined service established by Brandis was, and is, a quasi-military force, with khaki uniforms, caps, and barracks for officers and families. A consequence throughout Asia has been the perception that the forested landscapes, inhabited for millenia by swiddening minorities, have been usurped by "governments" and their forest departments. From the 18th century to the present, dozens of Adivasi (tribal communities) have fought for control of Indian forests, and they have been manipulated by political parties with very different aims.

Even where communities cultivated only the more fertile lower slopes, villagers often assumed that the upper forested ridges and slopes were their heritage. They depended on them for water, game, and non-timber goods, and saw them as the abode of their gods. Rajah Vyner Brooke, the third White Rajah of Sarawak, whose family founded and ruled the Kingdom of Sarawak from 1841 to 1946, refused to set limits on such land uses by the Iban tribes under his control.

Forestry in other colonies. The Dutch were mainly content to extract timber from existing teak stands, primarily in eastern Java, and were late in establishing a forest department. The French were even later in Indochina, as were the Spanish in the Philippines. Forestry practiced in the British colonies was by far the most advanced and effective.

Under ideal circumstances, the government should represent the nation, and uplanders should receive services, such as schools and medical facilities, in exchange for their contributions to the natural economy and their acceptance of limits to their activities. Timber harvesting brought roads to the uplands, and with them came opportunities for uplanders to share more equally in the benefits of the national economy. However, social and economic equity remained the remotest of ideals in most places. In the worst cases, benefits from logging failed to trickle down to anyone beyond the concessionaires and their political patrons. Even where they did, urban and lowland populations benefited at the cost of the forest uplanders. In short, there has been a massive transfer of wealth from the poor to the rich.

Environmental consequences of forest conversion. The introduction of tea cultivation did not by itself lead to the intrusion of commodity agriculture into the traditional lands and forests of the upland tribes. Their lands rarely extended as high as the lower montane forests where tea was planted, because such cool climates yield few lowland tropical crops. The environmental consequences for mountain forests became an increasing concern, however, because they affected both water supply and plant communities. J. D. Hooker, director of the Royal Botanic Gardens, Kew (fig. 16.3) and an expert on the Indian flora, used his influence against deforestation in British dependencies and campaigned for the establishment of a formal Forest Service. As early as 1874, he initiated an inquiry into evidence for changes in rainfall patterns from montane forest clearance. His recommendation that forest conversion should be confined below 1,600 m, and slopes of less than

45% in India and Sri Lanka, was heeded. It continues in some measure to the present day in Peninsular Malaysia and elsewhere, where slopes greater than 30% are protected from conversion.

Sustainable Forest Management

For former British India—a vast swath of tropical Asia from Pakistan to Burma plus Peninsular Malaysia and North Borneo—British foresters, particularly H. G. Champion and J. Wyatt-Smith[4] (p. 170), completed important work on silviculture.

Teak and sal. Colonial forestry focused at first on commercial timber production for metropolitan home markets, but by the late 19th century, it increasingly served local markets. Attempts at management of tropical forests started with the less species-rich deciduous forests. European foresters had been trained to manage their species-poor forests, and their interest in tropical India was overwhelmingly in the sustainable management of the two commercially important, durable hardwood species, teak (*Tectona grandis*) and sal (*Shorea robusta*, p. 184). Teak was in international demand; sal was used locally for railway sleepers (ties) and construction.

Fire. The first priorities for management, following European forestry tradition, were the exclusion of fire and the reduction or elimination of browsing and trampling by domestic cattle (which alienated local farming communities). Fire killed young seedlings and reduced older seedlings to re-sprouting root stumps, but damage was insignificant after a few years in the drier forests (p. 161). However, fire protection inhibited regeneration in the higher-quality forests in climates with shorter dry seasons. Fire was eventually developed as a management tool in Burmese moist teak forest. It eliminated the evergreen species competing with teak and was required for teak germination. Sal soils (p. 183) are highly prone to trampling, especially in drier climates. The resulting management protocols aimed to increase stocking rates of one of these two trees. Success in achieving that goal almost inevitably came at the cost of other species, many of them important in the rural economy.

Pests. Inevitably, host-specific pests and pathogens increased with management. Notorious among these was the longicorn beetle of sal, which also attacked other trees in its range. Similar problems arose with teak. The best management results, both for sal and teak, were often achieved in mixed stands where the species of interest was not strongly dominant. Plantation forestry was established wherever natural regeneration was poor.

Shelterwood systems. Attempts to manage mixed evergreen forests came late, and they were at first similarly focused on individual commercial species or genera. The lower species diversity of seasonal evergreen forests, combined with greater regeneration capacity after catastrophe, made these forests easier to manage than the MDF of western Malesia. Management focused on dipterocarp regeneration, especially the *Dipterocarpus* species used for railway ties. Where natural regeneration of focal species was reliable, *shelterwood* methods of silvicultural management were attempted. These vary in pattern and timing of canopy removal, but all have several common characteristics. First, they emulate an episodic natural disturbance, so that there is always one dominant cohort of recruits that develops, matures, and is regenerated again. Second, canopy is removed (the "disturbance") over a large area of similar floristics and structure (that is, in foresters' terms, a *stand*, p. 117) to homogenize tree development and thereby create a more economically manageable unit of timber, known as a *compartment*. Third, success of the shelterwood system depends on regeneration at the time of final felling; this may require prior subcanopy removal and site treatments to facilitate the initial germination and establishment of canopy tree seedlings. All silvicultural manipulation aimed at releasing a well-stocked, fast-growing future stand of trees occurs at the final timber harvest of canopy trees. It involves partial or total removal of unwanted species and competitors of well-spaced, superior individuals of the species desired. Where it could be adopted, as in northeast India, enrichment planting was also successful.

The Malayan experience. The most comprehensive attempts to manage natural mixed evergreen forest for sustained timber production were in Peninsular Malaysia.[5] The emergent canopy of mixed dipterocarp forest in the everwet Sunda climate is rich in red meranti (*Shorea* spp.) and keruing (*Dipterocarpus* spp.). These yield general-purpose light hardwoods with favorable properties, including ease of planing and resistance to warping. By the 1930s, timber was sold on local markets, and exports were restricted by law to ensure a sustainable supply. Though they are light-demanding and fast-growing as juveniles, they occupy a late successional stage in the forest cycle. On other continents, individual species such as mahoganies have similar silvicultural characteristics. This silvicultural program has been exhaustively documented, and the sociopolitical context of Peninsular Malaysian forestry has been well-reviewed.[6] Attempts to establish sustainable commercial forestry in Malaysian MDF must ultimately be viewed as a succession of responses to rapidly changing markets: changes in emphasis from single species to many, and from climax species to successional.

The social and economic background. The colonial administration of Malaya (now Peninsular Malaysia) reduced local hostilities, stabilized government, and introduced medical services that heralded the control of malaria. Such governance (combined with immigration from denser populations, particularly China) promoted population growth. As towns grew and commodity plantations became widely established, populations migrated inland. Early forest product export markets were in *gutta percha*. Stocks were rapidly depleted, and there was limited success in planting within the forest following canopy thinning. Highest demand within the country at first was for heavy, durable hardwoods, particularly *cengal* (the dipterocarp *Neobalanocarpus heimii*) and the rosewood *merbau* (*Intsia palembanica*). These are shade-tolerant, grow slowly, and produce abundant seedlings; neither of them, however, responded well to initial pre-felling canopy opening. Still, there was an optimal canopy gap diameter for growth.

By the early 1920s, the growing Malayan population increased demand for fuelwood, creating a market for "improvement fellings" of the pole-sized trees removed to enhance growth of the shade-tolerant, heavy hardwoods then in demand. This reduced management costs early in the felling cycle, improving the economic viability of sustainable management. Shelterwood silvicultural systems were consequently adopted, following Indian practice. The canopy was manipulated by intermediate fellings to encourage regeneration prior to the main cut. But by the mid-1930s, kerosene and other cheap, petroleum-based fuels reduced the demand for fuelwood and charcoal in an increasingly prosperous economy. Improvement fellings continued at a reduced rate, with removal by girdling and poisoning.

In the 1930s, the trees were felled with axes and saws, and logs were extracted by buffalo, minimally damaging regeneration (fig. 16.8). The main felling itself often produced a satisfactory increase in regeneration. Pre-felling thinning was therefore finally abandoned, though thinning continued post-logging and in primary forest.

Economically viable management, which had become difficult during the depression of the 1930s, returned later in that decade, this time thanks to the mechanization of milling. Hand pit-sawing gave way to small sawmills, which sprang up wherever markets were expanding, and this led to demand for a wider range of timbers. Plans were interrupted by the war; management ceased, and forests were destroyed both by unmanaged timber exploitation and forest clearing for food production. Much of the forest that had been silviculturally treated, both prior to and following logging,

Fig. 16.8 Pit sawing of logs in forest, Sirumalai Hills, South India (D. L., 2000).

was lost. Following the war, labor costs increased rapidly. Before the forest service could thoroughly assess the condition of the forest estate and regain full control, the communist-inspired civil war, or Emergency, had started. This lasted for more than a decade, when much of the forest was inaccessible to all but the military. At the same time, world timber prices greatly increased, as did local demand, and research focused increasingly on intensive management of hardwoods.

The Malayan Uniform System. This difficult period nevertheless led to the only successful method yet devised for sustaining timber production in a biodiverse, tropical mixed hardwood rainforest: the Malayan Uniform System (MUS). The MUS aimed at sustained production of light hardwood red and yellow merantis (*Shorea* spp.). In Sundaland, these are generally abundant in the lowland MDF on zonal udult clay loam and sandy clay soils. Pressure treatment of timber with preservatives added value to light hardwoods, and a wider range of species could now be marketed.

The MUS was a simple shelterwood system, relying upon the regeneration beneath the canopy. The removal of the overstory in a single cutting released the regeneration. A felling cycle of 70 years was predicted. Forest was demarcated through area regulation, and1/70 of the production forest within a reserve would be logged and regenerated each year in perpetuity. These coups would in effect still simulate large natural canopy gaps, and the resulting single-aged cohorts (or "stands") would constitute the basic silvicultural unit of the management system. Under the MUS, the first logging was heavy, and individuals of noncommercial species were poisoned, opening the forest. Freed from competition and with their crowns in full sun, the early regeneration responded vigorously—if it survived the logging. Mechanized logging had by mid-century become general, so careful planning and oversight of the operation was critical.

MDF that had been felled and treated in this manner (or by the earlier shelterwood system of the 1930s, which promoted similar regeneration) produced spectacular stands of light red meranti whose volume exceeded that in the original forest. However, the MUS did reduce tree species richness drastically where unwanted juveniles were poisoned. It would likely lead to gradual elimination of shade-tolerant species over successive felling cycles because many would fail to reach reproductive age within a cycle. It may seem surprising that Wyatt-Smith, a keen naturalist and a founder of the Malayan Nature Society, did not express concern about this. The view was that MDF was floristically chaotic but overall uniform, and therefore the few large forests conserved in the national park and wildlife sanctuaries would be sufficient.

Selective logging systems. Forests of everwet climates, with high species diversity, were the most vulnerable to over-exploitation. The MUS had scarcely been implemented before peace returned in 1959, and Malaysia became independent. It was highly successful in achieving its primary purpose for managing MDF: increasing the yield of light dipterocarp hardwoods on a 70-year rotation. Tragically, the final introduction of MUS in Peninsular Malaysia coincided with the policy decision to convert all lowland forest on flat or undulating land—except the peat swamps—to commodity plantations, notably oil palm.

Selective systems of management had been recommended in the Philippines, where an almost continuous emergent canopy of light red meranti provided light for dense, pole-sized meranti regeneration. In true selection systems, only canopy individuals within the mature stands (comprising perhaps one-tenth of the area of the floristic association under management) are harvested. The residual stand is then further modified to create conditions for optimal growth of promising pole-sized juveniles of the required species (here merantis) as the next crop, which are marked and monitored. Growth rates are subsequently enhanced by periodic "regeneration thinnings," requiring a permanent road system for access. Such selection systems are therefore an irregular type of shelterwood system. They are particularly appropriate where stocking of emergent dipterocarp species is relatively low, and where more shade-tolerant, heavier hardwood *Shorea,* dark red meranti, and selangan batu (balau) prevail. Felling cycles must be of 20–30 years to be economic. Selection systems are demanding both of labor and the silvicultural skill to monitor individuals of varying growth rates.

Selective logging was optimistically introduced in Malaysia and Indonesia; it promised to shorten the felling cycle by relying on advance regeneration at or above a prescribed minimum felling diameter. The major disadvantage of such selective logging systems is ecological. Medium-sized individuals of a diversity of economic species are indeed often found in considerable densities (albeit few of the desired merantis), but most of these individuals are in fact suppressed, dying slowly. Owing to their malformed crowns, they cannot recover high growth rates even when released. Individuals of good crown form may maintain diameter growth at ~1cm/y, but in forests of calm climates, such trees are confined to natural gaps of less than 20% of the forest area. This problem can only be overcome by the pre- and post-harvest liberation thinnings that are fundamental to true selection systems—but such treatments require skilled, salaried field workers representing the owners. Unfortunately, the regional forest services did not establish and monitor long-term plots to examine the response of stands to selection systems.

The postwar timber boom was accompanied by the introduction of crawler tractors for road building and for timber extraction. This was cheaper, but greatly increased soil compression and scarification, killing advance regeneration. Tractor damage to soil in lowland forest on undulating land affected 20–80% of the logged area. Tractor and felling damage to the residual stands, by debarking and branch and crown snap, was estimated at 40–50%. The highest density and basal area of timber in the hill forests has been along the ridges where, without consideration of subsequent crops, the extraction roads are most easily located. High-lead logging produced even greater damage than tractor logging, both to seedlings and crowns in the residual stands on slopes.

Reduced-impact logging systems. Reducing damage, for instance by climber cutting, and restoration by enrichment with planted timber saplings were studied. Several protocols were suggested for reduced-impact logging (RIL), reducing damage to soil and stand by lowering logging intensity and maintaining partial canopy continuity.

In the Far East, such "counsels of perfection" did not reach policymakers, and for obvious reasons; their interest was in maximizing financial returns rather than achieving sustainable economic value of a renewable resource. A 35-year felling cycle promised higher and earlier profits, and perhaps even increased timber yield, but the evidence was to the contrary. Forest departments had not been funded to meet the need for oversight of larger and more remote areas under exploitation. Oversight was abandoned, leaving the fate of the forest in the hands of the concessionaire. But the prevalence of licenses of shorter duration than any felling cycle deprived that same concessionaire of any incentive to protect the residual stand—while it motivated the corrupt politician, acting as renter, to take maximum and most frequent rent.

Sustainability and optimal forestry. The MUS simulated the regeneration characteristics of the quite large windfall gaps prevailing on the udult clay and sandy clay soils in Peninsular Malaysia, which are the principal soils throughout lowland Sundaland. True selection systems, and RIL in principle, more closely simulate the small gaps that prevail on freely draining, deeper rooting but less fertile humult yellow soils of MDF on high ridges and sandstone. Either system, *if strictly implemented,* could be effective and optimize conservation of the tree flora and dependent organisms. Selection systems rely on the commitment, skill, and experience of the forester in charge. Reliance on the goodwill of concessionaires and loggers has been naïve at best, leading to the triumph of personal greed over the public interest. This sylvicultural history set the stage for the widespread deforestation that followed independence.

Notes

1. This chapter was condensed from *OTFTA*, chap. 8, 509–22; Freedman, Paul, *Out of the East: Spices and the Medieval Imagination* (New Haven, CT: Yale University Press, 2009).

2. Dove, Michael R., "Smallholder rubber and swidden agriculture in Borneo: A sustainable adaptation to the ecology and economy of the tropical forest," *Economic Botany* 47 (1993): 136–47; Dove, Michael R., "Transition from native forest rubbers to *Hevea brasiliensis* (Euphorbiaceae) among tribal smallholders in Borneo," *Economic Botany* 48 (1994): 382–96; Dove, Michael R., *The Banana Tree at the Gate: A History of Marginal*

Peoples and Global Markets in Borneo (New Haven, CT: Yale University Press, 2011); Peluso, Nancy Lee, "Rubber erasures, rubber producing rights: Making racialized territories in West Kalimantan, Indonesia," *Development and Change* 40 (2009): 47–80.

3. Doolittle, Amity, "Colliding discourses: Western land laws and native customary rights in North Borneo, 1881–1918," *Journal of Southeast Asian Studies* 34 (2005): 97–126; Doolittle, Amity A., "Stories and maps, images and archives: Multimethod approach to the political ecology of native property rights and natural resource management in Sabah, Malaysia," *Environmental Management* 45 (2010): 67–81; Brosius, J. Peter, "Green dots, pink hearts: Displacing politics from the Malaysian rainforest," *American Anthropologist* 101 (1999): 37–57; Peluso, Nancy Lee, and Peter Vandergeest, "Genealogies of the political forest and customary rights in Indonesia, Malaysia, and Thailand," *Journal of Asian Studies* 60 (2001): 761–812; Roseman, Monica, "Singers in the landscape: Song, history and property rights in the Malaysian rainforest," in *Culture and the Question of Rights. Forests, Courts and Seas in Southeast Asia*, ed. Charles Zerner (Durham, NC: Duke University Press, 2003), 109–35.

4. Wyatt-Smith, J., *Manual of Malayan Silviculture for Inland Forests*, Malayan Forest Records 23 (Kepong: Forest Research Institute, 1963).

5. Ashton, Mark S., and Matthew J. Kelty, *The Practice of Sylviculture: Applied Forest Ecology*, 10th ed. (Hoboken, NJ: John Wiley, 2018).

6. Kathirathamby-Wells, Jeyamala, *Nature and Nation: Forests and Development in Peninsular Malaysia* (Copenhagen: Nias, 2005).

As I was applying for a new job in an overseas aid agency, one interviewer challenged me: "Why is it that, having assisted in converting forest to oil palm plantation as a source of future stable revenue to the state, the local villagers had not responded to our employment offers, so that Indonesian labor had to be imported instead?' It dawned on me that this official had not the remotest understanding of those villagers' minds. Their forest had been destroyed, and their rich ancestral heritage, knowledge, and skills permitting their continued use of a diverse natural resource, were about to be lost with it. Was it beyond the capability of contemporary science and economics to foster the enrichment of ancient traditions while promoting a modestly prosperous, mixed rural economy? I realized then that I was applying for the wrong job. — P. A.*

17

FOREST HISTORY: INDEPENDENCE

After the end of World War II, the benchmark for the beginning of global environmental problems, colonies in tropical Asia became independent. Thailand, Nepal, Brunei, and Bhutan had not been colonized. Indonesia, after a five-year struggle with the Dutch, became independent in 1950. The British granted independence to India and Pakistan in 1947, after a century of rebellion and the brutal partition; to Myanmar (Burma) in 1948; Sri Lanka in 1948; and Malaysia in 1957, after ending the communist Emergency. France granted independence to Laos (1949), Cambodia (1953), and Vietnam (1945). The Philippines became independent from Spain in 1898, but was controlled by the United States until 1946. Bangladesh broke away from Pakistan in 1971. Papua New Guinea became independent from Australia in 1975.

In 1950 tropical Asia had a population of 542 million. By 2020, it had increased to 2,099 million, even though the rate of population increase had fallen by half. A population that had been 11% of the world total increased to 27% by 2020. These countries inherited much from their ruling countries: institutions (including forestry schools and departments), infrastructure (roads and railroads for transporting commodities), bureaucracy (*red tape* took a special meaning to us when we saw the bundles of old documents stored on shelves!), and economic ties to the parent countries. Yet, there were opportunities to pursue new policies and directions. Despite the deforestation during the colonial period, each country had a substantial forest estate.

Three categories of still-forested land persisted in Asia.

Upland forests. In the dry seasonal tropics, deciduous forests had long been disturbed by pastoralists and by upland rice and millet cultivators. The wet seasonal tropics and the everwet tropics, where malaria was rife, remained widely under primeval forest into the 20th century.

Montane forests. Some lower montane forests were not cleared for commodity crops, because they were located on steep slopes or were otherwise inaccessible. Much upper montane forest persisted.

Lowland forests. Floodplain, swamp, and peat forests, a fifth of lowland forest, persisted. Lowland MDF remained in isolated areas; some was protected, and much was converted to rubber and some to oil palm. In the seasonal tropics, forests remained despite disturbance by fire and grazing, and conversion to agriculture.

Mercantilism Triumphs

After World War II, the world was economically devastated, and rapid economic development was pushed. The engines were cheap energy, cheap primary resources, and the technology from military-industrial research. The world's forests seemed to exceed any predicted demand. Large-scale logging started in the rainforests of the Far East in the 1960s. Within four decades, commercial forest resources were exhausted in several countries.[1]

It seems extraordinary that newly independent countries exploited and converted their forests at a rate never approached by former colonial regimes, and it is easy to identify political corruption and short-term, self-serving commercial interests as the causes (to discount them would be naïve). The colonizers did not prepare their subjects for independence by promoting national over personal interests, long-term versus short-term goals, and a sharing of wealth among all citizens. The new leaders' priority was to provide work and better livelihoods for rapidly increasing populations, and the necessary infrastructure and capital for industrialization and economic development. What values did the forested lowlands have for a developing economy, other than as timber and land ideal for commodity plantations?

Forest-clearing technology (military in origin), cheap fuel to run it, and cheap credit to buy it—especially for the international conglomerates—promoted logging. Over the last half-century, a renewable resource was exploited as if it were a non-renewable mineral, to great financial profit for the few and immense long-term cost to the majority. The demands of free global trade have been blamed, but were this the root problem, the selling price of timber should have increased as the commodity declined in availability, providing an incentive for reinvestment in forest management. Instead, perverse national policies, including under-taxation (low royalties) and promoting intercontinental trade, prevented market signals from operating. Above all, selling unprocessed timber at low prices gave the importing nations opportunity to add value by processing. This discouraged forest-owning nations from developing the rural forest-based industries that could have transformed the future of upland peoples. Royalties remained low. The timber was thus sold at a lower price than the cost of managing the subsequent crop, especially so when the high discount rates of less developed economies are taken into account.

Deforestation

With independence, deforestation became widespread throughout the humid tropics of Asia, varying in causes and extents among different countries.

Malaysia. Oil palm overtook *Hevea* rubber in profitability by the 1980s, as the global markets for processed foods and biofuels exploded. The remaining lowland MDF and, more recently, even the peat swamp forests are being converted to oil palm plantations, leaving only a few small patches of research forest and national parks. Thus, in the last half-century, a forest formation 23 million years old has diminished from perhaps 75% of its original area to less than 10%. Most of this remnant outside the modest conserved areas has itself been logged and has an uncertain future (fig. 17.1).

Peninsular Malaysia. This region is unique in the everwet Far East for its long history of commercial forestry and research, and its early, well-developed home market.[2] Due to control of forests and licenses by state governments, timber companies remained local and focused on that market. After silvicultural treatment was abandoned (p. 367), sustainability rested on the care of timber extraction. Concessions were generally too small and too brief for sustainable management. Taxes on timber often failed to increase with inflation, or decreased, or were avoided. Timber exploitation with weak government intervention has long been irresistible to commercial enterprises and their political patrons.

Sarawak (Malaysian Borneo). In Sarawak the Forest Department originally prescribed a selection felling system, cautiously employing a 70-year cycle owing to the steepness of the inland hills and patchiness of advance re-

Fig. 17.1 Forestry in postcolonial West Malaysia. *Left*, clear-cut in Terengannu, 1973, in preparation for oil palm plantation (D. L.); *right*, cutting from hill-forest ridge in Perak, 1991 (D. L.).

generation. Concessions were increasingly distributed under political pressure, and often for disastrously short tenure, overriding stated Forest Department policy. The International Tropical Timber Organization (ITTO) recommended a reduction in logging and greater emphasis on in-country processing if production were to be sustained. The government later mandated a 35-year cycle selection system to increase timber production for export and to satisfy local demand, on the mistaken assumption that a selection system could yield crops on a rotation of that length. Timber production continued to decline. Forests are now re-entered at ever shorter intervals, pole-sized trees are now exploited, and the forests degraded. Land ownership among many ethnicities had been locally established in Sarawak. The Forest Department mapped the forest under state control for management planning, then disregarded traditional claims.

Once the logging is over, promises by concessions of improvements to education, health, transportation, and employment are met to varying degrees. The land, once partly owned or at least controlled by villagers, is converted to company-owned oil palm plantations. The upland farmer, with generations of rural knowledge, becomes a laborer on what was once his own land. This he has often declined. The companies then import foreign labor to harvest their estates. In contrast, in those fewer, long-settled communities with fruit gardens and permanent agriculture, as in west Sarawak, land has been secured by law.

With timber all but exhausted, the Sarawak Forest Department became partly privatized as a corporation charged with management but dependent on the declining revenue of a degraded resource. Little primary, unlogged forest remains outside the national parks and wildlife sanctuaries. Some of these too have been illegally logged, and rafts descend the rivers, loaded with undersized logs (fig. 17.2). Now the corporation is again owned by the government.

These degraded and vanished forests, along with those in neighboring

Fig. 17.2 Deforestation in Borneo. *Top left*, illegal log removal from Gunung Palung National Park, West Kalimantan (© Tim Laman); *top right*, peat swamp forest fires on Borneo, 2002 (Wikimedia); *bottom left*, lack of dipterocarp regeneration in abandoned logging area, Danum, Sabah (H. H.); *bottom right*, clearing peat swamp forest for oil plantation (Google Earth).

West Kalimantan and Brunei, were the most biodiverse in the Old World (p. 296). The profits from logging have been invested in oil palm or in private homes overseas in Vancouver or London. In Indonesia and Sarawak, logging companies with global ambition have arisen, funded by the enormous profits extracted from concessions at home. To their new contracts in Brazil and New Guinea, they bring their own workers, among them the Iban Dayaks, prized for their courage and ability, the youths who traditionally went on a personal expedition, or *bejalai*, to qualify for manhood.

The permanent forest estate has retreated to steep lands and mountains, demanding for sustainable forestry. Even into the 1970s, we were fortunate to witness the primeval forests of Malaya and of Borneo in their magnificence, as Beccari and Wallace had seen them. Now, the lowland primary forests have all vanished, and the hills are ravaged by uncontrolled logging (fig. 17.3).

Indonesia. This large and populous country once contained the greatest forest resources in tropical Asia (table 17.1). Under colonial rule, all forested land became property of the government. It was tightly controlled in Java, and left to local rulers elsewhere. Later attempts at more centralized control were only partly successful. After independence, the government passed the Basic Forestry Law of 1967, which placed the entire forest estate of 143 million hectares (more than three-fourths of total land area) under state control. National development objectives overrode traditional local claims of forest use. Under President Suharto, large forest concessions were granted to family and friends (often senior military officers); local communities did not benefit from this exploitation, except from some small-scale timber extraction. This system came to a sudden halt after Suharto was removed from power in 1998.[3] Reforms transferred economic controls (including control of timber extraction) to local districts and municipalities, but some concessions bypassed provincial control. This same structure promoted oil palm cultivation, replacing the degraded forest estate on a massive scale, particularly in the peat swamp forests of Sumatra and Borneo. Abetting this activity were the transmigration schemes, which moved whole villages from overpopulated areas, particularly in Java, to the outlying islands. Consequently, Indonesia became the greatest exporter of tropical timber (some 40% of it illegal) and had the highest rates of deforestation (table 17.1, fig. 17.4), with only 5% of the original forest remaining.

Plantation economics. The trends of clearing forest for plantations continued the earlier colonial policies established in much of tropical Asia. At first,

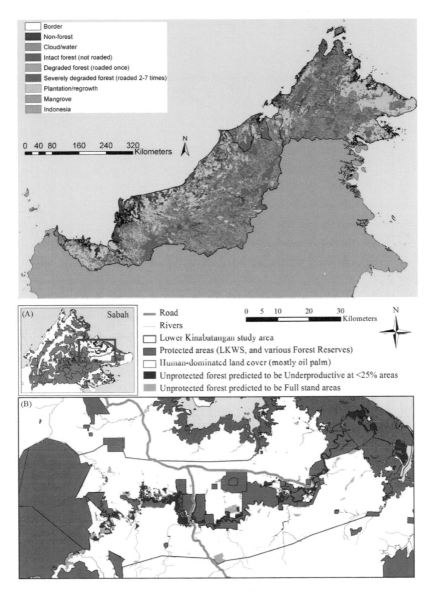

Fig. 17.3 Land use changes in Sarawak and Sabah (Malaysian Borneo), and Brunei (courtesy of *PLoS*). *Top*, note the extent of deforestation, except in Brunei (from satellite imagery analysis, up to 2009; see J. E. Bryan et al., "Extreme differences in forest degradation in Borneo: Comparing practices in Sarawak, Sabah, and Brunei," *PLoS ONE* 8 [2013]: e69679; doi:10.1371/journal.pone.0069679); *bottom*, land use changes in northeast Sabah; grey is protected forest, white is mainly oil palm (2009–2011, Google Earth images; see N. K. Abram et al., "Synergies for improving oil palm production and forest conservation in floodplain landscapes," *PLoS ONE* 9 [2014]: e95388; doi:10.1371/journal.pone.0095388).

Table 17.1 Land and forest area in tropical Asian countries

Country	Land area	1970 Forest	1970 % Forest	1990 Forest	2015 Forest	2015 % Forest	2000–12 % Loss	2014 % Protected
Bangladesh	130.2	NA	NA	14.9	14.3	11	0.4	14.6
Bhutan	38.4	NA	78	25.1	27.6	72	0.3	47.3
Brunei	5.3	4.4	92	4.1	3.8	72	3	44.1
Cambodia	176.7	133.3	43	129.4	94.6	53	7.1	26
East Timor	14.2	NA	NA	9.7	6.9	49	1.2	8.7
India	2,973.20	689.6	13	639.4	706.8	24	0.3	5.4
Indonesia	1,811.60	1,212	67	118.5	91	5	8.4	14.7
Laos	230.8	150	43	176.4	187.6	81	5.3	16.7
Malaysia	329.6	240.3	81	223.8	222	67	14.4	18.4
Myanmar	653.5	452.7	60	392.2	290.4	44	2.3	7.2
Papua New Guinea	452.9	364.2	85	336.3	335.6	74	1.4	3.1
Philippines	298.2	119.1	53	65.6	80.4	27	2.1	11
Singapore	7	0.02	< 1	0.2	0.2	< 1	0	10.9
Sri Lanka	62.2	29	57	22.8	20.7	33	1.5	23.2
Thailand	510.9	326	64	140	164	32	2.4	18.8
Viet Nam	310.2	133.4	43	93.6	147.7	48	2.8	6.5

Note: Forest/land area in thousands of km². Forest areas are from FAO Forestry Yearbooks of 1960 and 2015; protected areas from World Bank, 2015; net forest reduction from analysis of satellite imagery, 2000–2012.

Source: M. C. Hansen et al., "High-resolution global maps of 21st-century forest cover change," *Science* 342 (2013): 850–53.

these were established by American/European companies (like Sime Darby and Guthries) and then by more local ones, and even by a few nationalized foreign companies. In the 1970s and later, oil palm plantations began to dominate the forest clearing. In Malaysia and Indonesia, government agencies established huge plantations in attempts to provide more opportunities for the rural poor, including schools and medical dispensaries. In Malaysia, FELDA (the Federal Land Development Agency, or Lembaga Kemajuan Tanah Persekutuan) was the principal organization. In Indonesia, the central government PIR (Persatuan Indonesia Raya) organized large plantation

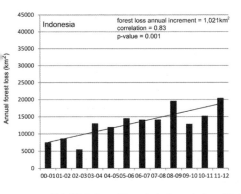
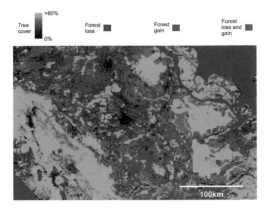

Fig. 17.4 Forest loss in Indonesia (see M. C. Hansen et al., "High-resolution global maps of 21st-century forest cover change," *Science* 342 [2013]: 850–53; reproduced by permission of *Science*). *Left*, Forest loss estimations in entire country, from analysis of satellite imagery 2000–2012; *right*, land use changes across central Sumatra (Pekanbaru is the large black spot in center); forest loss in red (2012).

transmigration schemes of impoverished people from densely populated areas. In both countries, smallholder agriculture was supported along with the plantations (in Indonesia, by the Plasma Small Holdings).[4]

The continuing need for natural rubber (partly for radial and aviation tires, and condoms) still is a cause for deforestation in the Indo-Chinese countries, such as Thailand and Cambodia, and in Yunnan, South China.[5]

Surviving lowland forests. We can still see the various lowland forest types in those nations that have established and protected forest preserves for nature conservation. Superb protected forested landscapes remain in India, Sri Lanka, and New Guinea. Protected forested landscapes survive, mostly on more limited scale, in Indo-Burma and Malaysia. That any seasonal lowland forest survives at all—often in degraded condition—is due to the persistent dependence of rural communities on the forest, combined with the difficulty of growing any of the major exotic commodity crops in the more seasonal lowlands.

Forests in the seasonal tropics. The forests and woodlands of seasonal Asia, adapted to climates prone to large-scale climatic catastrophes—drought and hurricane—are also better adapted to regenerate following human intervention than those of the everwet regions.

The weakening of traditional rule in India following independence, and a forest service under relentless criticism from both rural communities and the politically motivated public, made Forest Reserves into common land.

Forests suffered degradation, as such lands have elsewhere, as population and demand for their resources increased. Catchments were deforested. Although the legislated forest area was increased, specifically for fuelwood plantations, forest degradation also continued to increase. In the rain-forested northeast, expansion of *jhum* (swidden), together with corruption-abetted illegal logging, decimated the forest. In response, the policy of joint forest management was introduced. Eventually, a national moratorium on logging was established and continues to the present. It is not just logging that has degraded Indian deciduous forests, but the increasing demands of an ever-expanding population, rural and urban.

The cow was traditionally venerated, and milk and curd are important sources of protein in the South Asian diet. Cattle browsing eliminates both seedling and coppice regeneration. Browsing and trampling harm regeneration and compact the soil, but they are restricted during the monsoon, allowing hay cutting and storage for fodder during the long dry season. Cattle have made wastelands of much of India's former deciduous forest.

Consequences of Forest Loss

Overharvesting of timber has altered the structure and ecology of remaining forests, and it has led to the ultimate removal of forests and conversion to agriculture or fire-prone, coarse *Imperata* (cogon, lalang) grassland. Forest changes have had ancillary effects.

Fire. Fire is a natural phenomenon in the seasonal tropics. Humans have traditionally used fire to encourage grass for cattle and to attract wild ungulates (p. 161). For swiddeners, originally hill tribals but now also landless migrants, fire is a fundamental instrument in cultivating marginal land. Fire frequency has increased markedly in many places over the last half-century, including national parks, such as Mudumalai Wildlife Sanctuary in India (p. 155).

During the last 50 years, forests of wet seasonal and everwet climates, in or close to logged forests, have burned, associated with ENSO-induced droughts. Fires of 1982–1983 killed 16–20% of understory trees and up to 70% of canopy trees in the primary forest of Kutei National Park (East Kalimantan). Logging increases the intensity of droughts locally, damaging the canopy. Fallen branches greatly enhance proneness to fire. One in Sabah killed 19–27% of canopy trees in an unlogged forest tract and 38–94% in a logged tract. More than 80% of juveniles and understory trees in both tracts

Fig. 17.5 Conversion of kerapa forest to oil palm plantation in Murum, Sarawak, in 2016 (P. C.).

died. Recovery proceeds slowly, with sharply reduced species richness. Dipterocarps are notably patchy or absent in the regeneration, which is at first saturated with *Macaranga* and other pioneers. Successive episodes of burning produce *Imperata* grasslands, the true "green desert."

Particularly worrying has been the conversion of peat swamp forests throughout the Far East into oil palm plantations (fig. 17.5). Not only is a unique ecosystem being destroyed wholesale, but the burning of the cut forest and the peat—to a depth increased by the construction of drainage canals—contributes hugely to atmospheric carbon. The resulting pervasive regional haze may kill vulnerable residents and reduce tree growth by as much as half (fig. 17.2).

Fire also converts upper montane forest to grassland, which tribal cowherders initiated many centuries ago in South Asia and New Guinea (p. 349). Forest burning severely affects the water yield of upper montane forests. When healthy—their surface area much amplified by cryptogamic epiphytes—they remove water from foggy monsoon winds, which irrigates the paddies far below.

Wildlife. People have lived with wildlife and shared its resources for millennia in the deciduous forest regions. This continues only in the wildlife parks. In rainforests, certain mammals persist and even increase in numbers in logged and disturbed primary forest, in secondary forest and close to villages where hunting has not intensified. Logged forests promote hunting for some game species, often large vertebrates, such as wild cattle and elephants, which are now in steep decline there. Arboreal vertebrates eventually return to regenerating selectively logged forest as the canopy closes. If deprived of canopy corridors and fruit trees, some species decline. Small mammals and birds decline in logged forest even where overall richness may increase, owing to the intrusion of forest margin and open country species.

Fragmentation and edges. The remaining primary lowland forests in Asia are now fragments (figs. 17.4 and 17.6). In some, such as the Bukit Timah Sanctuary in Singapore, primary forest may cover less than 100 ha (p. 229). Others may cover a few thousand, such as the 6,800 ha Lambir National Park (Sarawak) or the Danum Forest area in Sabah. Patches of primary for-

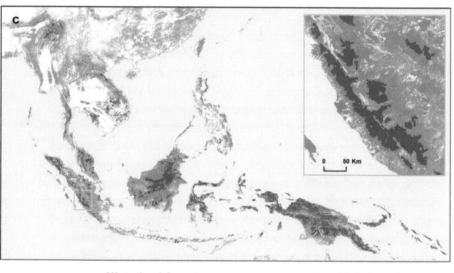

Hinterland forests   **Non-hinterland forests**

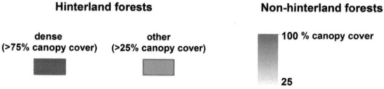

dense   **other**   100 % canopy cover
(>75% canopy cover)   (>25% canopy cover)

                                                          25

Fig. 17.6 Hinterland forests 2007, Southeast Asia: inset, coast of Sumatra, Indonesia. These are residual forests with high foliage density, based on analysis of satellite imagery, most noticeable in New Guinea and remote central Borneo (see Tyukavina, Alexandra, M. C. Hansen, P. V. Potapov, A. M. Krylovm, and S. J. Goetz, "Pan-tropical hinterland forests: Mapping minimally disturbed forests," *Global Ecology and Biogeography* 25 [2016]: 151– 63; reproduced by permission of *Global Ecology and Biogeography*).

est survive within larger conserved areas—and these are often mountainous. Examples include Gunung Palung and Kutei National Parks in Indonesia; Gunung Gading, Gunung Kinabalu, Gunung Mulu, and Taman Negara in Malaysia. Most other patches of true primary forest persist as small or tiny relict fragments.

Edges dry out and brighten the interiors of rainforest fragments, introducing open-country vertebrates into the understory, and they may reduce biomass well into the fragment, as when surrounded by oil palm.[6] The effect is most severe in smaller irregular fragments with more edge relative to overall area, and in seasonally windy regions.

Preservation plots, more than 300 of which have been initiated since 1905, promote conservation in India. Generally covering only a few hectares, they are controls for silvicultural treatments, refuges for biota threatened by logging operations, and examples of major forest types for staff training. In 1959, the Forest Department of Peninsular Malaysia set aside Virgin Jungle Reserves (VJR) throughout the production forest reserve system, inspired by the Indian example. Most of these stands survive embedded in logged forest, and sometimes even now isolated in a sea of oil palm, no longer used for their original purpose.

Invasive plants. Innumerable exotic plants have been introduced in the tropics. Noxious Neotropical weeds have become pervasive in deciduous and seasonal evergreen forests of Asia. Very few weeds invade primary everwet rainforest, but the fecund berry-bearing shrub *Clidemia hirta*, from the Neotropics, is now widespread in Asia.

Gene pools. Farmers in ancient China and Japan, and later in Europe, gradually improved the quality of fruits and other products of trees introduced from the forests. This practice has hardly developed in those places where the opportunity remains greatest—the lands of the hyper-diverse MDF. Neglect of the genetic diversity of these potential crops will impede future selection and improvement of them.

The Social and Political Environments of Forests

Political liberation released pent-up expectations in every country of tropical Asia. Forest-poor countries, already with dense and impoverished rural populations, capitalized on their forests' socioeconomic value. Governments of countries well-endowed with forests, still with small populations, logged and converted many to commodity crops, not so much to provide land for the poor as to raise capital, ostensibly for industrial development and infrastructure—which in turn increased local demand for timber. Most of the private profit was invested elsewhere and lost to the forest-based economy, a loss to forest communities.

Rural populations and urban interests. Growing rural communities contributed to forest degradation and deforestation, via fuelwood collection and cattle browsing in dry regions, and shortened swidden fallow periods in wet regions. Increase of rural populations has also accompanied rising urban expectations and the intrusions of broad-based market forces into fragile rural economies. Traditional constraints on overexploitation forsaken, these trends early became apparent in regions in the seasonal tropics that had long been densely populated, and where cropland was repeatedly subdivided as families expanded beyond their capacity to support themselves.

Joint forest management (JFM). The village republics of the ancient Bali-Hindu tradition, still partially surviving in Bali, and the longhouse communities of the Bornean Iban retained considerable independence. However, even the most isolated rural economies are now enmeshed in relations with wider economies, with administrative agencies, and with surrounding ecological patterns of change, all of which both create incentives and constrain actions.[7] Under former conditions of low and fluctuating population densities, long-standing patterns of resource use could be maintained in some places, such as in the biodiverse forests of West Kalimantan. There, community forests, enriched with tengkawang fruit trees (*Shorea* spp.), hardly differed in their composition in commoner species from the nearby primary forest. The same applied elsewhere in tribal societies where traditions of local decision-making prevailed. Now, with increasingly intensive land use, it is unsurprising that the most promising examples of participatory forest management in general have been reported from the less-diverse deciduous forests, where uncontrolled grazing had compromised forest regeneration.

Some of the most successful examples are from India. JFM schemes have often failed to cede practical decision-making powers to forest users, especially the Adivasi. Sometimes, owing to divisions of caste and social status from those they serve or poor management, forest department staff have found it difficult to relax authority where that is essential for success. JFM has been most successful where the land has the greatest capacity to respond resiliently, with less success on laterite and other marginal sites where it is often most needed.

Regional trade. Nations that have imposed logging moratoria must import on a large scale to meet home demand. These include Thailand, India, and Sri Lanka; Sri Lanka is a laudable exception, with plantations and extensive forests remaining in its double-monsoon northern and eastern sectors. These imports are now mostly coming from countries which either never had strong forestry oversight, or in which the overseeing agencies have been politically overridden and demoralized. Nowhere has this been more tragic than in Myanmar. There, the world's finest teak forests, excellently managed for over a century, have been rapidly decimated for timber exports to China, India, and even Thailand, in exchange for foreign currency to prop up the still military-influenced government. The upland minorities are effectively reduced to servitude, often without legal title to land and cowed by forced labor (*corvée*). The taungya system of forest management, formalized by the British there (p. 364), has forced thousands of uplanders and lowland Burmese migrants into temporary labor, farming government forest lands in exchange for replanting teak and other species, but without land of their own and often without schools or health services. Laos and Cambodia are suffering similar, albeit less socially harsh, exploitation. Illegal logging of rosewood (*Dalbergia*) for the Chinese market persists in Thailand, moratorium notwithstanding.

The expansion of China's mercantile interests in Southeast Asia has a long but intermittent history. In the postcolonial era, aided by modern communications, the overseas Chinese community in the Far East has been able to reassert its long-established influence in trade. Between 1997 and 2007 the value of wood imports into China tripled, making it the world's leading importer (fig. 17.7). Wood imports to China may double again in the coming decade. Asian and Pacific countries, including Russia, have supplied ~70% of demand, so far. At the same time, China has become the major exporter of finished wood products. It makes a third of the world's furniture, with exports rising from 91 million individual pieces in 2000 to 248 million in 2006. China has also become the second largest producer of paper and cardboard. Between 1997 and 2005, exports of wood products from China to Western nations rose 700%. Clearly, the West can point no accusing finger!

In effect, a new commercial imperialism has arisen. Burma has become an economic vassal of China, recalling the earlier conquest of inland southern China by the Han. Besides timber, products illegal in international trade, including tiger parts and ivory, are being exported to China and Vietnam. Chinese planners should encourage trade policies that might sustain that flow of goods and capital emanating from the tropical Far East. Such ex-

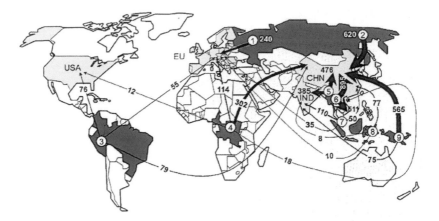

Fig. 17.7 Legal and illegal flows of timber and logs from sources to destinations in millions of USD, also indicated by the thickness of the arrows, in 2014 (see Jianbang Gan et al., "Quantifying illegal logging and related timber trade," in *Illegal Logging and Related Timber Trade—Dimensions, Drivers, Impacts and Responses: A Global Scientific Rapid Response Assessment Report*, ed. Daniela Kleinschmit, Stephanie Mansourian, Christoph Wildburger, and Andre Purret [Vienna: IUFRO World Series, 2016], 35:38–59; reproduced by permission of IUFRO World Series). Tropical Asian sources are (5) Myanmar; (6) Laos; (7) Malaysia; (8) Indonesia; and (9) Papua New Guinea. China dominates the trade.

ploitation of neighbor states represents an extreme case of the financial interests of mercantilism being allowed to override the longer-term economic interest of nations.

This seemingly irresistible vacuuming up of forest-based products into the insatiable maw of growing and industrializing economies is clearly just one element in the overall pattern of long-armed national and international resource exploitation presently promoted by the combination of cheap transport, rapid communications, hyper-mobile capital, and half-hearted oversight that we know as "globalization." Modern globalization is itself the logical outcome of previous cycles of economic imperialism, including the industrialization of the West in the 19th century.

Exploitation by commercial interests serving one community against the interests of another (usually a forest minority) happens within individual national economies too. The social and economic consequences of such exploitation have often been disastrous. A poignant example is the Ifugao, who (living in narrow, rain-shadow valleys in northern Luzon, p. 343) patiently carved and sculpted their near-perpendicular slopes to irrigate rice. They cultivated rice throughout the year, thanks to perennial springs dependent on precipitation from fog in mossy upper montane forest along the high ridges during the dry northeast monsoon. Farmers from the lowlands converted many to potato and temperate vegetable cultivation. The streams dried up during the dry season, and Ifugao farmers suffered.

Large-scale population movements. Migration from areas of high population into the forested hinterlands, whether organized in settlement schemes or illegal, has caused deforestation. Migrants are seldom familiar with the intricacies of the terrain and the soils on which they settle, or with the peoples whose traditional forest lands they usurp. These lands had often remained forested over time because they are agriculturally marginal.

Two exceptions. Long-term planning and policy can work. Two small Asian nations have preserved the greater part of a pristine forest estate. In the Sultanate of Brunei, oil had deferred the exploitation of forest capital, but now formal legislation of a conservation policy and conserved forest areas is being implemented (fig. 17.3). In Bhutan, the annual estimate of Gross National Happiness is preferred to Gross Domestic Product. The Bhutanese royal government values the combining of intensified agriculture on fertile sites with conservation of ancient forest landscapes for "happiness"— tourism and export of hydropower. The king is a constitutional monarch and has wisely retained authority over land. Bhutan's rural population has yet to outgrow its arable land. Bhutan's farmers have legal title to their land. Forests near villages have been given legal community ownership. The "Achilles heel" in this system remains the cattle, which require fencing and planting to regenerate forests.

Knowledge and Identity

Dramatically altering forests in which hundreds of tribes have lived for thousands of years is bound to affect their well-being (fig. 17.8). Their culture and language are bound up in the forest, and taking that away is disorienting. Their knowledge is also valuable to the global community, because of the value of minor forest products, particularly medicines.[8] With the destruction of most of the lowland forests during the past half century, we can only imagine the depths of the suffering of these forest people. Best known among them are perhaps the Penan, hunter-gatherers of the interior of Borneo (fig. 17.8). They were a wandering people, speaking their own distinct dialects, who only began to settle under the influence of missionaries in the mid20th century. They felt the brunt and injustice of logging, which even-

Fig. 17.8 Indigenous forest inhabitants in tropical Asia. *Left*, New Guinea villagers dressed for a festival (© Tim Laman); *right*, nomadic Penan family in Mulu National Park, Sarawak in 1958 (P. A.)—now excluded from the forest.

tually altered the forests to the extent that they could no longer live in them. The Penan initially resisted the logging efforts (with public support from environmental NGOs), and were subsequently persecuted by the logging companies and governments.[9]

Such people are further alienated from their forest heritage by the universal use of "screens:" TVs, computers, and smart phones substituting for their original forest culture. Now it is nearly too late. A vast body of indigenous knowledge, much of it profoundly place-specific, is fading away.

Non-timber forest products. These products, known as NTFP, remain critical components in the household economies of the poorest regions, which (with some exceptions) are those with seasonally dry climates. Harvesting levels are increasing with population, but some also because of growing expectations and increasing pressure from national urban and even international markets.

The seasonal tropics of Asia are rich in NTFP derived from palms and bamboo. Among palms, the palmyra (*Borassus flabellifer*), from northeast Africa in ancient times, has been widely grown from India to the Lesser Sundas (fig. 17.9). Use of the palmyra is recorded in a Tamil Nadu sutra that extols its 800 uses.

Some products other than timber increase mightily following logging and disturbance. The most influential for the future trajectory of the forest is bamboo (p. 106), a source of fiber for producing containers, paper, food,

Fig. 17.9 Non timber forest products (NTFP). *Top left*, bamboo basket makers, Selangor, West Malaysia; *top center*, rattan (*Calamus* sp.) Xishuangbanna, SW China; *top right*, Soppina Betta forest, Karnataka, India, with betel palm in understory. *Bottom*, NTFP entrepreneurs, Ganeshpuri, Maharashtra, India. *Left*, preparing delicious palmyra nuts for sale; *right*, selling betelnut and bidis in a typical village shop (all D. L.).

and other products. In low-wage economies such as those in drier rural India, other NTFP are important. Tendu, the leaves of *Diospyros melanoxylon*, is the covering of the bidi (an Indian cigarette, fig. 17.9). The Soppina Betta of the Western Ghats are natural patches of forest set aside by hamlets specializing in betelnut (from the indigenous palm *Areca catechu*, fig. 17.9). Betel and bidis are sold together in innumerable shops throughout tropical Asia, satisfying the addictions of hundreds of millions of users.

In wetter climates of Southeast Asia, forest minorities forage for NTFP in the farming off-season, and sell them to increase their income. These include *Fritillaria cirrhosa* bulbs sold to traders from Yunnan; tengkawang or engkabang (p. 104) for its valuable oil; fragrant gaharu, or eaglewood

(p. 351); and rattans (some 400 species of viny palms), exploited through-out tropical Asia, and raw materials for a huge furniture industry. Collection for distant markets has led to their overexploitation and decline.

Whereas the rural poor of less-developed economies continue to de-pend on forests for grass and browse, game, fuelwood, timber, and medici-nal herbs, forest use is dramatically changing in advancing economies (no-tably that of Peninsular Malaysia). Prized products of former times, such as Borneo camphor, from *Dryobalanops aromatica*, or the oleaginous resins of Indo-Burmese *Dipterocarpus*, are neglected. The latex gutta percha was an early victim of the destructive exploitation by Western mercantile interests (p. 357); presently it is only used in dentistry.

Tropical Asia is home to an unparalleled diversity of sophisticated me-dicinal traditions, studied by the British and Dutch in colonial times. The two most highly developed traditional medicinal systems are the Hindu-Buddhist ayurvedic tradition and the Chinese Confucian system; both are humoral in theory and rely on herbs from surrounding forests (fig. 17.10). Tribes vary in their medicinal knowledge, and the plants used by them are often unique to the indigenous floras where they live. Lowland people de-pended for many medicinal preparations on the uplanders, who collected from the forest. Indigenous drug production is economically important in India and China, and these products sell in the health stores of Europe and North America. In spite of this, many medicinal species suffer over-exploitation, and conversion to commercial cultivation remains the ex-ception.

Stirrings of a Conservation Movement?

The idea of conserving areas for the wildlife of tropical Asia originated among European colonial officials in the late 19th century, with the first Indian Wildlife Sanctuary in 1898. National parks were then established in South Asia, often from game reserves. Wildlife reserves were added else-where in the interwar years, including in the Dutch East Indies. Taman Ne-gara (National Park) was established in Peninsular Malaysia in 1938 and in-cludes the peninsula's highest mountain. What remains of primary MDF in Peninsular Malaysia is almost entirely confined to the national park system, part of which has been illegally logged. After World War II and the end of local insurrections, the establishment of national parks accelerated, notably in Thailand, Indonesia, the Philippines, India, and Sri Lanka. The objectives

Fig. 17.10 Medicinal NTFP. *Top left*, producing ayurvedic medicines in Gandhigram, South India; raw materials are collected by Adivasi from the Sirumalai Hills in background; *top right*, ayurvedic physicians from Gandhigram, next to a vasaka plant (*Justicia adhatoda*, Acanthaceae, including inset), source of a potent cough and asthma medicine. *Bottom left to right*, dragon's blood resin, from *Dracaena cochinchinensis* (Asparagaceae), now cultivated in Xishuangbanna and sold as a medicine (all D. L.).

of these parks are, in most cases, to protect extensive examples of natural landscapes, together with their native vegetation and wildlife, usually where at least one charismatic, "flagship" vertebrate has survived.

In Sarawak, Bako National Park (established in 1957) protects a landscape of unique forest associations. In succeeding years, more, including the large Mulu National Park, were inaugurated, preserving lowland MDF. Parks are susceptible to degradation by illegal logging (fig. 17.2), especially in the Philippines and Indonesia. Thailand's park system incorporates the principal landscape features and geology of this diverse country. Sri Lanka has expanded its conservation areas to more completely represent overall species diversity.

Even though they provide a broad range of benefits (economic, cultural, environmental), the future of tropical Asia's forests and the people who depend on them seems dire. In the final chapter we discuss some developments that give room for cautious optimism.

Notes

1. This chapter condenses and updates parts of chapter 8 (pp. 523–51) in *OTFTA*, with assistance from Reinmar Seidler.

2. Kathirathamby-Wells, Jeyamala, *Nature and Nation: Forests and Development in Peninsular Malaysia* (Copenhagen: Nias, 2005).

3. Day, Tony, ed., *Identifying with Freedom: Indonesia after Suharto* (New York: Berhahn Books, 2007); Murray Li, Tania, "Articulating indigenous identity in Indonesia: Resource politics and the tribal slot," *Comparative Studies in Society and History* 42 (2000): 149–79; Obidzinski, Krystof, and Koen Kusters, "Formalizing the logging sector in Indonesia: Historical dynamics and lessons for current policy initiatives," *Society & Natural Resources* 28 (2015): 530–42; Miettinen, Jukka, Aljosja Hooijer, Jianjun Wang, Chenghua Shi, and Soo Chin Liew, "Peatland degradation and conversion sequences and interrelations in Sumatra," *Regional Environmental Change* 12 (2012): 729–37; Miettinen, Jukka, Chenghua Shi, and Soo Chin Liew, "Two decades of destruction in Southeast Asia's peat swamp forests," *Frontiers in Ecology & the Environment* 10 (2012): 124–28; Linkie, Matthew, Sean Sloan, Rahmad Kasia, Dedy Kiswayadi, and Wahdi Azmi, "Breaking the vicious circle of illegal logging in Indonesia," *Conservation Biology* 28 (2014): 1023–33; Carlson, Kimberly M., et al., "Committed carbon emissions, deforestation, and community land conversion from oil palm plantation expansion in West Kalimantan, Indonesia," *Proceedings of the National Academy of Sciences U.S.* 109 (2012): 7559–64; Gatto, Marcel, Meike Wollni, and Matin Qaim, "Oil palm boom and land-use dynamics in Indonesia: The role of policies and socioeconomic factors," *Land Use Policy* 46 (2015): 292–303.

4. Gaveau, David L. A., et al., "Rapid conversions and avoided deforestation: Examining four decades of industrial plantation expansion in Borneo," *Scientific Reports* 6 (2016): article no. 32017. Good statistics on deforestation and plantations for the region.

5. Dove, Michael R., "Rubber versus forest on contested Asian land," *Nature Plants* 4 (2018): 321–22; Grogan, Kenneth, Dirk Pflugmacher, Patrick Hostert, Ole Mertz, and Rasmus Fensholt, "Unravelling the link between global rubber price and tropical deforestation in Cambodia," *Nature Plants* 5 (2019): 47–53; Zeng, Zhenzhong, Drew B. Gower, and Eric F. Wood, "Accelerating forest loss in Southeast Asian Massif in the 21st century: A case study in Nan Province, Thailand," *Global Change Biology* 24 (2018): 4682–95.

6. Ordway, Elsa M., and Gregory P. Asner, "Carbon declines along tropical forest edges correspond to heterogeneous effects on canopy structure and function," *Proceedings of the National Academy of Sciences, U.S.* 117 (2000): 7863–70.

7. Cramb, Rob A., "Explaining variations in Bornean land tenure: The Iban case," *Ethnology* 28 (1989): 277–300; Cramb, Rob, and Patrick S. Sujang, "'Shifting ground':

Renegotiating land rights and rural livelihoods in Sarawak, Malaysia," *Asia Pacific Viewpoint* 52 (2011): 136–47; Peluso, Nancy Lee, "Territorializing local struggles for resource control: A look at environmental discussion and politics in Indonesia," in *Nature in the Global South. Environmental Projects in South and Southeast Asia*, ed. Paul Greenough and Anna Lowenhaupt Tsing (Durham, NC: Duke University Press, 2003), 231–52; Peluso, "Rubber erasures, rubber producing rights: Making racialized territories in West Kalimantan, Indonesia," *Development and Change* 40 (2009): 47–80.

8. Ashton, Mark S., I. A. U. N. Gunatilleke, C. V. S. Gunatilleke, K. U. Tennakoob, and P. S. Ashton, "Use and cultivation of plants that yield products other than timber from South Asian tropical forests, and their potential in forest restoration," *Forest Ecology and Management* 329 (2014): 360–74.

9. Brosius, J. Peter, "Green dots, pink hearts: Displacing politics from the Malaysian rainforest," *American Anthropologist* 101 (1999): 37–57; Brosius, J. Peter, "Local knowledges, global claims: on the significance of indigenous ecologies in Sarawak, East Malaysia," in John A. Grim, ed.: *Indigenous Traditions and Ecology* (Cambridge, MA: Harvard University Press, 2001), pp. 125–157; Colchester, Marcus, *Pirates, Squatters, and Poachers: The Political Ecology of Dispossession of the Native Peoples of Sarawak* (Petaling Jaya, Malaysia: Institute of Social Analysis, 1989); Colchester, Marcus, "Pirates, squatters and poachers: the political ecology of dispossession of the native peoples of Sarawak," *Global Ecology and Biogeography Letters* (The Political Ecology of Southeast Asian Forests: Transdisciplinary Discourses) 3 (1993): 158–179.

After Sri Vikrama Rajasinha, the last Kandyan king, was deposed by the British in 1815, his memory was consecrated by the sacred forest of Udavattekelle, below which is nestled the celebrated Sri Dalada Maligawa, sanctuary of the Tooth Relic. That forest remains, at the old royal capital of Kandy, near the center of Sri Lanka. For several decades, I have learned from professors Savitri and Nimal Gunatilleke, of the nearby University of Peradeniya, about the ecology of nearby examples of the Asian tradition of multi-species, planted forests known as home gardens. Land is used according to its capability: by secure land tenure; irrigated padi in the floodplains and lower slopes; roads, and home gardens with vegetables; spices and medicinals close to the residences mid-slope; and swiddening patches as insurance against failed harvests on upper slopes, below surviving indigenous forest patches along the ridges. The whole is a deeply spiritual realization of the unity of humanity and nature in a developed yet sustainable landscape. A strong conservation ethic, born of a traditional sense of unity with nature, persists in Sri Lanka. Commercial logging of indigenous forest is banned. Very few herbs, and no tree species, have been lost from Lanka's rich flora during the last two centuries, though only 5% of indigenous forest remains. — P. A.

18

FUTURE FORESTS

The burning of the great library at Alexandria in the 7th century retarded the early return of learning to Europe. Produced by hand, long before the printing press, each book was unique and irreplaceable. Similarly, the genetic information enshrined in the vast array of species in tropical forests is unique and irreplaceable. Like the books, each species may yield important information or insight. The subtitle of this book, *Exploring Tapovan*, suggests studying a forest to obtain deep knowledge through intense spiritual practice and contemplation. Tropical Asian forests are a store of complex knowledge, to be understood by such study.

It is now up to us, generations of the 21st century, to decide the fate of much of the global biodiversity library. How can we meet our collective and individual responsibilities toward tropical biodiversity?[1]

The Dilemma

In the 60 plus years since independence, the forest estate of tropical Asia has almost vanished. Our surveillance methods and efforts need improvement, and the present estimates are approximate (table 17.1). However, in 1970 approximately 46% of the original forest cover had been lost, and that increased to 67% in 2015. These statistics include all types of forest, from deciduous to logged forests, and the planting and harvesting of plantations. We estimate that perhaps 5% of the original MDF remains. Deforestation has varied among countries. It has been most extensive in Indonesia and Malaysia, and high also in Laos and Cambodia; it is now low in the Philippines (not much left to remove), and low but increasing in Papua New Guinea, which has the greatest remaining area of everwet forest. India still has the greatest extent of tropical forest in Asia, mostly deciduous.

We have discussed the causes of deforestation in the previous two chapters. First is the unsustainable logging pressure, despite the earlier development of sustainable silvicultural systems that were never properly implemented. Forests have been damaged primarily by the impact of mechanized logging and by the too frequent cutting of remaining trees, especially of undersized trees. Second in importance is the conversion of forest to commodity plantations. This began under colonization with the cultivation of tea and coffee in lower montane forests and rubber in the lowlands, and it accelerated after independence in the lowlands with cultivation of oil palm. Third is the gradual damage to forests, particularly deciduous, by local people. Intense use by nearby villagers, who take firewood and fodder for animals, and whose browsing animals compact the soils, degrades the canopy and capacity for regeneration. Swidden, when practiced in short cycles, may contribute to the conversion of forest to wasteland.

These practices decrease the capacity of forests to harbor biodiversity. The theory of biogeography (p. 308) predicts the erosion of biodiversity in a nonlinear manner, with 95% loss of forest corresponding with a ~40% loss in species. That calculation is based on animals, many of which, being mobile, occupy whole forest regions. Rates of species erosion apply rather to the loss of ecological island habitats occupied by plants within the forest regions, which more closely correlates with the rate of forest loss.

Forest damage and removal—and species loss—must be seen in the context of population increases in tropical Asia (table 18.1). Population in the region increased from 137 million in 1600 (p. 352) to 598 million in 1950, at an annual growth rate over the 350 years of 0.4%, and a doubling time of 139 years. In just the 70 years between 1950 and 2020, population reached 2.189 billion (with an annual average growth rate of 1.9% and a doubling time of 36 years (table 18.1). In another 30 years it is expected to increase to 2.610 billion, at a declining rate (0.26%) less than that early growth. Population densities in the region will increase from 57/km^2 to an estimated 362/km^2. Countries vary widely in projected densities: Bangladesh, India and the Philippines are the highest; Papua New Guinea, the lowest and most rapidly increasing. To what extent will these demographics affect our ability to conserve forest and biodiversity (fig. 18.1)?

... Population in tropical Asian countries. Population data from Department... of Economic and Social Affairs of the United Nations Secretariat; urban estimates from UN Development Program.

	Population in 1000s				% Urban population				Population density (persons/km²)			% Population increase			
	1950	1990	2020	2050	1950	1990	2020	2050	1950	2020	2050	1950–55	1990–95	2020–25	2045–50
Tropical Asia	591,383	1,420,012	2,189,435	2,610,514	16.0	28.0	45.0	57.0	74	274	326	1.90	1.74	0.92	0.26
COUNTRY															
Bangladesh	37,895	107,386	169,566	201,948	4.3	19.8	38.0	55.7	291	825	1,551	2.58	2.20	0.96	0.24
Bhutan	177	536	322	980	2.1	16.4	42.2	55.0	5	14	26	2.48	1.02	0.97	0.25
Brunei	48	257	454	546	26.8	65.8	78.6	84.0	9	48.5	185	5.56	2.76	1.02	0.18
Cambodia	4,433	9,057	16,947	22,569	10.2	15.5	22.0	36.2	25	95.9	128	2.58	3.46	1.34	0.64
East Timor	433	751	1,286	2,087	9.9	20.8	35.8	48.3	30	90.6	147	1.26	2.85	1.95	1.34
India	376,325	868,891	1,353,305	1,620,051	17.0	25.5	34.8	50.3	127	455	545	1.66	1.73	0.94	0.29
Indonesia	72,592	178,633	269,413	321,377	12.4	30.6	57.2	70.9	40	149	177	1.73	1.66	0.91	0.26
Laos	1,683	4,245	7,651	10,579	7.2	15.4	43.5	60.8	7	33	46	2.32	2.76	1.52	0.70
Malaysia	6,110	18,211	32,358	42,113	20.4	49.8	77.7	85.9	18	100	128	2.77	2.59	1.24	0.51
Myanmar	17,527	42,123	56,125	58,645	16.2	24.6	36.9	54.9	27	86	90	1.91	1.47	0.54	-0.19
Papua New Guinea	1,413	3,683	7,260	10,111	2.0	16.9	15.4	20.5	3	16	22	1.81	2.67	1.7	0.76
Philippines	18,580	61,949	110,404	157,118	27.1	48.6	44.3	56.3	62	370	527	3.54	2.33	1.54	0.85
Singapore	1,022	3,016	6,057	7,065	99.4	100.	100.	100.	146	430	1,009	4.90	2.87	0.76	0.19
Sri Lanka	8,076	17,324	22,338	23,834	15.3	18.6	18.8	30.2	130	359	383	1.79	1.03	0.48	-0.02
Thailand	20,607	56,583	67,858	61,740	16.5	29.4	55.8	71.8	40	133	121	2.76	0.83	0.01	-0.66
Viet Nam	24,949	68,910	97,057	103,697	11.6	20.3	36.8	53.8	80	313	334	2.54	1.24	0.56	-0.12

Fig. 18.1 The importance of bringing our population into balance. The back end of an Indian truck says, "We two, ours one" (i.e., "We have only one child"; Ma. A.).

Economics

We should be able to conserve forests because they are valuable. The economic value of forests changes during national development. The eminent economist Simon Kuznets plotted inequality versus per capita income, and others have plotted Kuznets curves to describe environmental health, including forest value, against the path of economic development (fig. 18.2).[2] The high value assigned to forest by subsistence societies plummets with access to cash-based markets and employment offered by timber concessionaires. Technologies, including hunting guns and chainsaws, make life so much easier that future costs due to the draw-down of forest capital are ignored. However, if the capital thereby released as profits were attracted to the same national and local economy, rather than invested worldwide wherever returns are greater, more and better employment options would result. Such reinvestment depends on political stability, land security, and some means by which the work force can acquire new skills. In some countries, reinvestment has initially been in commodity plantation agriculture. The profits from this, when retained in-country, have in turn been invested in industry, which provides better-paid employment and demands a skilled workforce. But it is also drawing people off the land into the cities, degrading traditional, experience-based forest knowledge, and impoverishing what were mixed rural economies—turning proud freeholders into estate labor-

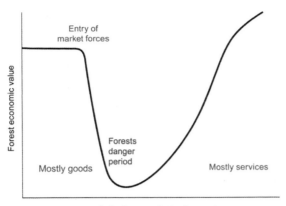

Fig. 18.2 A Kuznets curve, plotting the value of forest against the path of economic development (I. B.). Initially, the forest is heavily used at low cost, resources are depleted, and then the forest recovers with a much higher value.

ers. We are presently in the danger period of this curve. What can we do to restore the forest's value?

Multiple uses. First, the value of the principal current forest product, timber, can only be enhanced by growth or restoration of profitable, indigenous, timber based industries, including for construction, furniture, and other enterprises that create products for export. Second, values other than timber should be included. Ecosystems benefit many stakeholders, from forest-fringe villages to the global community. Vital goods and services come from forests, from medicinal plants and fuelwood to crop pollination, pest control, ecological benefits, soil conservation (and silt retention), weather amelioration, water conservation and quality, carbon sequestration, and biodiversity conservation. Unfortunately, these services are difficult to value, but there is no alternative to weigh their importance relative to that of timber or certain NTFP. The benefits, including income and compensatory payments from loggers, to elites who control forest fates and other beneficiaries in the wider world, should mostly be helping villagers who are living sustainably with forests.

Human capital. A growing, financially secure, and educated citizenry comes to appreciate that forests yield services essential to a stable prosperous economy, including dependable, high-quality water, equable weather, and space for recreation. People of the Asian tropics are becoming better educated and accessing the internet in greater numbers. Through their growing influence, policy can be reformed to secure a sustainable forest estate. If rural communities are better educated, they can better share in a successfully diversifying economy; demand for traditional products can increase the value of remaining forests. Security with prosperity—the final achievement of a stable, "developed" society—depends on sustaining forest services.

Reviving forest management. Sustainable management of MDF faces a host of challenges. Not the least of these is the presence of rural communities. Their needs for the goods and services of the forest, and their claims to the land are urgent and not easy to reconcile with sustainable timber harvesting. Countries in the Asian tropics are driving to build industrial economies. Will demand for non-exploitative products of forest diversity expand? As forest goods, such as long-lived hardwoods and wild fruit trees—or tigers—become rare through overexploitation, their value inevitably increases. Theoretically, this *scarcity value* should provide incentives for conservation, but only if ownership of the product is clear and secure. Successful economic development depends on retaining the full range of options for future use of the aboriginal forest estate—an asset that includes many species that are truly unique to each country and therefore an irreplaceable national heritage. Protecting the genetic library, which, in the Asian and American tropical lowland rainforests, far exceeds that of any other terrestrial ecosystem on Earth, is essential. Such forests must survive the stage of economic development when forests decline to their lowest perceived values.

Urban versus rural. From where might future capital investment—for environmental services in particular—be generated? Urban populations have grown disproportionately (table 18.1) and are better off, but they have lost awareness of the extent to which their wealth has arisen from forests, and those left behind in the inland valleys and hills have been disenfranchised. Modern tropical deforestation in Asia is in many ways a new manifestation of an age-old pattern: exploitation of isolated and poorly organized upland communities by lowlanders with access to information, resources, or political patronage (p. 350). This pattern is strengthened by modern communications and technologies.

In many parts of Asia, particularly in the drier seasonal regions, rural populations continue to live in subsistence economies, near the starting point of the Kuznets curve (fig. 18.2). There, multiple use is compromised by pressure, especially from gathering firewood and browsing. Wood remains the principal energy source in poorer agricultural communities, although improvements in stove technology and conversion to alternative energy sources are reducing demand. Unfortunately, we lack data from rigorous monitoring of both the forest resource and patterns of its use that would help us to recommend optimal management strategies.

Sustainable timber. Silvicultural techniques can potentially be applied to the remaining tropical forests. These are largely deciduous forests that have escaped conversion to commodity plantations. The remaining everwet forests are largely on steep slopes and ridges, dangerously prone to degradation and slips if cleared for oil palm or plantation forestry. Some remain on drought-prone, low fertility podzols and sandy humult ultisols where growth rates are slow. Much of that limited forest estate has been heavily harvested, and it will require careful management to provide future economic yields.

Semi-evergreen and deciduous forests. Social scientists' push for more equitable and sustainable products and services from tropical deciduous forests has rarely been balanced by estimates of forests' yield. The primary determinants of succession in these forests are the frequency and timing of timber harvest, fire intensity, biomass (fuelwood) removal, and browsing and trampling by domestic cattle. These deflect forest succession from more evergreen, floristic associations (richer and taller) toward shorter, deciduous ones (p. 153); semi-evergreen and tall deciduous forests are most susceptible. Regular but infrequent ground fires sustain tall deciduous and semi-evergreen forests that include valuable deciduous rosewoods (*Dalbergia, Pterocarpus, Afzelia, Xylia*), teak, and sal, by excluding evergreen species of little timber value.

Grazing and browsing pressures are reduced by rotation and fencing. Fencing, although expensive, may also permit tree regeneration, even from seed, if fire is controlled. Fuelwood rotation should also enhance forest productivity. These are excellent public investments to ameliorate rural poverty.

Evergreen tropical forests. Timber harvesting in indigenous forest still can be managed to maximize a sustainable out-turn of logs. Treated well, a logged forest may continue to provide many natural services, but this depends entirely on the details of how it is managed. In developing economies, where discount rates often remain comparatively high, it can be tough to make a commercial case for slow-growing heavy hardwoods. In the Sunda MDF, the light hardwood red and yellow merantis (p. 369) grow quickly, but because they mostly grow on land more profitable for oil palm, they can only be the backbone of a sustainable timber industry if policy gives them priority.

Active management will require experienced professional silviculturists, knowledgeable about the local trees and field operations. Regenerated stands will never reach the stature, clear bole length, or diameter of old growth, but

they will be comparable to the best stands in temperate and tropical deciduous forests. Future smaller logs will be extractable using lighter machinery,
with less damage to soil and better regeneration.

A major reason for the success of natural forest management in Peninsular Malaysia was the shift in responsibility for licensing and oversight of
logging operations from federal service headquarters to local forest offices
and the states. This, and the early strength of the home market, fostered a
processing industry of relatively small enterprises. Recent devolvement of
government to the provinces in Indonesia (p. 380) and the Philippines provides similar opportunities, but this shift further exacerbates deforestation
due to local corruption and lack of well-trained and experienced silviculturists. These need to apply and modify the systems researched and partially
implemented during the colonial period (p. 366). Without a new and active
program of field research and experimentation in rigorously applied, selective silvicultural management (p. 372), the future of the indigenous forest is
bleak. It will be important to tweak the systems to optimize the net present
value (NPV, from financial costs and benefits) of forest stands over time. As
trees grow, the NPV of timber harvests increases; they eventually decline
when the growth rates fall below discount rates. Unfortunately, all silvicultural systems eventually erode biodiversity.

Achievement of sustainable felling cycles of less than 50 years within a
polycyclic, selective management system will require skillful and stringent
management. The forest cannot recover unless returned to management by
well-trained, experienced, and dedicated forest service staff: servants—and
heroes—of the nation. Even then, the economics can only work if the ecological services and other values of the forest are considered. Always, forest
must be strictly protected for endangered species to survive.

Recovering natural forest. Nobody knows how long forest recovery will
take, but the less severe the modification, the less time is required for regeneration. Evergreen rainforests recover slowly, since few rainforest species
possess dormant seeds, seed dispersal distances are short, and fewer species
recover by coppicing. Furthermore, surface soil organic matter, which is the
main external source of readily available nutrients, is easily lost.

Asian tropical deciduous forests and the Himalayan lower montane forests, with a prevalence of coppicing species and lower species richness than
the rainforests, have survived the impacts of cattle grazing and periodic fire
for centuries. They appear amenable to rapid restoration. The Phnom Kulen
Plateau, north of the great medieval city of Angkor Thom, was deforested
more than 1,000 years ago for irrigation. Angkor was abandoned six centu-

Fig. 18.3 Forest recovery. *Left*, MDF forest at edge of an oil palm plantation; the forest is not degraded in this windless climate, but neither does it reinvade where soils have been degraded and leached (© Tim Laman); *right*, reintroduction of useful, indigenous MDF trees, palms, and herbs into restorative plantation, using the diffuse-crowned *Pinus caribaea* as nurse, in the buffer zone of the Sinharaja World Heritage forest (M. A.).

ries ago. Only now does the seasonal evergreen dipterocarp forest clothe the plateau, to all appearances primary but for the persistent lychee trees (*Litchi chinensis*), and only the scattered brick *prasat* (towers) betray its past.

A rainforest carefully and selectively logged may not return to anything close to its original composition and species abundances for at least two centuries. If the forest has been degraded, full recovery could take at least a millennium. Abandoned oil palm or rubber plantations, absent their original forest flora, would recover similarly, provided there is a seed source (fig. 18.3). The return of hyper-diverse rainforest to such areas cannot be expected to occur for thousands of years, if ever. Mixed indigenous species plantations may achieve similar timber productivity and carbon storage as aboriginal rainforests. They can accommodate non-timber producers, including medicinal products, rattan, and sugar palm, but they cannot accumulate its biodiversity. They are as prone to soil erosion on drought-prone soils and steep topography as are commodity plantations (fig. 18.3).

Climate Change

Primarily because of human activity, temperatures are steadily increasing globally, especially at high latitudes. The trends are well documented and understandable from the changes in energy balance due to increased infrared absorption by greenhouse gases. Temperature increases secondarily affect (1) regional climates and precipitation patterns; (2) the distribution of biomes; and (3) sea levels and the inundation of coastal forests.

Weather. In the tropics, the greatest increases in temperature are expected in continental and seasonal regions. Temperature may increase less than 2° C in the next 80 years in the humid tropics. Temperatures will increase more at greater distances from the equator, at least partly due to the expansion of the Hadley cells (p. 4).[3]

Rising temperatures should increase fluctuations in weather and more catastrophic events: drought, hurricanes, floods, and fires. In the absence of a tropical continental landmass, the steep declines in annual rainfall inland, such as those predicted for sub-Saharan Africa and the Amazon basin, are unlikely, though northern India is already beginning to lose monsoon rains and also to suffer extreme cyclonic storms. Increases in rainfall seasonality and drought, including penetration into currently everwet regions, are also predicted.

Greater rainfall seasonality in continental Asia promotes the replacement of semi-evergreen dipterocarp and other lowland forests by less species-rich and less productive deciduous forests. Frequency and intensity of ENSO-related droughts have increased during the last century,[4] apparently influenced by climate change.[5] So far, ENSO-correlated droughts and floods have hardly affected primary forests in everwet Asia, but their impact on degraded forests, through devastating fire, has increased. Deforestation and urbanization can alter local rainfall patterns. Total annual rainfall may not change, but the severity of storms will increase.

An apparent rise in the ecotone between upper montane and subalpine thicket and in the mortality of exposed canopy trees there are likely caused by climate change, as in the rising of zones on Chimborazo, in Ecuador.[6] The changing intertropical conversion zone (ITCZ) may also influence cloud formation, thereby shifting the boundaries between upper and lower montane forest.[7] The recent mortality of emergent or otherwise exposed canopy trees near the mountain tree line in equatorial Asia may be caused by drought, but freezing may also contribute (p. 210).[8]

Tree metabolism. Higher temperatures lead to increased metabolic rates and increased CO_2 (but results are inconsistent). Increased temperature, by changing the relative growth rates of co-occurring species, may alter their competitive relationships. It may also increase isoprene emissions by trees, with uncertain effects on climate. Drought may further reduce carbon uptake.[9]

Carbon. The growth of trees (and carbon fixation) is continuous, while their death is sudden and may reduce stand biomass. Stands are permanently in a state of recovery from periodic natural catastrophe. Major surges in mortality usually take place at long intervals in biodiverse equatorial rainforests. Predictions of globally increasing tropical rainforest biomass are debatable; local and regional trends may be driven by other factors. To date the most persuasive evidence points to a decline in mean growth rates with rising temperature, possibly caused by increased respiration rates from higher nocturnal temperatures.

Deforestation and atmospheric carbon. Deforestation, mostly tropical, presently releases carbon into the atmosphere at a rate equivalent to the world's automobile emissions, about 20% of all emissions, but the rate is increasing.[10] The problem has been exacerbated in the Far East by the felling and clearing of peat swamp; the carbon stored in the peat, up to 15 m deep, may exceed that in the standing biomass above, while peat accumulation at up to 3 mm per decade likely exceeds all other terrestrial carbon sinks. The burning of peat during the 1995–1996 ENSO-related drought released a quantity of carbon comparable to that released by the world's power plants and industry. The clearing of the majestic peat swamp forests, burning, draining and liming of their peat before planting with oil palm (which itself demands carbon for fertilization, protection, harvesting, refining, and transporting the oil) generates atmospheric carbon comparable to the melting of subarctic permafrost. Such efforts will come to naught when salt water invades with the rising sea, predicted within the next quarter-century. The reduction of deforestation rates along with plantation or natural regeneration are crucial parts of the solution to atmospheric carbon dioxide increase.

Stands of tall old trees contain the greatest carbon in their biomass. Logged evergreen dipterocarp forests will not approach their original biomass (therefore their carbon content) for centuries. Tropical evergreen forests therefore constitute an important carbon bank.[11] Carbon offsets can provide additional incentives for rainforest conservation where opportunity costs are low or where other values justify conservation.

The tropical lowland peat swamps of the Far East (p. 140) accumulate carbon more rapidly than anywhere else, worldwide. Their continuing destruction is a catastrophe, producing excessive CO_2 and methane.[12] Compensation for their protection on behalf of all humanity must be given, among all forests, the highest priority. Tropical deciduous forests have greater biomass below ground than above it, in part owing to their deep root systems. Sequestered carbon may partly be charcoal, a consequence of centuries, even millennia, of burning.

One of the paradoxes resulting from cheap, hydrocarbon-derived energy has been the almost complete change from timber to concrete residential houses in forest-rich tropical countries, in contrast to Scandinavia, Russia, and much of North America. Current energy prices, which favor cement manufacture (a huge CO_2 producer), are unlikely to continue into the medium and long term. A return to timber will both increase the value of the residual forest and the incentive to actively manage production forests. It will also increase these forests' capacity for carbon sequestration, by yielding timber that will be retained in semi-permanent structures.

At present, compensation to forest owners for conserving carbon in trees does not extend to unlogged primary forest. A United Nations initiative, Reduced Emissions from Deforestation and Degradation (REDD), is intended to correct that. Problems in obtaining reliable data on qualifications and from opportunities for cheating may be resolved with new techniques in remote sensing.[13]

Sea level rise. Current estimates, approaching a rise of 1 m over the 21st century, would make it extremely difficult for coastal forests (including mangrove) to migrate and reestablish without extinctions. Such a rise will erode the majority, perhaps all, of the coastal peat swamps of the Far East, which will release additional quantities of atmospheric carbon as they degrade. Most tree species will survive on those remaining upland plateau peat swamps and *kerangas*. But the major impact will be the inundation and increase in salinity of the great rice bowls of the Asian coastal and riverine plains. Many farmers made landless will likely migrate into the residual forests of the hills.

Conserving Biodiversity

The forests richest in biodiversity, which include the lowland MDF of Malesia, have fluctuated in area during periods of climate change, reducing the Sunda continent to islands over short geological periods (p. 66). But the fragmentation now occurring is beyond any that occurred during the last 23 million years, when our forest flora and fauna evolved. We must sustain as much as possible of that biodiversity, an integral requirement for a sustainable future.

Genetic libraries. Biodiversity is genetic information, a living library of gene sequences contained in populations that have been accruing over evolutionary (that is geological) time. Biodiversity maintains itself. The extraordinary chemical diversity of plants and microorganisms continues to yield novel, complex compounds and resistances to pathogens.

Most of the world's biodiversity is stored in lowland tropical rainforests. Within that biome, diversity is overwhelmingly concentrated below 400 m altitude and in those large, continuously wet equatorial land masses. The Sunda lands and New Guinea are two such major land masses. Southwest Sri Lanka is smaller but important; distinctly different gene pools are stored in each.

Diversity is overwhelmingly concentrated among the arthropods (especially insects) and microorganisms (including the fungi). The diversity of these groups directly or indirectly depends on the diversity of plants as primary producers, and tree species richness thus serves as a proxy for genetic diversity in biodiverse tropical forests.

Most of the tree species of Asian rainforests are represented in their chosen communities by less than one reproductive individual per hectare. Several patches of suitable habitat for commoner tree species of at least 200 ha and, for some species, a few patches of at least 10,000 ha, are therefore necessary for a region. A few large conservation areas might therefore appear optimal for conservation of both trees and animals, but many plants are confined to small habitat islands. Thus, full protection of plant diversity will require many small conservation areas in addition to the few large ones. Conservation of animals, especially large vertebrates, requires large, protected areas. These can best be located by first using geological maps to locate areas of exceptional surface geology (and therefore soils) and then evaluating their communities and flora on the ground. Plant conservation requires, at least for the foreseeable future, a network of preserves representative of the climate, geology, and landforms of their region. Protection of riparian forests (fig. 18.4) provides a network for migration of dispersers among them. These may be quite small, but they must be strictly protected from logging, and they are best in a forest matrix. Indian sacred forests and Malaysian virgin jungle reserves (p. 340) are good models for such reserves.

SLOSS. The debate about the relative merits of protecting "single large or several small" (SLOSS) areas has largely died down; conservationists

Fig. 18.4 Forests and conservation. Sabah has adopted the policy of retaining primary forest along the banks of the Kinabatangan and other major rivers, but it is a narrow band, representing only riparian forest and habitat; see also fig. 17.3 (© Tim Laman).

rarely enjoy the luxury of making such a choice. Protected area establishment in the "real world" has turned out to be largely a question of political and financial opportunity. Although large "charismatic" vertebrates benefit from protection in a few large preserves to contain their mobile populations, much plant diversity (including trees) can be conserved by setting aside forests on agriculturally marginal land, though the costs of protection will remain substantial. Much animal diversity can be maintained by retaining forest corridors between smaller conservation areas.

The criteria of the International Union for the Conservation of Nature (IUCN) for endangerment categories are principally designed for animals, for almost all mature rainforest tree species would thereby qualify as endangered, which cannot be practical policy advice. In the long term, we must accept these criteria as applicable to trees as well.

We are "five minutes before midnight" for the conservation of biodiversity in Asia's rainforests. Today, unlogged lowland MDF are almost gone throughout the region, except in those few cases that have been set aside for conservation or research. The less diverse hill and upper dipterocarp forests have greater hope of long-term survival. In Peninsular Malaysia, Sri Lanka and some other countries, legislation forbids commodity plantation and other cultivation on steep slopes and logging on mountains above 1,000 m, although these rules have been ignored in parts of Borneo. These upper dipterocarp forests also therefore serve as protection forests, while upper montane forests on ridges absorb and release moisture.

Success will depend on securing local support and commitment, active protection against overhunting, conservation of fauna crucial for pollination and seed dispersal, predators to control herbivores, and international subsidies for biodiversity conservation.

Ecotourism and biodiversity. Public interest worldwide has been attracted to the wonder of tropical forests and wildlife, partly by television programs of extraordinary beauty—particularly by BBC. Starting with a trickle of avid birdwatchers and other naturalists, often wealthy and preferring long stays in the tranquility of remote forest bungalows, such interest is expanding as a mass market develops both for local and international middle-income tourists. These are primarily attracted to picturesque landscapes and exotic means of transport and accommodation that allow them to experience adventure without danger or discomfort. Ecotourism may now comprise 20% of international travel.

With a major tract of its magnificent mixed dipterocarp forest, though previously logged, recovering as a corridor linking national parks, fortunate in its majestic Kinabalu and the Crocker range, and some of the globally richest coral reefs conserved in a pristine state, Sabah is now a major regional ecotourist destination in an increasingly prosperous economy. Much is due to one enlightened and committed director of forests, Datuk Sam Manan, and to a dynamic forestry research program. Sabah gives room for hope.

Unfortunately, ecotourism enterprises foster the unrealistic expectation that charismatic animals can easily be seen at close quarters in rainforests, and that they can be viewed without the discomfort of sticky climates and biting insects. The mass market is mostly attracted to parks where large animals can reliably be seen; commercial success is often determined by the attractions of the location rather than by its conservation value. Habitat degradation in such parks is a continuing threat, which must be counteracted by careful management and community involvement. Mass tourism also results in social problems and dislocation. Consequently, smaller, community-based tourism programs, such as opportunities for home-stays and participation in the life of remote villages, have multiplied (fig. 18.5). Many programs are run by locally based NGOs that often have external funding but aim to retain proceeds within the local economy. Empowering communities to bene-

Fig. 18.5 Ecotourism: a Semelai village visit in the Tasek Bera wetlands reserve in Peninsular Malaysia. *Left*, the homestay; *center*, tour of wetlands in dugout canoes; *right*, music from a Semelai elder (all D. L.).

fit from ecotourism depends on people's ability to manage a rather complex business with national and international links and, potentially, substantial cash flows. Ecotourism can probably only strengthen conservation where there is strong technical and regulatory oversight by public-minded government agencies or by resourceful, ethically committed and politically adroit NGOs. Ecotourism is least likely to lead to negative impacts on nature and on communities in societies that have already achieved a level of education and economic sophistication.

Indigenous people should be an important part of ecotourism enterprises. They have the deep knowledge of the forest to share with visitors.[14]

What to Do? A Dream for an Ideal World

First, we must save remaining areas. Perhaps counterintuitively, a review of the case for multiple use forestry, especially when focused on specific uses across the landscape, identifies conservation of forests of noted biodiversity as the only objective respected by both policymakers and exploiters.[15] Protecting 30% of remaining forest conserves 50% of its biodiversity. The method for finding such areas is straightforward, though not necessarily easily realized. First, identify areas that represent variations in climate and geology. Second, use satellite imagery and local inquiry to search for residual unlogged (or lightly logged) forest fragments (fig. 17.6). Third, explore candidate areas to assess their flora. Fourth, prioritize candidate areas according to the level of endangerment of their habitats and associations, their

Fig. 18.6 Preserving or enhancing the value of plants in forests of Sri Lanka. Kanneliya Conservation Forest, established to protect medicinal plants (Ma. A.).

species (as far as known), and their endemism (fig. 18.6). Fifth, negotiate the conservation of a comprehensive but parsimonious series of reserves. Sixth, seek to maximize forested corridors, whether logged, secondary, or otherwise, upland or riparian, between reserves (fig. 18.4). Seven, collaborate in gaining community and political interest and support, as well as legal protection (fig. 18.7).

Save productive forests. We can agree on some general principles to protect remaining forests and biodiversity. First, forest policy and management must be executed at the scale of whole landscapes, because optimization for more than one objective, notably timber production and biodiversity conservation, is unachievable within the same forest. Second, conservation areas such as national parks and nature reserves must never be logged. Third, inviolate reserves are essential to overall biodiversity conservation, even when they are too small to support many vertebrates. Opportunities for periodic immigration of dispersal agents and other organisms essential to ecosystem function, especially via forested corridors, must be found and

Fig. 18.7 States are beginning to foster conservation. *Left*, the sign at the entrance to the Sinharaja World Heritage forest ensures that everybody knows about it. Nimal Gunatilleke, right, and left, the ForestGEO plot manager, Gunadasa (P. A.); *top right*, a protest is proclaimed by a hill tribe, New Guinea (S. D.); *bottom right*, villagers, supported by a local conservation movement, post pleas for the retention of a majestic *Dipterocarpus turbinatus* grove threatened by road widening, Chiang Dao, northern Thailand (H. H.).

protected. Fourth, representative tracts of primary forest within timber production forests are refuges for animals, from which they can reinvade regenerating logged stands, as well as controls with which to monitor management. Fifth, conservation of species with broad ecological tolerances, which include many vertebrates, can best be achieved by sustaining habitat connectivity in a landscape divided into fragments, with differing uses and management. Sixth, the necessity for managing hunting becomes ever more urgent as forests become more accessible.

Improve management. Forests must be managed correctly, whether for timber production or conservation. This often involves conflicts between top-down versus bottom-up management.

The conservation of resources and their sustainable use are often most successful in the hands of individuals and organizations with long-term interests, who therefore have incentives to protect—if these incentives come with good information, good communication, and clearly defined powers of decision-making within a legal framework. Every tropical forest has a multitude of stakeholders. These range from the local inhabitants who harvest products directly to those whose health and welfare depend on ecosystem services, and to the regional and global communities, increasingly urban, that benefit from forests' amelioration of climate, carbon sequestration, and the biodiversity unique to each forest. Institutional reform is necessary, so that all stakeholders may participate in planning the future of the forest. Even then, options for changes in land-use must be defined by the law, overseen by an enlightened forest service, and mediated by an independent judiciary.

To preserve forest, we need to combine the approaches described above as governmental with those from below, based on village ownership. Restoration of authority to the forestry service (previously usurped by politicians) is essential if forest goods and services are to be sustained. *Local* rights and franchise must be respected, and the training and development of forest service staff must be appropriate and adequate.

The era of the large timber corporation, logging with minimum reinvestment in the forest resource, is over in Asia. Their activities have left us with the need for highly skilled staff and site-specific management, especially where the forest has been degraded. The added costs will fall on those governments that had invested in the management of first logging operations, these costs now passed on to consumers and taxpayers. Local, smaller enterprises, as in India and Malaysia, can better guide the fate of a forest. The interests of forest-dwelling communities have been appallingly neglected by authoritarian or paternalistic governments; these are already a major cause of conflict in most nations that retain tropical forest worldwide.

Gain international support. In our present globally linked economy, forests benefit not just the adjacent rural communities, but inhabitants of the watershed, the nation, and the wider world. Different benefits accrue to each group. Values of some services, especially biodiversity maintenance, benefit in the long term. The sense of urgency to "save the rainforest" among the urban middle classes of the advanced temperate economies cannot by itself greatly influence the future fate of the forest (although their past influence can hardly be ignored, and their future financial subsidy should be expected). The mission of international NGOs is always to address forest conservation in pursuit of one or a limited range of objectives. Their role will thus always be limited and has often contradicted those of major local stakeholders.

The rich world has played an ambiguous role in this regard, for it has often been responsible for offering economic incentives for overexploitation. In future, in addition to carbon sequestration, strictly protected areas for wildlife and biodiversity will need to be subsidized by the global community. In 2010 at Nagoya (Japan), representatives of more than half the world's nations agreed to a Convention on Biological Diversity. This set an objective of permanently protecting 17% of remaining forest land.

Promote public action. In every country in the Asian tropics, there is a stirring of interest and pride in indigenous wildlife and in its conservation, but so far this is happening mostly among the urban educated elite. Urban conservation movements in the tropics have been stimulated in part by the media. Just as in countries outside the tropics, interest is mostly expressed as sentiment and political opinion, although a lively growth of interest in natural history is also contributing important advances in knowledge.

In Peninsular Malaysia, and especially in Thailand, the growing urban middle class and professional population are gaining in influence for sound resource management and conservation of forests. The Malaysian Nature Society (MNS), for example, is now among the largest associations for amateur naturalists in the tropics.

At present, this upswing of interest and support for conservation is primarily an urban phenomenon, but it will move into rural areas with time, aided by growing literacy and access to the internet (fig. 18.8).

The future of Asian tropical forests depends on a shared understanding of its multiple values. Unless timber-importing nations ensure that those imports come from well-managed forests, the forest will cease to exist as a productive resource. Trade in endangered species must be controlled. The collection and consumption of endangered plants and animals for unproven medicinal or restorative purposes and as a spurious mark of prestige are surely pathetic admissions of barbarity, and commit future generations to an impoverished patrimony. From a similar perspective, can a state claim to be civilized while it allows the rampant exploitation of the resources of a less powerful neighbor, harvesting their renewable assets to destruction?

Overriding ethics. That ancient recognition and understanding of our interdependence with nature, which achieved an unequaled sophistication in Asia, persists but urgently needs reinvigoration and communication worldwide. Some religious leaders are at last beginning to recognize, once again,

Fig. 18.8 Friends and foes of biodiversity conservation. *Left,* Logging allows entry of outsiders to hunt: logging company employees, replanting during the day, go back at night to hunt, starting fires to flush game and regenerate the grass for wild ungulates, Ma Da Forest, Dong Nai Province, Vietnam, 1982 (P. A.); *right,* villagers of the Santal tribes, living in Similipal National Park, Orissa, set up their own conservation corps with the encouragement of the park director (H. H.).

the moral urgency behind the conservation of nature, and in Thailand, they are participating in practical efforts. Without such reinvigoration, the collective memory of our ancient forest heritage will surely fade among our grandchildren. Although they may never realize it, it will have mattered, and we shall have caused them to suffer for it.

An aspiring dream. The beginnings of beneficial changes in the rural and forest economy are visible in some countries that have achieved middle-income status, where a diversified landscape reminiscent of Japan or even French Provence (industrialized countries successful in retaining a diversified, traditional rural economy and landscape) still prospers and welcomes the unfamiliar visitor. Sri Lanka, for example, is now a such a country. Evergreen lowland rainforest, originally restricted to less than 10,000 km² of the island's southwest, has retreated to a mere 5% of that area, nearly all of which has been logged once. But there is a strong conservation ethic, deeply held by all sections of the rural as well as urban communities. Although factions still campaign for intrusive new roads, and land-hungry farmers still nibble the forest edge, public approbation usually wins out. A moratorium on logging was introduced and is strictly upheld, while larger fragments of the residual lowland rainforest are conserved in a natural state as medicinal plant reserves, and as a UNESCO World Heritage Forest, the Sinharaja

wilderness of ancient fame and respect. The landscape of the densely populated southwest is vegetated following tradition (p. 398). This landscape
continues to support communities in moderate prosperity, with a quality
of life equal to any. Collection of forest produce is declining as community
wealth increases, while hunting, never strong, is now illegal. Communities
are currently further enriched by the opportunities they offer for ecotourism: Visitors from far afield come to learn traditional methods of health en-
hancement and herbal remedies, and to explore nature with knowledgeable
villagers while staying in home accommodations within communities, or in
attractive rural hotels of moderate size and often glorious location that are
the renovated rest houses of colonial times.

The everwet Far East has enormous potential for such a transformation.
Smallholder oil palm would occupy land that, in seasonal regions and most
fertile soils, would support irrigated rice, providing the principal family cash
income. Ancient knowledge of nature, seen in the extraordinary diversity of
tree species still cultivated in home gardens by the Kedayan people of Brunei, themselves inheritors of the ancient Hindu-Buddhist Javan tradition,
and in the lesser but still rich home gardens of other settled Sunda forest
ethnicities, will revive. And the remaining forest, conserving wildlife and
flora as well as water, and offering protection from landslides, can lead a rejuvenated and sustained forest economy, with a local industry born of a joint
approach to management.

An Epilogue

The Silent Valley, or Sairandhri's Forest, is in the Western Ghats of the Indian state of Kerala. It is thickly forested, watered with some 6 m of yearly
rainfall from the two monsoons, and it is the home of tigers, gaur, elephants,
and four primates—including the endangered lion-tailed macaque and
the Nilgiri langur. Legend describes the valley as the home of Sairandhri
(Draupadi), the wife of the Pandavas in the epic of the Mahabharata, thus
a sacred forest. The valley was threatened by inundation for a hydroelectric
project in the 1970s, but it was saved by Prime Minister Indira Gandhi in
September of 1984 (two months before her assassination), when she made
the valley a national park (fig. 18.9). Her sympathy for such a place was undoubtedly influenced by her travels with her father, Jawaharlal Nehru, to
natural areas in the Himalayas, and her studies under Rabindranath Tagore,
at Shantiniketan. Tagore established this place of learning as a forest ashram

Fig. 18.9 Silent Valley. *Top left,* Indira Gandhi, who established the national park in 1984, at age 7 with Mahatmas Gandhi during his 1924 fast for Muslim-Hindu Unity (Wikimedia); *bottom left*, lion-tailed macaque, an endangered monkey whose largest population is within the valley (Wikimedia: Aaron Logan); *right*, view toward the north of the valley from Pechapary (D. L.).

(as in the tradition of the Mahabharata) where studies and nature were combined under strict discipline—*tapovan.*

Silent Valley National Park was the catalyst for the establishment of the Nilgiris Biosphere Preserve in 1986, which was designated as a UNESCO World Heritage Site in 2012. It was more than that, though. It became a model for biodiversity conservation in India and elsewhere (there are now 18 such preserves in India, 11 recognized by UNESCO). The Preserve also became a magnet for the establishment of nongovernmental organizations (NGOs) whose purposes range from fostering basic research and developing conservation policies to advocating for the rights of indigenous people (Adivasi). The earliest was the Nilgiri Game Association, established by British colonials in 1877 to control the game harvest, and recast as the Nilgiri Wildlife and Environmental Association in 1979, an activist protector of Nilgiri biodiversity. Kamal Bawa (p. 226) helped establish the Ashoka Trust for Research in Ecology and the Environment (ATREE), in 1996; it has developed a global reputation.

The park continues to receive threats from hydroelectric development, fire, marijuana cultivation, and Maoist guerrilla activity—and now government neglect during a pandemic, but it is still the best example of peninsular lower montane forest. Its history illustrates the hope for the future of forests in the Asian tropics, as indigenous conservation movements unite the rural poor (even forest dwellers) with prosperous urban groups, overcoming traditional mistrust and converging on a vision of healthy sustainable forests.

Notes

1. *OTFTA*, chap. 9, 552–96, with assistance of Reinmar Seidler.

2. Panayotou, T., and P. S. Ashton, *Not by Timber Alone: Economics and Ecology for Sustaining Tropical Forests* (Washington, DC: Island Press, 1992).

3. Hefferman, Olive, "The mystery of the expanding tropics: Cause—and consequences," *Nature* 530 (2016): 20–22.

4. Villafuerte, Marcelino Q., and Jun Matsumoto, "Significant influences of global mean temperature and ENSO on extreme rainfall in Southeast Asia," *Journal of Climate* 28 (2015): 1905–19.

5. Cai, Wenju, et al., "Increasing frequency of extreme El Niño events due to greenhouse warming," *Nature Climate Change* 4 (2014): 111–16.

6. Morueta-Holme, Naia, Kristine Engemann, Pablo Sandoval-Acuña, Jeremy D. Jonas, R. Max Segnitz, and Jens-Christian Svenning, "Strong upslope shifts in Chimborazo's vegetation over two centuries since Humboldt," *Proceedings of the National Academy of Sciences U.S.* 112 (2015): 12741–745.

7. Crausbay, Shelley D., Patrick H. Martin, and Eugene F. Kelly, "Tropical montane vegetation dynamics near the upper cloud belt strongly associated with a shifting ITCZ and fire," *Journal of Ecology* 103 (2015): 891–903.

8. Rehm, Evan M., and Kenneth J. Feeley, "Freezing temperatures as a limit to forest recruitment above tropical Andean tree lines," *Ecology* 96 (2015): 1856–65.

9. Qie, Lan, et al., "Long-term carbon sink in Borneo's forests halted by drought and vulnerable to edge effects," *Nature Communications* 8 (2017): DOI: 10.1038/s41467-017-01997-0; Popkin, Gabriel, "How much can forests fight climate change? Trees are supposed to slow global warming, but growing evidence suggests they might not always be climate saviours," *Nature* 565 (2019): 280–82; Taylor, Tyeen C., et al., "Isoprene emission structures tropical tree biogeography and community assembly responses to climate," *New Phytologist* 220 (2018): 435–46; Ribeiro, Kelly, et al., "Tropical peatlands and their contribution to the global carbon cycle and climate change," *Global Change Biology* 27 (2021): 489–505

10. Yu, Kailiang, et al., "Pervasive decreases in living vegetation carbon turnover time across forest climate zones," *Proceedings of the National Academy of Sciences, U.S.* 116 (2019): 24662–667; Wijedasa, Lahiru S., et al., "Carbon emissions from Southeast Asian peatlands will increase despite emission-reduction schemes," *Global Change Biology* 24 (2018): 4598–4613.

11. Harris, Nancy L., et al., "Baseline map of carbon emissions from deforestation in tropical regions," *Science* 336 (2012): 1573–76; Baccini, A., W. Walker, L. Carvalho, M.

Farina, D. Sulla-Menashe and R. A. Houghton, "Tropical forests are a net carbon source based on aboveground measurements of gain and loss," *Science* 358 (2017): 230–34; Smith, Jesse, "Measuring Earth's carbon cycle," *Science* 358 (2017): 186–87; Sakabe, Ayaka, Masayuki Itoh, Takashi Hirano, and Kitso Kusin, "Ecosystem-scale methane flux in tropical peat swamp forest in Indonesia," *Global Change Biology* 24 (2018): 5123–36.

12. Wijedasa et al., "Carbon emissions from Southeast Asian peatlands"; Ayaka, et al., "Ecosystem-scale methane flux in tropical peat swamp forest in Indonesia."

13. Sun, Y., et al., "OCO-2 advances photosynthesis observation from space via solar induced chlorophyll fluorescence," *Science* 358 (2017): DOI: 10.1126/science.aam5747; Nechita-Banda, N., et al. "Monitoring emissions from the 2015 Indonesian fires using CO satellite data," *Philosophical Transactions of the Royal Society B* 373 (2018): 20170307. http://dx.doi.org/10.1098/rstb.2017.0307; Baldeck, Claire A., Gregory P. Asner, et al., "Operational tree species mapping in a diverse tropical forest with airborne imaging spectroscopy," *PLoS ONE* 10 (2015): e0118403. doi:10.1371/journal.

14. Cuong, Chu Van, Peter Dart, Nigel Dudley, and Marc Hockings, "Building stakeholder awareness and engagement strategy to enhance biosphere reserve performance and sustainability: The case of Kien Giang, Vietnam," *Environmental Management* 62 (2018): 877–91; Paulson, Nels, Ann Laudati, Amity Doolittle, Meredith Welsh-Devine, and Pablo Pena, "Indigenous peoples' participation in global conservation: Looking beyond Headdresses and face paint," *Environmental Values* 2 (2012): 255–76; Ellen, Roy, ed., *Modern Crises and Traditional strategies: Local Ecological Knowledge in Island Southeast Asia* (New York: Berghahn Books, 2007); Brosius, J. Peter, "Local knowledges, global claims: On the significance of indigenous ecologies in Sarawak, East Malaysia," in *Indigenous Traditions and Ecology*, ed. John A. Grim (Cambridge, MA: Harvard University Press, 2001), 125–57.

15. Knoke, Thomas, et al., "Accounting for multiple ecosystem services in a simulation of land-use decisions: Does it reduce tropical deforestation?," *Global Change Biology* 26 (2020): 2403–20; Hannah, Lee, et al., "30% land conservation and climate action reduces tropical extinction risk by more than 50%," *Ecography* 43 (2020): 1–11.

Acknowledgments

The many experts who shared their expertise in *OTFTA* have indirectly aided in the preparation of this book, particularly Reinmar Seidler, who assisted in writing the final two chapters of *OTFTA* that were reorganized as the final four chapters of this book. Ian Baillie was an indispensable aid in the production of graphics, for both the original and present books. Bob Morley reviewed chapters 2, 3, and 13, and he added new figures for the latter chapter. Jeff Blossom revised the two important maps on climate and forest distributions, and Carrie Caverley produced the soils diagram in chapter 4. We also thank the reviewers for their criticism of the manuscript and the staff at the University of Chicago Press for helping make this book a reality. Our thanks go to Christie Henry for early support and encouragement. Rachel Kelly-Unger, associate editor for life sciences, rescued this project mid-stream and made it a reality, along with Joseph Calamia, executive editor. Nick Murray, our copy editor, improved the manuscript greatly by suggesting textual changes and finding errors small and large, and he did this all with patience and good humor. Michaela Luckey, editorial assistant, solved many problems, and Mary Corrado, then Tamara Ghattas, in the manuscript editing department, saw this project to completion.

Appendix A

Geological Time Line

This simple time scale provides an equivalence between time in millions of years ago (MYA in the figure), and eras, periods, and epochs. The three horizontal bars are at different scales, top to bottom: 540 million years to present, 66 million to present, and 2.5 million to present (used with permission of Steve Lougheed).

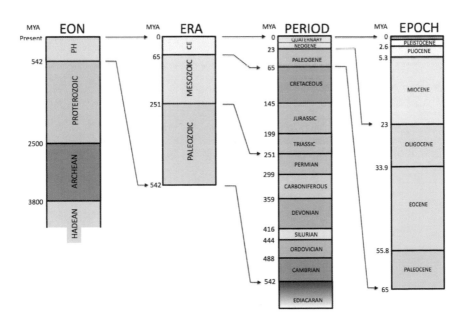

Appendix B
ForestGEO

This important organization and network of forest plots was established as a result of a conversation between Peter Ashton and Steve Hubbell (p. 000). Its first research sites were at Barro Colorado Island, Panama, followed by one at Pasoh, in West Malaysia. Other forest sites were established in subsequent years, particularly in Asia (fig. B.1). For a time, the headquarters were at the Arnold Arboretum, with support from the Smithsonian Tropical Research Institute (STRI), but they were moved to the Smithsonian in 2012. These sites collectively became a powerful resource for the scientific study of tropical forests because of the large plot size (optimally of 50 Ha with some at 25 Ha) and the standard methods used for collecting, storing, and manipulating data—making comparison among sites possible and creating even larger data sets by pooling information among sites.[1] In each plot, all free-standing trees with a trunk diameter at breast height (DBH) of at least 1 cm are tagged, measured, and identified to species. The trees are recensused about every 5 years.

In 2004, the large plot network was expanded to include temperate sites. This made comparisons of forests in latitudinal gradients feasible.[2] The expanded network became known as CTFS-ForestGEO, and then just ForestGEO. This network now comprises 63 forest plots in 24 countries, with 6 million trees from 10,000 species. Thirty of these sites are in Asia, particularly in China. These encompass a latitudinal range of 51.8° N to 5.2° S. Many of the forests described in this book are part of this network, as Mudumalai in India, Sinharaja in Sri Lanka, Doi Inthanon, Huai Kha Khaeng and

Fig. B.1 Global map of the present 30 ForestGEO sites (www.forestgeo.si.edu.).

Khao Chong in Thailand, Pasoh and Lambir in Malaysia, Xishuangbanna in China, and Kuala Belalong in Brunei.

Notes

1. Anderson-Texeira, Kristine, Stuart J. Davies, et al., "CTFS-ForestGEO: A worldwide network monitoring forests in an era of global change," *Global Change Biology* 21 (2015): 528–49; Muller-Landau, Helene, Matteo Detto, Ryan A. Chisholm, Stephen P. Hubbell, and Richard Condit, "Detecting and projecting changes in forest biomass from plot data," in *Forests and Global Change*, ed. David A. Coomes, David F. R. P. Burslem, and William D. Simonson (Cambridge: Cambridge University Press, 2014), 381–415; Davies, Stuart J., et al. (153 co-authors), "ForestGEO: Understanding forest diversity and dynamics through a global observatory network," *Biological Conservation* 253 (2021): 108907.

2. LaManna, Joseph A., et al., "Plant diversity increases with the strength of negative density dependence at the global scale," *Science* 356 (2017): 1389–92; Usinowicz, Jacob, et al., "Temporal coexistence mechanisms contribute to the latitudinal gradient in forest diversity," *Nature* 550 (2017): 105–8; Schepaschenko, Dmitry, et al., "The Forest Observation System, building a global reference dataset for remote sensing of forest biomass," *Scientific Data* 6 (2019): 198 | (https://doi.org/10.1038/s41597-019-0196-1).

Appendix C
An Ecotourism Guide to Tropical Asian Forests

Where can these forests best be visited? Despite deforestation, magnificent examples of the forests described in this book remain. Many we recommend are those to which our colleagues in the region have taken us. Bear in mind that many of the best are not on the usual tourist trail, and these include several national parks. It is wise, when first arriving in a country, to seek advice and assistance from the forest service. If you are a foreign visitor planning research, it is our experience that you will be best served—and in fact will immeasurably gain—by first establishing collaboration with an in-country researcher or institution.

Some will be disappointed that there is a roughly inverse relationship between the ease of seeing wildlife, and the magnificence of the vegetation and richness of flora. Don't be misled by what you see on television; it is hard to see wildlife in rainforests, except in those few places where iconic primates have been captured from wildlife thieves and reintroduced. And remember, Asian rainforests cannot compare, in diversity or density of wildlife, with those of the New World. But the great wildlife parks of seasonal tropical Asia, Nagarhole, Kaziranga, and others, especially in India, with their open grasslands and forests leafless in the cool winter dry season, are unmatched for their large terrestrial mammals and their birds—while tigers and elephants are currently recovering their numbers in India and Thailand.

Here we first suggest a limited number of forest tours that can capture examples of all major forest formations in tropical Asia.

Access and accommodations are sometimes commented on in our forest tours, but accommodations change and may already be out of date. We recommend the *Lonely Planet* tourist guidebooks for the best access and accommodations for explorers unfamiliar with living in the regions, but a search on Google can provide a wider choice, especially for the camper and student on a limited budget.

Some of the best accessible forests. Bhutan and the adjacent lowland "Duars" of India, India's Western Ghats mountains and adjacent western lowlands, Sri Lanka, Brunei, Malaysia, and Thailand have the best record for biodiverse evergreen forest and wildlife conservation, but parks in the deciduous forest regions of India are also magnificent.

Malesia

The Sultanate of Brunei Darussalam shares with the Kingdom of Bhutan the accolade of being the only remaining well-forested nations of tropical Asia, thanks to enlightened governance by traditional rulers (although Thailand is not far behind). The landscapes of both are substantially more than half-clothed in magnificent indigenous forest harboring fauna and flora of unsurpassed diversity and richness. Brunei includes superb and accessible examples of most of the major inland lowland evergreen forest types of the Sunda region; in addition, it has the richest and most extensive mangrove in the region and therefore the world. The hyperdiverse *Anisoptera grossivenia* MDF on humult yellow sands is conserved in a 300 ha patch, a site of Peter's early research; it is now opposite the Forest Research Station at Sungei Liang in the coastal Andulau hills (fig. 7.1). A drive along this road northeast beyond Labi village leads to the east slope of Bukit Teraja, where a hike up its scarp through MDF on clay loams reveals, near the hill crest, that the western dip slope instead supports *Anisoptera grossivenia* MDF. Another outstanding location, at Kuala Belalong (fig. 7.2), where Peter documented *Shorea parvifolia* MDF on dramatic steep terrain beside the picturesque Temburong torrent, now hosts a field studies center of the university and a tourist hotel. A good day's hike there and back along the narrow ridge dividing the Temburong and Belalong valleys takes you to the Bukit Belalong summit. Just under 1,000 m, its upper slopes are clothed in a fine, ferny island of lower montane oak-laurel forest. Tramping these high ridges can still recall that least forgettable of Borneo experiences: the rush of the rapids echoing up from below in the cool, upwelling orographic breeze, the ghostly cackle of the helmeted hornbill summoning from a distant hilltop. *Kerangas* is well represented on the sandstone plateaus of Patoi and adjacent Peradayan hills in the lower part of the same district. Although partly damaged by fire, a fine stand of *Agathis kerangas* survives on the raised Pleistocene sea beach of Badas forest, Belait District. Here the only remaining complete, raised peat swamp (fig. 2.7), with all phasic communities largely intact, survives between the Belait river and the Sarawak frontier to the west—beyond which it has been destroyed, and the peat will be oxidized through drainage and submerged in the rising sea within the next half-century. On either side of the Belait river (fig. 2.7), also at Loagan Bunut National Park on the Tinjar tributary of Sarawak's Baram River, are curious tributary floodplains where shallow flood water backed up from the main rivers can stand for weeks, becoming hot and anoxic, and resulting in unique equatorial savannas when the water has drained—recalling similar woodland savannas in the Wyanaad and elsewhere just east of the Western Ghats.

The adjacent Malaysian state of Sarawak witnessed the sad fate of what was the richest terrestrial landscape in the world, under a political regime dominated by forest corruption. Patches remain in the several national parks, themselves periodically invaded by illegal loggers. Perhaps the richest forest tree communities now surviving are the two main MDF types, and hilltop *kerangas*, notionally protected in Lambir National Park, just west of the Brunei frontier (figs., 2.2, 7.3, 9.4, 9.7); but even the small arboreal wildlife has mostly been hunted out, and the future will be bleak unless seed dispersers are reintroduced. Also, just west and south of Brunei's Belait District, is the large Mulu National Park, which has a tourist hotel and camping. The park includes the finest limestone karst in the region, reaching almost 2,000 m, as a range fronting sandstone Mulu mountain, 2,371 m, with its lower montane *kerangas* and stunted, densely mossy upper montane "cloud" forest (figs. 2.8, 10.9). The mountain is well furnished with overnight camping huts. In the west, Kuching, the capital, is well situated for excursions to several forests of interest. Those fragments that remain are among the richest in arboreal endemism in northern Borneo. The once ubiquitous mixed dipterocarp forest of the extreme northwest of Borneo on siliceous humult lowland sandy clays is now restricted to a tiny patch at the Semengoh forest and wildlife sanctuary, but a more extensive, albeit floristically distinct, dipterocarp forest clothes the granite slopes in Gunung Gading National Park, rising above the town of Lundu, while patches survive on the lower slopes of Gunung Matang in Kobah National Park, a day's hike to the summit from Kuching. Neither mountain exceeds 1,000 m, but the upper slopes and summits—where not disturbed by trigonometric points or former army posts, support distinct lower montane *kerangas*. Gunung Santubong is an easily scaled, coastal sandstone mountain on which frequent diurnal cloud, and therefore a dry form of upper montane forest near its summit, completely truncates any lower montane forest. On a peninsula jutting north into the South China sea, it is accessible by road leading to the resort along its western shoreline. Well worth the climb, the path first follows a scarp up to a porphyry dike, where the majestic *Pterygota horsfieldii* towers above the canopy. This is the only known stand in the region; its closest known locality otherwise is in southeast Borneo. Above, the climb steepens, passing from coastal hill dipterocarp forest with abundant *Shorea bracteolata* into *kerangas*, in which some lower montane elements appear higher up. Finally, after scaling sandstone rock, the summit is reached, which has abundant moss, the fern *Matonia pectinata*, the tree *Leptospermum flavescens*, and other upper montane elements. On the far side of the bay to Santubong's east is Bako National Park, which has some guesthouses. The park supports unique coastal *kerangas*, partly reduced by fire in the past, on its sandstone plateau and mixed dipterocarp forests along its western coastal scarp and southern ridge (figs. 2.7, 7.1, 13.9). Fine limestone woods can be explored at Bau, though the summits have often experienced fire. All are short car journeys from Kuching.

The State of Sabah, to Brunei's east and unlike Sarawak, has become a model of wise conservation and sensitive ecotourism unmatched elsewhere in equatorial Asia. Kinabalu National Park is a major attraction, but the magnificent *Dryobalanops lanceolata* MDF of Sepilok Forest Reserve, where there is an orangutan and rhinoceros rehabilitation sanctuary, is well worth a visit, and the similar Danum MDF (figs. 2.2, 7.1, 7.2), where there is an ecotourism lodge, and a ForestGEO plot. The tallest forest stands in the tropics are in Sabah, at Tawau River Forest Reserve and the Imbai Gorge National Park. A recent, enlightened director of forests has successfully negotiated the uniting of these primary forests in a vast matrix of logged but regenerating forest, creating arguably the richest lowland wildlife preserve in all the Sunda lands. Forests on ultramafic rocks can be examined by climbing the coastal Mount Silam, near Lahad Datu, where there is a commercial airport. The jewel in Sabah's crown, of course, is Mount Kinabalu, highest peak between the (increasingly intermittently) snowy mountains of New Guinea, also equatorial, and the Himalayas at the tropical margin (figs. 10.1, 10.5, 10.9, 10.10, 10.12). Easily accessible from the excellent hotels in and around the capital, Kota Kinabalu, comfortable accommodation can be booked at the two rest houses (with helipad) 1,000 m below the summit. The climb starts at ~1,000 m, in mixed lower montane oak-laurel forest and lower montane *kerangas*. This is the drier flank of the massif and, although ferns and other lower montane ground and epiphytic flora increase with altitude, they are not as rich along the trail as on more humid slopes—which can be better seen where trails branch off to side destinations. As the forest shortens, mossy sleeves on branches herald the ecotone to upper montane forest. Higher up, upper montane forest prevails, and with it the widespread dominance of *Leptospermum javanicum*. A further change, from this species to *L. recurvum*, with smaller spoon-shaped leaves, and other species with similar foliage, proclaims arrival of subalpine thicket. The rest houses are eventually sighted, backed by the bare granite face of the Kinabalu summit dome. From its foot can be seen the only natural forest and *tree-shrub altitudinal limit* in the Asian tropics. On the New Guinea mountains, and even the Himalayas, this limit has been modified and lowered by domestic cattle and fire since ancient times.

Peninsular Malaysia's once-dominant red meranti-keruing MDF is now reduced to patches, in Taman Negara, the major national park primarily established to conserve the highest peninsular peak, Tahan mountain, 2,187 m, where there is *Shorea curtisii*–dominated hill dipterocarp forest. A more picturesque landscape is experienced in the Royal Belum National Park, south of the Thai frontier, at the northern end of the Tasek Temenggor reservoir damming the upper Perak River. There, mixed and associated hill dipterocarp forests are approaching their northern climatic limits. Distinct MDF communities that include the Sumatran camphor tree *Dryobalanops aromatica* cover the sandstone Endau-Rompin National Park. MDF with distinct Riau Pocket flora covers the hills to the west of the lower Perak river, notably the Segari-Melintang Forest Reserve. Fine *S. curtisii*–dominated coastal MDF can be seen at Mukah Head, Penang (fig. 2.2), while a noble stand survives within a great city, Singapore, at Bukit Timah, still replete with seed-dispersing birds and primates—a treasure for the citizenry and a tribute to their respect for nature. A visit to the northwest must not ignore the dramatic karst Langkawi archipelago, now a resort, where craggy peaks are clothed in the southernmost stands of *Shorea siamensis*–dominated deciduous dipterocarp forest, while lower scree slopes support the southernmost elements of seasonal evergreen dipterocarp forest. The most accessible altitudinal zonation of forests in the peninsula is up the trail to the Gunung Tahan summit. Fine lower montane forests line the paths descending the summit ridges of Fraser's Hill, while short, densely mossy upper montane forest can still be visited along the ridge beyond the resort of Genting Highlands (figs. 10.6, 10.9). The Cameron Highlands also bear less mossy examples of upper montane forest.

Indonesia is currently experiencing more deforestation than any other tropical country, mainly for oil palm, but also because of mining. Many national parks have been logged or otherwise degraded. Tall deciduous teak forests, intensively managed by replanting, still remain in East Java. But mountain forests remain more widely: in Borneo much of Gunung Palung National Park, between Pontianak and Sukadana in the west, still survives with its 1,100 m coastal granodiorite mountain, MDF, and lower montane forest. Undisturbed montane forests (figs. 7.2, 7.3), now more widespread than in the lowland forests, can be accessed in several locations. Korinci in Sumatra, which still possesses upper dipterocarp forest, and Gede, West Java, which is climbed from the Cibodas mountain botanic garden at 1,400 m, offer excellent examples of the more gradual zonation on equatorial volcanoes and of equatorial lower montane oak-laurel forest.

Little remains of lowland forest in the Philippines either, even in the national parks, but a patch of MDF survives around the ForestGEO plot at Palanan, on the east coast of Luzon, requiring flight by domestic light aircraft, grazing the summit ridge of the Sierra Madre. Small patches do survive elsewhere, including *Dipterocarpus grandiflorus* seasonal evergreen dipterocarp forest in Quezon National Park east of Manila, with fine karst, and MDF in the prison compound outside Puerto Princesa, Palawan.

East of Wallace's Line, forest exploration requires more knowledge of local opportunities, and contacts, as well as means of transportation and accommodation. Notable forests do survive on all the main islands, including Lombok (also Bali to the west, fig. 10.6); Sulawesi. where Lore Lindu National Park offers opportunities to subalpine elevations; Ceram, where little remains of the great groves of the Kayu Bapa, *Shorea celanica*, the only red meranti east of Borneo; and New Guinea, where few lowland forests have never experienced clearing for swidden agriculture in the many thousands of years of agriculture. The daring visitor will gain unique experience along the trails up and beyond the tree line on mounts Trikora, Wilhelm, (fig. 10.10) and Albert Edward (fig. 10.9).

Karst woodland occurs as ecological islands covering the archipelagos of limestone karst that dot Malesia, from northern Peninsular Malaysia to New Guinea. Viewed from afar, karst is magnificent, but the cliffs deter all but the most agile. For the most pristine, we recommend scaling the trail up to the summit pinnacles, resulting from rain solution of dolomite, on the karst front range at everwet Mulu National Park, Sarawak (fig. 2.8).

Indo-Burma

Thailand offers the most convenient access to most lowland forest types of seasonal regions. Southern seasonal evergreen dipterocarp forest of the *Parashorea stellata* type can be viewed at Khao Chong National Park, near Trang in the peninsula, with a ForestGEO-associated plot, and the *Dipterocarpus costatus* coastal MDF on Tarutao island. Southern semi-evergreen forest of the *Hopea odorata* subtype is best viewed in the Huai Kha Khaeng–Tung Yai Naresuan National Park and wildlife sanctuary complex on the Burmese frontier (2.4, 8.2, 8.3), the largest conserved area in Indo-Burma, with another ForestGEO plot. Fine examples of deciduous dipterocarp forest, and a diversity of tall deciduous forests are widely present. A circular tour leads west from Chiangmai to Mae Sariang, where a turn to the south leads to Mae Ping National Park, with fine stands of tall *Dipterocarpus tuberculatus* deciduous dipterocarp forest on a sandy plain, and with *Shorea siamensis* on low karst. A stand of teak that is reaching maturity following logging one century ago, with associated tall deciduous forest subcanopy flora, is worth seeing. The road then strikes north to Mae Hon Song, north and west to Mae Pai, then back to Chiang Mai via Mae Chiang Dao. It offers fine deciduous dipterocarp forest and tall deciduous forest at Salawin (Salween) National Park; there is extensive deciduous dipterocarp forest, often with *Pinus merkusii*, along the road to Mae Hong Son, while much deciduous forest, with *Pinus kesiya* silhouetted along the dramatic high ridges, follows the road toward Pai, where there is massive karst. Descending to the valley of Chiang Dao, one moves from sandy to loam soils, where patches of majestic *Dipterocarpus turbinatus* semi-evergreen forest survives. The lower slopes of Doi (mount) Chiang Dao, 2,175 m, support fine stands of *D. costatus* upper dipterocarp forest. Doi Chiang Dao has karst around its summit bearing grassy slopes with temperate herbs, a result of periodic frost. Excellent lower montane oak-laurel forests, which include tropical Asia's only montane ForestGEO plot, can be visited by the road to the summit of Thailand's highest mountain, Doi Inthanon, 2,590m, as they decline in stature and increase in canopy smoothness and mossiness—more than 500 m above the ecotone to upper montane forest on ever-cloudy equatorial peaks (figs. 2.4,10.5,

15.10). There, a short descent into a narrow dell reveals frost-hardy temperate shrubs, including the magnificent *Rhododendron arboreum delaveyanum*. Phang Nga National Park, on a plateau at ~1,200 m near the central Thai town of Phetchabun, provides a pleasant diversion to examine short-stature lower montane forest in a distinctly seasonal climate, *D. costatus*-dominated upper dipterocarp forest, and a remarkable plain clothed with a *Pinus kesiya* savanna, which suffered dramatic mutilation at the hands of a recent southwest monsoon cyclone.

Cambodia also had fine examples of these same lowland forest types but no karst, plus rare seasonal *kerangas* on raised beaches along the road following coastal lowlands south of the Cardamom mountains up Andoung Tuk; however, we have not visited since the extensive logging and deforestation of recent years. Further east and toward the end of the Elephant Chain is the steep road winding up through seasonal evergreen and then lower montane forest to the ruined hill station sitting on the scarp: Bokor (or Popork Vul: "the clouds turn"). To its north, the gentle dip slope supports a fine stand of *Dacrydium elatum* and *Dacrycarpus imbricatus*—dominated lower montane *kerangas*. The tourist-saturated ruins around Angkor Thom remain embedded in the ghost of former *Dipterocarpus alatus*–dominated southern semi-evergreen dipterocarp forest. To the north, the Phnom Kulen plateau, source of the Siem Riep river that once fed the great tanks that supplied and sanctified the ancient Khmer capital, is now clothed in fine forest of this type.

The tragically war-damaged relic forests of southern Vietnam do include an astonishing continuum of increasing rainfall seasonality in now partially conserved Ma Da forest, Dong Nai Province, north-east of Ho Chi Minh city, which transitions from southern seasonal evergreen dipterocarp forest through southern semi-evergreen forest to tall deciduous forest and deciduous dipterocarp forest. Although degraded by agent orange sorties to varying extent, and by fire during and since the war, fine stands do survive. Most are on sandy soil, but patches of loam over basalt extrusions toward the moist eastern end support both seasonal evergreen and semi-evergreen stands.

In the north, northern seasonal evergreen dipterocarp forests are now reduced to several small patches. Those most easily visited include patches among the karst crags of Cuc Phuong National Park down the coast south from Hanoi, and the magnificent, albeit refugial, stands of *Parashorea chinensis* in Mengla District, near Xishuangbanna, Yunnan, where there is a ForestGEO-associated plot (figs. 2.3, 10.5). Northern semi-evergreen forests, often dominated by the beautiful *Terminalia myriocarpa*, remain widespread, although often logged or ancient secondary. A visit to Dalat, a hill station at the southern end of the Annamite range, provides access to patches of lower montane forest. Along the summit ridges are conifer forests, including fire-maintained savanna of the endemic *Pinus dalatensis*, and closed, broadleafed forest with the extraordinary *P. kremfii*, itself broad-leafed. Stands of several pine species survive at lower montane forest elevations. Patches of montane forests along the northern tropical margin, as along the road up Fan Si-Pan (though here much destroyed), make an interesting comparison with those still extensively surviving in Bhutan toward the western end of the wet tropical Himalayas.

Short deciduous forest is now perhaps entirely degraded in Indo-Burma except in northern Laos; but it remains extensive, though also degraded, in Burma, on the upper Irrawaddy plains west of the river. There it is dominated by deciduous dipterocarps, notably *Shorea siamensis*, and also includes notable endemic tree species (see chapter 13). We hope that Burma may eventually allow interested visitors to come to her forests, many of which still survive where not devastated by Chinese and Thai interests in the northeast and east.

Karst limestone occurs as ecological archipelagos up the western side of the Thai-Burmese peninsula, in northern Thailand and in northern Vietnam. It occurs more widely in adjacent southern China and in several locations throughout Malesia.

South Asia and the Himalayas

Sri Lanka lost most of its lowland forests in colonial times but, such is the universal spiritual respect for nature among its people, the remaining 5% is as well protected as any in the Asian tropics. The lowland MDF is conserved in several small patches, mostly logged, but one superb patch with its ForestGEO plot remains toward the western end of the Sinharaja UNESCO World Heritage Forest (figs. 2.2, 9.3), approached from the northern side, where the outstanding naturalist Martin Wijesinghe might still welcome you at his forest lodge. There are other small country hotels and home stays around Kudawa on the north side, and a gorgeous modern resort in a tea estate embedded in majestic rata dun lower montane forest on the south side at Deniyaya: a great place for birding. The Kanneliya Conservation Forest is also to the south; though once logged, it is also worth a visit, owing to its distinct flora on more siliceous metamorphic rocks. Other, smaller MDF forests are worth a visit, notably at Kottawe, along the roadside between Galle and Kanneliya-Nellowe. The north and east of the island, with its two dry seasons, still widely supports semi-evergreen notophyll forest, mostly degraded but with fine surviving areas around the base of the isolated mountain Ritigala, itself worth a climb through forests that do not easily conform with those of their region. At the attractive reservoir location of Giritale, there is an excellent government rest house. Further south, and to the east of the mountains, patches of southern semi-evergreen forest survive in Moneragala District, with species rare in Sri Lanka, such as *Alstonia scholaris*. The once extensive lower montane forests of Sri Lanka's central mountain massif, unique for their dipterocarp richness and dominated by *Shorea gardneri*, with twisted bole and writhing branches bearing diffuse microphyll crowns, that tower over single-species populations of point-endemic *Stemonoporus*, were all but replaced by tea one and a half centuries ago. Magnificent stands do survive in the Peak Sanctuary (figs. 10.4, 14.5), especially along the Balangoda-Bogawantalawa road. Adam's Peak, 2,243 m, least rainy in April, if scaled (or descended) along the long southern trail to Carney Estate, Ratnapura Province, provides the best sequence of forest zones on an everwet mountain subject to the full force of monsoon winds. Below the estate is the lowland MDF of Gilimale Forest Reserve. Higher up the trail, the full force of the southwest monsoon is associated with a thicket of upper montane structure but lower montane flora. At the eastern end of the Peak Sanctuary is Horton Plains, approachable by car from Nuwara Eliya. At 2,100 m, it bears forest with the character and flora of an upper montane forest—subalpine thicket ecotone (dense twigginess, some concave leaves; see fig. 10.9). It has been much converted to grassland by fire and agriculture, in which the sole frost-hardy tree is the *Rhododendron arboreum* subspecies *zeylanicum*. When it is in flower, which is often, keen observers will enjoy watching pollination by the endemic Ceylon White-eye, congener of the pollinators of Japan's Camellias and other red-flowered Himalayan rhododendrons. Horton Plains experiences a short dry season and light winter frost, and the lightly mossy woodland abounds with *Usnea* lichen. Other, drier, montane forests can be observed on the slope to the peak above Hakgala montane botanic garden, where survivors from the original, mid-19th-century consignment of tea saplings from Assam are worth a visit.

In India, seasonal evergreen dipterocarp forests, confined to the western slopes of the southern part of the Western Ghats, have mostly been logged or converted. A good example of the richest, southernmost forests can still be viewed at Periyar National Park, while toward their northern limit, lofty primary examples of *Dipterocarpus indicus*–dominated stands exist in sacred groves, including Pilarkan and the Katlekan Dark Forest on the lower slopes of the Karnataka Ghats. North again, sacred groves still conserve refuge stands of southern semi-evergreen forest. Northern seasonal evergreen forests reach their western limits in eastern Bhutan, but the best example is at Lakhimpur, north of the Brahmaputra in Assam, where it is dominated by *Dipterocarpus retusus*. Northern semi-evergreen forests may be viewed along the riverbanks passing through the duars raised beaches in Buxa Tiger Reserve, at Manas National Park in Bhutan, and west to central Nepal, where they merge with tall deciduous sal forest. They extend up adjacent slopes of the Himalayan front ranges to the base of adjacent lower montane oak-laurel forest, within which *Terminalia myriocarpa* remains abundant. These same rivers, gushing out from the mountains during the monsoon, laden with rock and gravel, flood the plain of the Brahmaputra to the south with shallow water that may heat and become deoxygenated, so that tree seeds fail to germinate, except near the banks, where khair sissoo woodlands dominate. The floodplains are thereby covered by grassland savanna, supporting a celebrated large mammal fauna, including rhinoceros and tiger, as at Kaziranga, Manas, and other wildlife parks along these northern plains.

The full range of deciduous forests is best seen in India. There, the continuum from tall to short deciduous teak forests, although extensively logged 50 and more years ago, is still superbly represented in the great lunette of hills conserved in the southern Karnataka and the north-west Tamil Nadu Deccan. Here, the largest remaining herd of elephants migrates from the thorn woodlands and short deciduous forests of Bandipur National Park and the upper Moyar valley (figs. 2.5, 2.6, 8.6, 8.7), where they feed in the wet monsoon, to the moist, tall deciduous and semi-evergreen forests abutting the base of the ghats at Mudumalai Wildlife Sanctuary (figs. 2.5, 2.6, 8.4), and north to Nagarhole National Park and the teak forests of the Wayanaad in the dry season. Tall sal forests, including their shorter, high-hill manifestation, remain intact and extensive in Similipal National Park, Odisha, the former domain of the Maharaja of Mayurbhanj, where small areas of short deciduous sal also occur on slopes sheltered from the northeast monsoon. Tall stands, more or less degraded by overgrazing, illegal logging, or fuelwood collection, do remain in what were the great sal forests of Singbhum and around the hill town of Ranchi, in Bihar state.

Thorn woodlands are also confined to South Asia except for an area on the Burmese northern Irrawaddy plain. As the vegetation most defiant to human intrusion, much of it remains. The "Saharo-Sindian" elements swathe the hills and plains of southern Rajasthan. Those least influenced, in the south, may be seen along the Moyar Gorge (fig. 2.6) or in the lowlands around and west of Tiruchchirappalli, Tamil Nadu, where the striking, emergent umbrella crowns of *Acacia planifrons* betray their presence.

Peninsular Indian montane evergreen forests are confined to the Western Ghats, though elements are to be found in the upper valleys at Similipal Tiger Reserve and elsewhere in the Eastern Ghats. They are reduced to fragments, except in the Nilgiri Silent Valley (fig. 18.9), which extends far into the Nilgiri massif from the west. But for sheer grandeur and beauty, one must visit the wet eastern Himalayas, especially Bhutan. Entering by road at Phuntsholing, lower montane forest with locally abundant *Castanopsis indica* is encountered at 850 m, along with fine examples of the tropical-warm temperate ecotone, on the left hand, upslope side of the road at Gedu, 2,100 m. The road tracks eastward from the capital, Thimphu, to more than 3,000 m at the several mountain passes between which, as in the woods near Punakha, the forest transition at the tropical margin can be examined (fig. 10.2). It proceeds upward through warm, then cool temperate, broadleaved forests, and dense stands of different conifers throughout, according to temperature and altitude. Fragments of northern semi-evergreen forest, with abundant *Terminalia myriocarpa,* can be seen on the road down from Trongsa to Tingtibi, and from Punakha down the Sankosh valley, with its noble stands of chir, *Pinus roxburghii,* but the lowland forests at the base are off-limits.

There is no karst in South Asia, but much to see still there!

Illustration Credits and Abbreviations

Most of the photographs and almost all of the diagrams were used in *OT-FTA*. Permission for their use, mainly from former students of Peter's and colleagues of both of us, was obtained for this book as well. A preponderance of photographs was taken by Hans Hazebroek (H. H.), who traveled to the Asian tropics with Peter and Mary, taking photographs for that book, but many were taken by the authors. Most diagrams were obtained from the publications of colleagues and former students, and they were redrawn by Ian Baillie (I. B.). For chapter 17, several diagrams were taken from recent journal issues, and use was granted by the journals indicated in figure captions.

Abbreviations of contributors to illustrations. A. C. = Alexander Cobb; A. F. = A. Farjon; A. S. = Alan Smith; B. M. = Robert (Bob) Morley; C. Lo. = Cynthia Lobato; D. L. = David Lee; D. W. = Dena Willmore; H. H.= Hans Hazebroek; H. L. = H. S. Lee; I. B. = Ian Baillie; J. B. = Jeff Blossom; J. G. = Jaboury Ghazoul; O. = Ong Jin Eong; K. A. = K. Abu Salim; L. E. = Lilly Eluvathingal; L. S. = Lucy Smith; M. A. = Mark Ashton; Ma. A. = Mary Ashton; M. D. = Dharmalingan Mohandas; N. G. = Nimal Gunatilleke; N. K. = Naveen Kadalaveni; P. A. = Peter Ashton; P. C. = Paul Chai; P. W. = Paddy (Paul) Woods; P. H. = Pamela Hall; R. H. = Rhett Harrison; R. S. = Raman Sukumar; S. D. = Stuart Davies; S. Dr. = Soejatmi Dransfield; S. H. = Sean Hoyland; S. L. = Saw Leng Guan; S. Le = Su See Lee; S. R. = Sabrina Russo; T. I. = Takao Itino; T. S. = T. R. Shankar Raman; T. W. = Tim Whitmore; U. G. = Uromi Manage Goodale; W. K. = Wong Khoon Meng; W. L. = Wang Luan Keng; Z. H. = Zhu Hua.

Index

Page numbers in boldface indicate illustrations (photographs, maps, diagrams).